Discrete Mathematics

Discrete Mathematics

Edited by
Desmond Ballard

Larsen & Keller
www.larsen-keller.com

Discrete Mathematics
Edited by Desmond Ballard
ISBN: 978-1-63549-088-6 (Hardback)

Larsen & Keller
Published by Larsen and Keller Education,
5 Penn Plaza,
19th Floor,
New York, NY 10001, USA

Cataloging-in-Publication Data

Discrete mathematics / edited by Desmond Ballard.
p. cm.
Includes bibliographical references and index.
ISBN 978-1-63549-088-6
1. Computer science--Mathematics. 2. Mathematics. I. Ballard, Desmond.
QA76.9.M35 D57 2017
004.015 1--dc23

Printed and bound in China.

For more information regarding Larsen and Keller Education and its products, please visit the publisher's website www.larsen-keller.com

Table of Contents

Preface

This book is a compilation of chapters that discuss the most vital concepts in the field of discrete mathematics. It discusses some of the most fundamental theories of this field in detail. Discrete mathematics refers to the study of discrete mathematical structures. It includes the study of logic statements, integers, and graphs, which have separate values. This book is a valuable compilation of topics, ranging from the basic to the most complex theories and principles in this field. This text attempts to understand the multiple branches that fall under this discipline and how such concepts have practical applications. The topics covered in this extensive book deal with the core subjects of discrete mathematics. It will be of great help to students studying industrial mathematics, control systems and game theory.

A short introduction to every chapter is written below to provide an overview of the content of the book:

Chapter 1 - Discrete mathematics is a branch of mathematics that is fundamentally discrete rather than continuous. The subjects concerned within discrete mathematics are integers, graphs and statements in logic. This chapter helps in understanding the basic concept of discrete mathematics and provides the reader with an understanding on the subject matter; **Chapter 2** - Arithmetic is the oldest branch of mathematics. It studies numbers; arithmetic is an elementary part of number theory. Some of the aspects of arithmetic discussed in this text are decimal, binary number, indeterminate form, greatest common divisor and least common multiple. The section on arithmetic offers an insightful focus, keeping in minds the aspects of arithmetic; **Chapter 3** - Probability is the possibility of an event happening. It can be measured between 0 to 1, where 0 is uncertainty and 1 signifies certainty. Sample space, event, random variable, expected value and conditional probability are some of the topics elucidated in the text. This chapter elucidates the crucial theories and principles of probability; **Chapter 4** - Sets are well- defined collections of objects. For example, numbers like 2 or 6 are distinct objects. This section also focuses on topics such as elements, Venn diagram, subset, intersection, complement and Cartesian product. This text will provide an integrated understanding of sets; **Chapter 5** - Number theory is a branch of pure mathematics. The number theory studies prime numbers and also the properties of integers. The topics explained in this section are analytic number theory, algebraic number theory, Baker's theorem and pentagonal number theorem. The chapter is an overview of the subject matter incorporating the major aspects of number theory; **Chapter 6** - The study of graphs is known as the graph theory. A graph is made up of points, nodes and lines. They are the main objects in discrete mathematics. The following text is an overview of the subject matter incorporating all the major aspects of graph theory; **Chapter 7** - Topography studies a collection of subsets that turn the given set into a topological space. Topology has a number of properties; some of these are connectedness and compactness. The essential aspects of topology that are dealt within this chapter are general topology, algebraic topology, differential topology and geometric topology.

Finally, I would like to thank my fellow scholars who gave constructive feedback and my family members who supported me at every step.

Editor

1

Introduction to Discrete Mathematics

Discrete mathematics is a branch of mathematics that is fundamentally discrete rather than continuous. The subjects concerned within discrete mathematics are integers, graphs and statements in logic. This chapter helps in understanding the basic concept of discrete mathematics and provides the reader with an understanding on the subject matter.

Discrete Mathematics

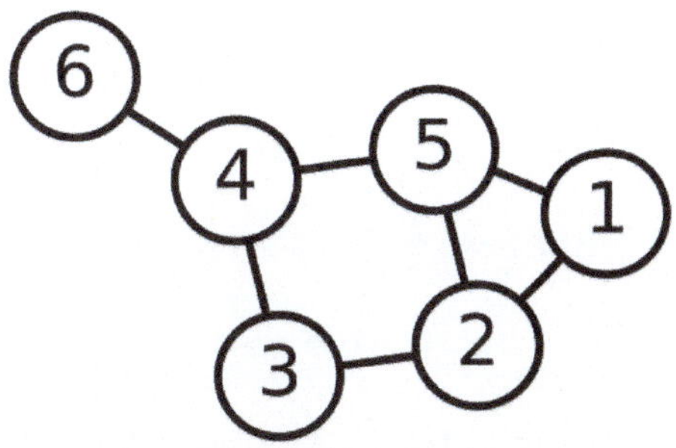

Graphs like this are among the objects studied by discrete mathematics, for their interesting mathematical properties, their usefulness as models of real-world problems, and their importance in developing computer algorithms.

Discrete mathematics is the study of mathematical structures that are fundamentally discrete rather than continuous. In contrast to real numbers that have the property of varying "smoothly", the objects studied in discrete mathematics – such as integers, graphs, and statements in logic – do not vary smoothly in this way, but have distinct, separated values. Discrete mathematics therefore excludes topics in "continuous mathematics" such as calculus and analysis. Discrete objects can often be enumerated by integers. More formally, discrete mathematics has been characterized as the branch of mathematics dealing with countable sets (sets that have the same cardinality as subsets of the natural numbers, including rational numbers but not real numbers). However, there is no exact definition of the term "discrete mathematics." Indeed, discrete mathematics is described less by what is included than by what is excluded: continuously varying quantities and related notions.

The set of objects studied in discrete mathematics can be finite or infinite. The term finite mathematics is sometimes applied to parts of the field of discrete mathematics that deals with finite sets, particularly those areas relevant to business.

Research in discrete mathematics increased in the latter half of the twentieth century partly due to the development of digital computers which operate in discrete steps and store data in discrete bits. Concepts and notations from discrete mathematics are useful in studying and describing objects and problems in branches of computer science, such as computer algorithms, programming languages, cryptography, automated theorem proving, and software development. Conversely, computer implementations are significant in applying ideas from discrete mathematics to real-world problems, such as in operations research.

Although the main objects of study in discrete mathematics are discrete objects, analytic methods from continuous mathematics are often employed as well.

In university curricula, “Discrete Mathematics” appeared in the 1980s, initially as a computer science support course; its contents were somewhat haphazard at the time. The curriculum has thereafter developed in conjunction with efforts by ACM and MAA into a course that is basically intended to develop mathematical maturity in freshmen; as such it is nowadays a prerequisite for mathematics majors in some universities as well. Some high-school-level discrete mathematics textbooks have appeared as well. At this level, discrete mathematics is sometimes seen as a preparatory course, not unlike precalculus in this respect.

The Fulkerson Prize is awarded for outstanding papers in discrete mathematics.

Grand Challenges, Past and Present

Much research in graph theory was motivated by attempts to prove that all maps, like this one, could be colored using only four colors so that no areas of the same color touched. Kenneth Appel and Wolfgang Haken proved this in 1976.

The history of discrete mathematics has involved a number of challenging problems which have focused attention within areas of the field. In graph theory, much research was motivated by attempts to prove the four color theorem, first stated in 1852, but not proved until 1976 (by Kenneth Appel and Wolfgang Haken, using substantial computer assistance).

In logic, the second problem on David Hilbert’s list of open problems presented in 1900 was to prove that the axioms of arithmetic are consistent. Gödel’s second incompleteness theorem, proved in 1931, showed that this was not possible – at least not within arithmetic itself. Hilbert’s tenth problem was to determine whether a given polynomial Diophantine equation with integer coefficients has an integer solution. In 1970, Yuri Matiyasevich proved that this could not be done.

The need to break German codes in World War II led to advances in cryptography and theoretical computer science, with the first programmable digital electronic computer being developed at England’s Bletchley Park with the guidance of Alan Turing and his seminal work, On Computable Numbers. At the same time, military requirements motivated advances in operations research. The Cold War meant that cryptography remained important, with fundamental advances such as public-key cryptography being developed in the following decades. Operations research remained important as a tool in business and project management, with the critical path method being developed in the 1950s. The telecommunication industry has also motivated advances in discrete mathematics, particularly in graph theory and information theory. Formal verification of statements in logic has been necessary for software development of safety-critical systems, and advances in automated theorem proving have been driven by this need.

Computational geometry has been an important part of the computer graphics incorporated into modern video games and computer-aided design tools.

Several fields of discrete mathematics, particularly theoretical computer science, graph theory, and combinatorics, are important in addressing the challenging bioinformatics problems associated with understanding the tree of life.

Currently, one of the most famous open problems in theoretical computer science is the P = NP problem, which involves the relationship between the complexity classes P and NP. The Clay Mathematics Institute has offered a $1 million USD prize for the first correct proof, along with prizes for six other mathematical problems.

Topics in Discrete Mathematics

Theoretical Computer Science

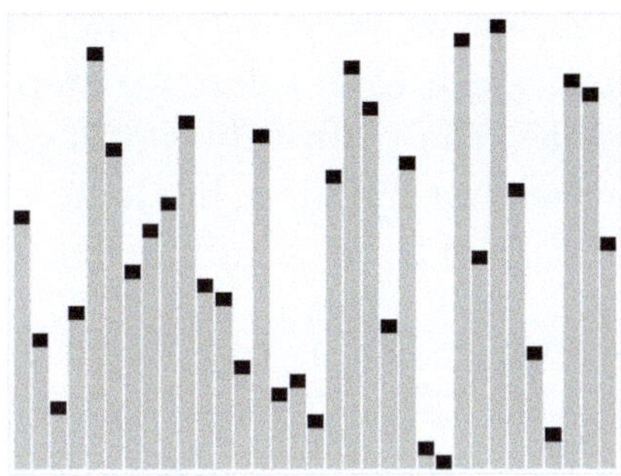

Complexity studies the time taken by algorithms, such as this sorting routine.

Theoretical computer science includes areas of discrete mathematics relevant to computing. It draws heavily on graph theory and mathematical logic. Included within theoretical computer science is the study of algorithms for computing mathematical results. Computability studies what can be computed in principle, and has close ties to logic, while complexity studies the time taken by computations. Automata theory and formal language theory are closely related to computability. Petri nets and process algebras are used to model computer systems, and methods from discrete mathematics are used in analyzing VLSI electronic circuits. Computational geometry applies algorithms to geometrical problems, while computer image analysis applies them to representations of images. Theoretical computer science also includes the study of various continuous computational topics.

Information Theory

Information theory involves the quantification of information. Closely related is coding theory which is used to design efficient and reliable data transmission and storage methods. Information theory also includes continuous topics such as: analog signals, analog coding, analog encryption.

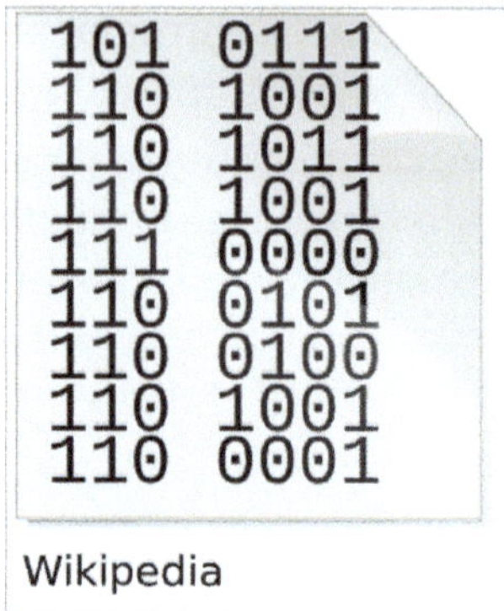

The ASCII codes for the word "Wikipedia", given here in binary, provide a way of representing the word in information theory, as well as for information-processing algorithms.

Logic

Logic is the study of the principles of valid reasoning and inference, as well as of consistency, soundness, and completeness. For example, in most systems of logic (but not in intuitionistic logic) Peirce's law $(((P \to Q) \to P) \to P)$ is a theorem. For classical logic, it can be easily verified with a truth table. The study of mathematical proof is particularly important in logic, and has applications to automated theorem proving and formal verification of software.

Logical formulas are discrete structures, as are proofs, which form finite trees or, more generally, directed acyclic graph structures (with each inference step combining one or more premise branches to give a single conclusion). The truth values of logical formulas usually form a finite set, generally restricted to two values: *true* and *false*, but logic can also be continuous-valued, e.g., fuzzy logic. Concepts such as infinite proof trees or infinite derivation trees have also been studied, e.g. infinitary logic.

Set Theory

Set theory is the branch of mathematics that studies sets, which are collections of objects, such as {blue, white, red} or the (infinite) set of all prime numbers. Partially ordered sets and sets with other relations have applications in several areas.

In discrete mathematics, countable sets (including finite sets) are the main focus. The beginning of set theory as a branch of mathematics is usually marked by Georg Cantor's work distinguishing between different kinds of infinite set, motivated by the study of trigonometric series, and further development of the theory of infinite sets is outside the scope of discrete mathematics. Indeed, contemporary work in descriptive set theory makes extensive use of traditional continuous mathematics.

Combinatorics

Combinatorics studies the way in which discrete structures can be combined or arranged. Enumerative combinatorics concentrates on counting the number of certain combinatorial objects - e.g. the twelvefold way provides a unified framework for counting permutations, combinations and partitions. Analytic combinatorics concerns the enumeration (i.e., determining the number) of combinatorial structures using tools from complex analysis and probability theory. In contrast with enumerative combinatorics which uses explicit combinatorial formulae and generating functions to describe the results, analytic combinatorics aims at obtaining asymptotic formulae. Design theory is a study of combinatorial designs, which are collections of subsets with certain intersection properties. Partition theory studies various enumeration and asymptotic problems related to integer partitions, and is closely related to q-series, special functions and orthogonal polynomials. Originally a part of number theory and analysis, partition theory is now considered a part of combinatorics or an independent field. Order theory is the study of partially ordered sets, both finite and infinite.

Graph Theory

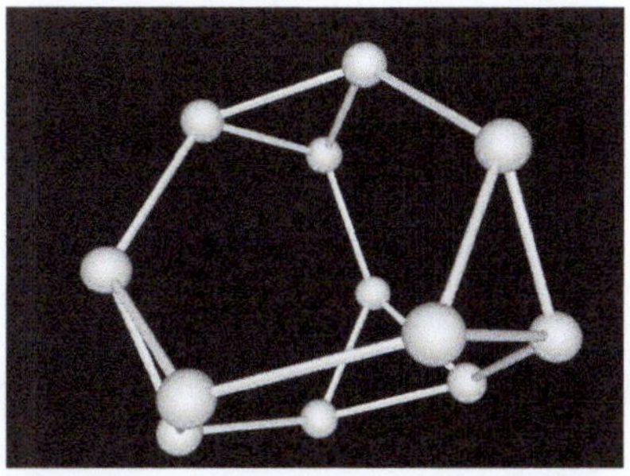

Graph theory has close links to group theory. This truncated tetrahedron graph is related to the alternating group A_4.

Graph theory, the study of graphs and networks, is often considered part of combinatorics, but has grown large enough and distinct enough, with its own kind of problems, to be regarded as a subject in its own right. Graphs are one of the prime objects of study in discrete mathematics. They are among the most ubiquitous models of both natural and human-made structures. They can model many types of relations and process dynamics in physical, biological and social systems. In computer science, they can represent networks of communication, data organization, computational devices, the flow of computation, etc. In mathematics, they are useful in geometry and certain parts of topology, e.g. knot theory. Algebraic graph theory has close links with group theory. There are also continuous graphs, however for the most part research in graph theory falls within the domain of discrete mathematics.

Probability

Discrete probability theory deals with events that occur in countable sample spaces. For example, count observations such as the numbers of birds in flocks comprise only natural number values {0, 1, 2, ...}. On the other hand, continuous observations such as the weights of birds comprise real number values and would typically be modeled by a continuous probability distribution such as the normal. Discrete probability distributions can be used to approximate continuous ones and vice versa. For highly constrained situations such as throwing dice or experiments with decks of cards, calculating the probability of events is basically enumerative combinatorics.

Number Theory

Number theory is concerned with the properties of numbers in general, particularly integers. It has applications to cryptography, cryptanalysis, and cryptology, particularly with regard to modular arithmetic, diophantine equations, linear and quadratic congruences, prime numbers and primality testing. Other discrete aspects of number theory include geometry of numbers. In analytic number theory, techniques from continuous mathematics are also used. Topics that go beyond discrete objects include transcendental numbers, diophantine approximation, p-adic analysis and function fields.

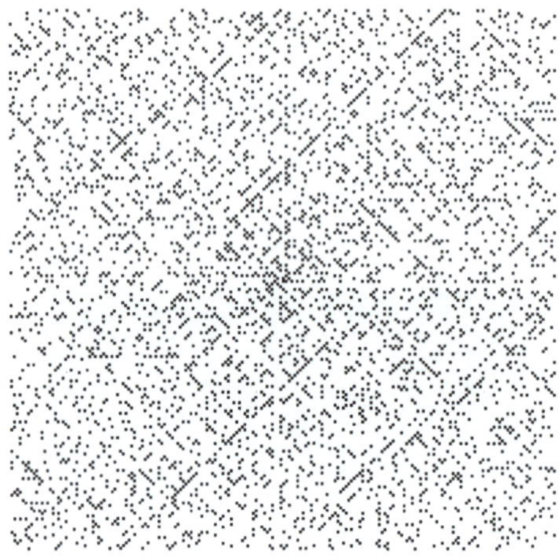

The Ulam spiral of numbers, with black pixels showing prime numbers. This diagram hints at patterns in the distribution of prime numbers.

Algebra

Algebraic structures occur as both discrete examples and continuous examples. Discrete algebras include: boolean algebra used in logic gates and programming; relational algebra used in databases; discrete and finite versions of groups, rings and fields are important in algebraic coding theory; discrete semigroups and monoids appear in the theory of formal languages.

Calculus of Finite Differences, Discrete Calculus or Discrete Analysis

A function defined on an interval of the integers is usually called a sequence. A sequence could be a finite sequence from a data source or an infinite sequence from a discrete dynamical system. Such a discrete function could be defined explicitly by a list (if its domain is finite), or by a formula for its general term, or it could be given implicitly by a recurrence relation or difference equation. Difference equations are similar to a differential equations, but replace differentiation by taking the difference between adjacent terms; they can be used to approximate differential equations or (more often) studied in their own right. Many questions and methods concerning differential equations have counterparts for difference equations. For instance, where there are integral transforms in harmonic analysis for studying continuous functions or analogue signals, there are discrete transforms for discrete functions or digital signals. As well as the discrete metric there are more general discrete or finite metric spaces and finite topological spaces.

Geometry

Discrete geometry and combinatorial geometry are about combinatorial properties of *discrete collections* of geometrical objects. A long-standing topic in discrete geometry is tiling of the plane. Computational geometry applies algorithms to geometrical problems.

Computational geometry applies computer algorithms to representations of geometrical objects.

Topology

Although topology is the field of mathematics that formalizes and generalizes the intuitive notion of "continuous deformation" of objects, it gives rise to many discrete topics; this can be attributed in part to the focus on topological invariants, which themselves usually take discrete values. See combinatorial topology, topological graph theory, topological combinatorics, computational topology, discrete topological space, finite topological space, topology (chemistry).

Operations Research

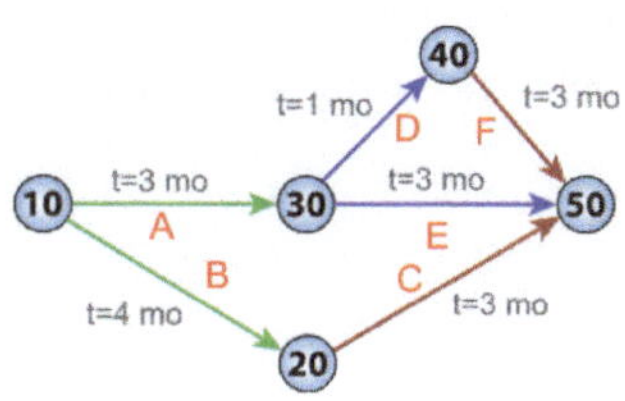

PERT charts like this provide a project management technique based on graph theory.

Operations research provides techniques for solving practical problems in engineering, business, and other fields — problems such as allocating resources to maximize profit, or scheduling project activities to minimize risk. Operations research techniques include linear programming and other areas of optimization, queuing theory, scheduling theory, network theory. Operations research also includes continuous topics such as continuous-time Markov process, continuous-time martingales, process optimization, and continuous and hybrid control theory.

Game Theory, Decision Theory, Utility Theory, Social Choice Theory

	Cooperate	Defect
Cooperate	−1, −1	−10, 0
Defect	0, −10	−5, −5

Payoff matrix for the Prisoner's dilemma, a common example in game theory. One player chooses a row, the other a column; the resulting pair gives their payoffs

Decision theory is concerned with identifying the values, uncertainties and other issues relevant in a given decision, its rationality, and the resulting optimal decision.

Utility theory is about measures of the relative economic satisfaction from, or desirability of, consumption of various goods and services.

Social choice theory is about voting. A more puzzle-based approach to voting is ballot theory.

Game theory deals with situations where success depends on the choices of others, which makes choosing the best course of action more complex. Topics include auction theory and fair division.

Discretization

Discretization concerns the process of transferring continuous models and equations into discrete counterparts, often for the purposes of making calculations easier by using approximations. Numerical analysis provides an important example.

Discrete Analogues of Continuous Mathematics

There are many concepts in continuous mathematics which have discrete versions, such as discrete calculus, discrete probability distributions, discrete Fourier transforms, discrete geometry, discrete logarithms, discrete differential geometry, discrete exterior calculus, discrete Morse theory, difference equations, discrete dynamical systems, and discrete vector measures.

In applied mathematics, discrete modelling is the discrete analogue of continuous modelling. In discrete modelling, discrete formulae are fit to data. A common method in this form of modelling is to use recurrence relation.

Hybrid Discrete and Continuous Mathematics

The time scale calculus is a unification of the theory of difference equations with that of differential equations, which has applications to fields requiring simultaneous modelling of discrete and continuous data. Another way of modeling such a situation is the notion of hybrid dynamical system.

References

- Albert Geoffrey Howson, ed. (1988). Mathematics as a Service Subject. Cambridge University Press. pp. 77–78. ISBN 978-0-521-35395-3.
- Joseph G. Rosenstein. Discrete Mathematics in the Schools. American Mathematical Soc. p. 323. ISBN 978-0-8218-8578-9.
- A. S. Troelstra; H. Schwichtenberg (2000-07-27). Basic Proof Theory. Cambridge University Press. p. 186. ISBN 978-0-521-77911-1.

2

Arithmetic

Arithmetic is the oldest branch of mathematics. It studies numbers; arithmetic is an elementary part of number theory. Some of the aspects of arithmetic discussed in this text are decimal, binary number, indeterminate form, greatest common divisor and least common multiple. The section on arithmetic offers an insightful focus, keeping in minds the aspects of arithmetic.

Arithmetic

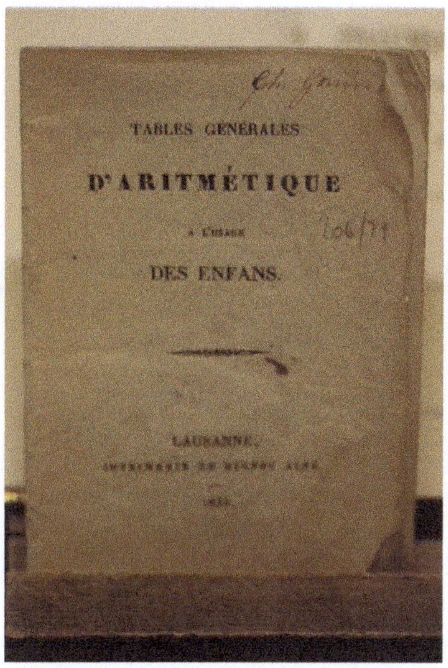

Arithmetic tables for children, Lausanne, 1835

Arithmetic or arithmetics is the oldest and most elementary branch of mathematics. It consists of the study of numbers, especially the properties of the traditional operations between them—addition, subtraction, multiplication and division. Arithmetic is an elementary part of number theory, and number theory is considered to be one of the top-level divisions of modern mathematics, along with algebra, geometry, and analysis. The terms *arithmetic* and *higher arithmetic* were used until the beginning of the 20th century as synonyms for *number theory* and are sometimes still used to refer to a wider part of number theory.

History

The prehistory of arithmetic is limited to a small number of artifacts which may indicate the conception of addition and subtraction, the best-known being the Ishango bone from central Africa, dating from somewhere between 20,000 and 18,000 BC, although its interpretation is disputed.

The earliest written records indicate the Egyptians and Babylonians used all the elementary arithmetic operations as early as 2000 BC. These artifacts do not always reveal the specific process used for solving problems, but the characteristics of the particular numeral system strongly influence the complexity of the methods. The hieroglyphic system for Egyptian numerals, like the later Roman numerals, descended from tally marks used for counting. In both cases, this origin resulted in values that used a decimal base but did not include positional notation. Complex calculations with Roman numerals required the assistance of a counting board or the Roman abacus to obtain the results.

Early number systems that included positional notation were not decimal, including the sexagesimal (base 60) system for Babylonian numerals and the vigesimal (base 20) system that defined Maya numerals. Because of this place-value concept, the ability to reuse the same digits for different values contributed to simpler and more efficient methods of calculation.

The continuous historical development of modern arithmetic starts with the Hellenistic civilization of ancient Greece, although it originated much later than the Babylonian and Egyptian examples. Prior to the works of Euclid around 300 BC, Greek studies in mathematics overlapped with philosophical and mystical beliefs. For example, Nicomachus summarized the viewpoint of the earlier Pythagorean approach to numbers, and their relationships to each other, in his *Introduction to Arithmetic.*

Greek numerals were used by Archimedes, Diophantus and others in a positional notation not very different from ours. Because the ancient Greeks lacked a symbol for zero (until the Hellenistic period), they used three separate sets of symbols. One set for the unit's place, one for the ten's place, and one for the hundred's. Then for the thousand's place they would reuse the symbols for the unit's place, and so on. Their addition algorithm was identical to ours, and their multiplication algorithm was only very slightly different. Their long division algorithm was the same, and the square root algorithm that was once taught in school was known to Archimedes, who may have invented it. He preferred it to Hero's method of successive approximation because, once computed, a digit doesn't change, and the square roots of perfect squares, such as 7485696, terminate immediately as 2736. For numbers with a fractional part, such as 546.934, they used negative powers of 60 instead of negative powers of 10 for the fractional part 0.934. The ancient Chinese used a similar positional notation. Because they also lacked a symbol for zero, they had one set of symbols for the unit's place, and a second set for the ten's place. For the hundred's place they then reused the symbols for the unit's place, and so on. Their symbols were based on the ancient counting rods. It is a complicated question to determine exactly when the Chinese started calculating with positional representation, but it was definitely before 400 BC. The Bishop of Syria, Severus Sebokht (650 AD), "Indians possess a method of calculation that no word can praise enough. Their rational system of mathematics, or of their method of calculation. I mean the system using nine symbols."

Leonardo of Pisa (Fibonacci) in 1200 AD wrote in *Liber Abaci* "The method of the Indians (Modus Indoram) surpasses any known method to compute. It's a marvelous method. They do their computations using nine figures and symbol zero".

The gradual development of Hindu–Arabic numerals independently devised the place-value concept and positional notation, which combined the simpler methods for computations with a decimal base and the use of a digit representing 0. This allowed the system to consistently represent both large and small integers. This approach eventually replaced all other systems. In the early 6th century AD, the Indian mathematician Aryabhata incorporated an existing version of this system in his work, and experimented with different notations. In the 7th century, Brahmagupta established the use of 0 as a separate number and determined the results for multiplication, division, addition and subtraction of zero and all other numbers, except for the result of division by 0. His contemporary, the Syriac bishop Severus Sebokht described the excellence of this system as "... valuable methods of calculation which surpass description". The Arabs also learned this new method and called it *hesab*.

Leibniz's Stepped Reckoner was the first calculator that could perform all four arithmetic operations.

Although the Codex Vigilanus described an early form of Arabic numerals (omitting 0) by 976 AD, Fibonacci was primarily responsible for spreading their use throughout Europe after the publication of his book *Liber Abaci* in 1202. He considered the significance of this "new" representation of numbers, which he styled the "Method of the Indians" (Latin *Modus Indorum*), so fundamental that all related mathematical foundations, including the results of Pythagoras and the algorism describing the methods for performing actual calculations, were "almost a mistake" in comparison.

In the Middle Ages, arithmetic was one of the seven liberal arts taught in universities.

The flourishing of algebra in the medieval Islamic world and in Renaissance Europe was an outgrowth of the enormous simplification of computation through decimal notation.

Various types of tools have been invented and widely used to assist in numeric calculations. Before Renaissance, they were various types of abaci. More recent examples include slide rules, nomograms and mechanical calculators, such as Pascal's calculator. Presently, they have been supplanted by electronic calculators and computers.

Arithmetic Operations

The basic arithmetic operations are addition, subtraction, multiplication and division, although this subject also includes more advanced operations, such as manipulations of percentages, square roots, exponentiation, and logarithmic functions. Arithmetic is performed according to an order of operations. Any set of objects upon which all four arithmetic operations (except division by 0) can be performed, and where these four operations obey the usual laws, is called a field.

Addition (+)

Addition is the basic operation of arithmetic. In its simplest form, addition combines two numbers, the *addends* or *terms*, into a single number, the *sum* of the numbers (Such as $2 + 2 = 4$ or $3 + 5 = 8$).

Adding more than two numbers can be viewed as repeated addition; this procedure is known as summation and includes ways to add infinitely many numbers in an infinite series; repeated addition of the number 1 is the most basic form of counting.

Addition is commutative and associative so the order the terms are added in does not matter. The identity element of addition (the additive identity) is 0, that is, adding 0 to any number yields that same number. Also, the inverse element of addition (the additive inverse) is the opposite of any number, that is, adding the opposite of any number to the number itself yields the additive identity, 0. For example, the opposite of 7 is −7, so $7 + (-7) = 0$.

Addition can be given geometrically as in the following example:

> If we have two sticks of lengths 2 and 5, then if we place the sticks one after the other, the length of the stick thus formed is $2 + 5 = 7$.

Subtraction (−)

Subtraction is the inverse of addition. Subtraction finds the *difference* between two numbers, the *minuend* minus the *subtrahend*. If the minuend is larger than the subtrahend, the difference is positive; if the minuend is smaller than the subtrahend, the difference is negative; if they are equal, the difference is 0.

Subtraction is neither commutative nor associative. For that reason, it is often helpful to look at subtraction as addition of the minuend and the opposite of the subtrahend, that is $a - b = a + (-b)$. When written as a sum, all the properties of addition hold.

There are several methods for calculating results, some of which are particularly advantageous to machine calculation. For example, digital computers employ the method of two's complement. Of great importance is the counting up method by which change is made. Suppose an amount P is given to pay the required amount Q, with P greater than Q. Rather than performing the subtraction $P - Q$ and counting out that amount in change, money is counted out starting at Q and continuing until reaching P. Although the amount counted out must equal the result of the subtraction $P - Q$, the subtraction was never really done and the value of $P - Q$ might still be unknown to the change-maker.

Multiplication (× or · or *)

Multiplication is the second basic operation of arithmetic. Multiplication also combines two numbers into a single number, the *product*. The two original numbers are called the *multiplier* and the *multiplicand*, sometimes both simply called *factors*.

Multiplication may be viewed as a scaling operation. If the numbers are imagined as lying in a line, multiplication by a number, say x, greater than 1 is the same as stretching everything away from 0 uniformly, in such a way that the number 1 itself is stretched to where x was. Similarly, multiplying by a number less than 1 can be imagined as squeezing towards 0. (Again, in such a way that 1 goes to the multiplicand.)

Multiplication is commutative and associative; further it is distributive over addition and subtraction. The multiplicative identity is 1, that is, multiplying any number by 1 yields that same number.

Also, the multiplicative inverse is the reciprocal of any number (except 0; 0 is the only number without a multiplicative inverse), that is, multiplying the reciprocal of any number by the number itself yields the multiplicative identity.

The product of a and b is written as $a \times b$ or $a \cdot b$. When a or b are expressions not written simply with digits, it is also written by simple juxtaposition: ab. In computer programming languages and software packages in which one can only use characters normally found on a keyboard, it is often written with an asterisk: $a * b$.

Division (÷ or /)

Division is essentially the inverse of multiplication. Division finds the *quotient* of two numbers, the *dividend* divided by the *divisor*. Any dividend divided by 0 is undefined. For distinct positive numbers, if the dividend is larger than the divisor, the quotient is greater than 1, otherwise it is less than 1 (a similar rule applies for negative numbers). The quotient multiplied by the divisor always yields the dividend.

Division is neither commutative nor associative. As it is helpful to look at subtraction as addition, it is helpful to look at division as multiplication of the dividend times the reciprocal of the divisor, that is $a \div b = a \times {}^{1}\!/_{b}.$ When written as a product, it obeys all the properties of multiplication.

Decimal Arithmetic

Decimal representation refers exclusively, in common use, to the written numeral system employing arabic numerals as the digits for a radix 10 ("decimal") positional notation; however, any numeral system based on powers of 10, e.g., Greek, Cyrillic, Roman, or Chinese numerals may conceptually be described as "decimal notation" or "decimal representation".

Modern methods for four fundamental operations (addition, subtraction, multiplication and division) were first devised by Brahmagupta of India. This was known during medieval Europe as "Modus Indoram" or Method of the Indians. Positional notation (also known as "place-value notation") refers to the representation or encoding of numbers using the same symbol for the different orders of magnitude (e.g., the "ones place", "tens place", "hundreds place") and, with a radix point, using those same symbols to represent fractions (e.g., the "tenths place", "hundredths place"). For example, 507.36 denotes 5 hundreds (10^2), plus 0 tens (10^1), plus 7 units (10^0), plus 3 tenths (10^{-1}) plus 6 hundredths (10^{-2}).

The concept of 0 as a number comparable to the other basic digits is essential to this notation, as is the concept of 0's use as a placeholder, and as is the definition of multiplication and addition with 0. The use of 0 as a placeholder and, therefore, the use of a positional notation is first attested to in the Jain text from India entitled the *Lokavibhâga*, dated 458 AD and it was only in the early 13th century that these concepts, transmitted via the scholarship of the Arabic world, were introduced into Europe by Fibonacci using the Hindu–Arabic numeral system.

Algorism comprises all of the rules for performing arithmetic computations using this type of written numeral. For example, addition produces the sum of two arbitrary numbers. The result is calculated by the repeated addition of single digits from each number that occupies the same position, proceeding from right to left. An addition table with ten rows and ten columns displays all possible values for each sum. If an individual sum exceeds the value 9, the result is represented with two digits. The rightmost digit is the value for the current position, and the result for the subsequent addition of the digits to the left increases by the value of the second (leftmost) digit, which is always one. This adjustment is termed a *carry* of the value 1.

The process for multiplying two arbitrary numbers is similar to the process for addition. A multiplication table with ten rows and ten columns lists the results for each pair of digits. If an individual product of a pair of digits exceeds 9, the *carry* adjustment increases the result of any subsequent multiplication from digits to the left by a value equal to the second (leftmost) digit, which is any value from 1 to 8 ($9 \times 9 = 81$). Additional steps define the final result.

Similar techniques exist for subtraction and division.

The creation of a correct process for multiplication relies on the relationship between values of adjacent digits. The value for any single digit in a numeral depends on its position. Also, each position to the left represents a value ten times larger than the position to the right. In mathematical terms, the exponent for the radix (base) of 10 increases by 1 (to the left) or decreases by 1 (to the right). Therefore, the value for any arbitrary digit is multiplied by a value of the form 10^n with integer n. The list of values corresponding to all possible positions for a single digit is written as $\{\ldots, 10^2, 10, 1, 10^{-1}, 10^{-2}, \ldots\}$.

Repeated multiplication of any value in this list by 10 produces another value in the list. In mathematical terminology, this characteristic is defined as closure, and the previous list is described as closed under multiplication. It is the basis for correctly finding the results of multiplication using the previous technique. This outcome is one example of the uses of number theory.

Compound Unit Arithmetic

Compound unit arithmetic is the application of arithmetic operations to mixed radix quantities such as feet and inches, gallons and pints, pounds shillings and pence, and so on. Prior to the use of decimal-based systems of money and units of measure, the use of compound unit arithmetic formed a significant part of commerce and industry.

Basic Arithmetic Operations

The techniques used for compound unit arithmetic were developed over many centuries and are well-documented in many textbooks in many different languages. In addition to the basic arithmetic functions encountered in decimal arithmetic, compound unit arithmetic employs three more functions:

- Reduction where a compound quantity is reduced to a single quantity, for example conversion of a distance expressed in yards, feet and inches to one expressed in inches.
- Expansion, the inverse function to reduction, is the conversion of a quantity that is expressed as a single unit of measure to a compound unit, such as expanding 24 oz to 1 lb, 8 oz.
- Normalization is the conversion of a set of compound units to a standard form – for example rewriting "1 ft 13 in" as "2 ft 1 in".

Knowledge of the relationship between the various units of measure, their multiples and their sub-multiples forms an essential part of compound unit arithmetic.

Principles of Compound Unit Arithmetic

There are two basic approaches to compound unit arithmetic:

- Reduction–expansion method where all the compound unit variables are reduced to single unit variables, the calculation performed and the result expanded back to compound units. This approach is suited for automated calculations. A typical example is the handling of time by Microsoft Excel where all time intervals are processed internally as days and decimal fractions of a day.
- On-going normalization method in which each unit is treated separately and the problem is continuously normalized as the solution develops. This approach, which is widely described in classical texts, is best suited for manual calculations. An example of the ongoing normalization method as applied to addition is shown below.

	£	s	d
	4	3	8
	5	17	7
	3	6	10
Answer	13	8	1
Column total	12	26	25
Carried forward	1	2	-
Unnormalised total	13	28	25
Normalisation	-	28÷20	25÷12

The addition operation is carried out from right to left; in this case, pence are processed first, then shillings followed by pounds. The numbers below the "answer line" are intermediate results.

The total in the pence column is 25. Since there are 12 pennies in a shilling, 25 is divided by 12 to give 2 with a remainder of 1. The value "1" is then written to the answer row and the value "2" carried forward to the shillings column. This operation is repeated using the values in the shillings column, with the additional step of adding the value that was carried forward from the pennies column. The intermediate total is divided by 20 as there are 20 shillings in a pound. The pound column is then processed, but as pounds are the largest unit that is being considered, no values are carried forward from the pounds column.

It should be noted that for the sake of simplicity, the example chosen did not have farthings.

Operations in Practice

During the 19th and 20th centuries various aids were developed to aid the manipulation of compound units, particularly in commercial applications. The most common aids were mechanical tills which were adapted in countries such as the United Kingdom to accommodate pounds, shillings, pennies and farthings and "Ready Reckoners" – books aimed at traders that catalogued the results of various routine calculations such as the percentages or multiples of various sums of money. One typical booklet that ran to 150 pages tabulated multiples "from one to ten thousand at the various prices from one farthing to one pound".

A scale calibrated in imperial units with an associated cost display.

The cumbersome nature of compound unit arithmetic has been recognized for many years – in 1586, the Flemish mathematician Simon Stevin published a small pamphlet called *De Thiende* ("the tenth") in which he declared that the universal introduction of decimal coinage, measures, and weights to be merely a question of time while in the modern era, many conversion programs, such as that embedded in the calculator supplied as a standard part of the Microsoft Windows 7 operating system display compound units in a reduced decimal format rather than using an expanded format (i.e. "2.5 ft" is displayed rather than "2 ft 6 in").

Number Theory

Until the 19th century, *number theory* was a synonym of "arithmetic". The addressed problems were directly related to the basic operations and concerned primality, divisibility, and the solution of equations in integers, such as Fermat's last theorem. It appeared that most of these problems, although very elementary to state, are very difficult and may not be solved without very deep mathematics involving concepts and methods from many other branches of mathematics. This led to new branches of number theory such as analytic number theory, algebraic number theory, Diophantine geometry and arithmetic algebraic geometry. Wiles' proof of Fermat's Last Theorem is a typical example of the necessity of sophisticated methods, which go far beyond the classical methods of arithmetic, for solving problems that can be stated in elementary arithmetic.

Arithmetic in Education

Primary education in mathematics often places a strong focus on algorithms for the arithmetic of natural numbers, integers, fractions, and decimals (using the decimal place-value system). This study is sometimes known as algorism.

The difficulty and unmotivated appearance of these algorithms has long led educators to question this curriculum, advocating the early teaching of more central and intuitive mathematical ideas. One notable movement in this direction was the New Math of the 1960s and 1970s, which attempted to teach arithmetic in the spirit of axiomatic development from set theory, an echo of the prevailing trend in higher mathematics.

Also, arithmetic was used by Islamic Scholars in order to teach application of the rulings related to Zakat and Irth. This was done in a book entitled *The Best of Arithmetic* by Abd-al-Fattah-al-Dumyati.

The book begins with the foundations of mathematics and proceeds to its application in the later chapters.

Decimal

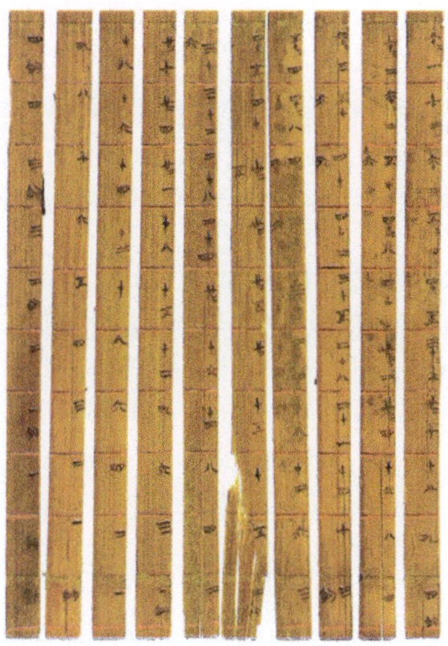

The world's earliest decimal multiplication table was made from bamboo slips, dating from 305 BC, during the Warring States period in China.

The decimal numeral system (also called base 10 or occasionally denary) has ten as its base. It is the numerical base most widely used by modern civilizations.

Decimal notation often refers to a base 10 positional notation such as the Hindu-Arabic numeral system or rod calculus; however, it can also be used more generally to refer to non-positional systems such as Roman or Chinese numerals which are also based on powers of ten.

A decimal number, or just decimal, refers to any number written in decimal notation, although it is more commonly used to refer to numbers that have a fractional part separated from the integer part with a decimal separator (e.g. 11.25).

A decimal may be a terminating decimal, which has a finite fractional part (e.g. 15.600); a repeating decimal, which has an infinite (non-terminating) fractional part made up of a repeating sequence of digits (e.g. 5.123144); or an infinite decimal, which has a fractional part that neither terminates nor has an infinitely repeating pattern (e.g. 3.14159265...). Decimal fractions have terminating decimal representations and other fractions have repeating decimal representations, whereas irrational numbers have infinite non-repeating decimal representations.

Decimal Notation

Decimal notation is the writing of numbers in a base 10 numeral system. Examples are Brahmi numerals, Greek numerals, Hebrew numerals, Roman numerals, and Chinese numerals, as well as the Hindu-Arabic numerals used by speakers of many European languages. Roman numerals have symbols for the decimal powers (1, 10, 100, 1000) and secondary symbols for half these values (5, 50, 500). Brahmi numerals have symbols for the nine numbers 1–9, the nine decades 10–90, plus a symbol for 100 and another for 1000. Chinese numerals have symbols for 1–9, and additional symbols for powers of 10, which in modern usage reach 10^{72}.

However, when people who use Hindu-Arabic numerals speak of decimal notation, they often mean not just decimal numeration, as above, but also decimal fractions, all conveyed as part of a positional system. Positional decimal systems include a zero and use symbols (called digits) for the ten values (0, 1, 2, 3, 4, 5, 6, 7, 8, and 9) to represent any number, no matter how large or how small. These digits are often used with a decimal separator which indicates the start of a fractional part, and with a symbol such as the plus sign + (for positive) or minus sign – (for negative) adjacent to the numeral to indicate whether it is greater or less than zero, respectively.

Positional notation uses positions for each power of ten: units, tens, hundreds, thousands, etc. The position of each digit within a number denotes the multiplier (power of ten) multiplied with that digit—each position has a value ten times that of the position to its right. There were at least two presumably independent sources of positional decimal systems in ancient civilization: the Chinese counting rod system and the Hindu-Arabic numeral system (the latter descended from Brahmi numerals).

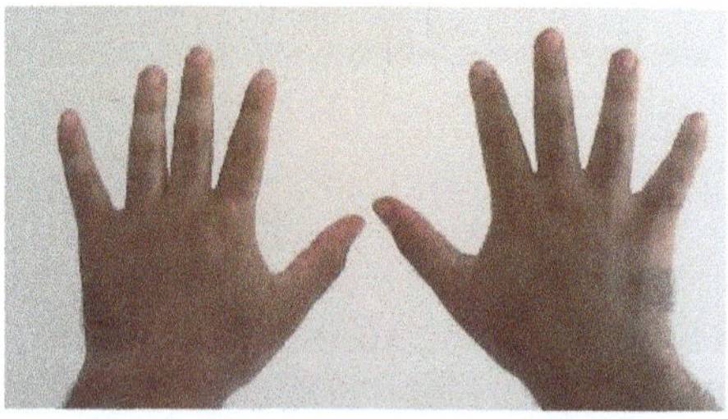

Ten fingers on two hands, the possible starting point of the decimal counting.

Ten is the number which is the count of fingers and thumbs on both hands (or toes on the feet). The English word digit as well as its translation in many languages is also the anatomical term for fingers and toes. In English, decimal (decimus < Lat.) means *tenth*, decimate means *reduce by a tenth*, and denary (denarius < Lat.) means *the unit of ten*.

The symbols for the digits in common use around the globe today are called Arabic numerals by Europeans and Indian numerals by Arabs, the two groups' terms both referring to the culture from which they learned the system. However, the symbols used in different areas are not identical; for instance, Western Arabic numerals (from which the European numerals are derived) differ from the forms used by other Arab cultures.

Decimal Fractions

A decimal fraction is a fraction the denominator of which is a power of ten.

Decimal fractions are commonly expressed in decimal notation rather than fraction notation by discarding the denominator and inserting the decimal separator into the numerator at the position from the right corresponding to the power of ten of the denominator and filling the gap with leading zeros if needed, e.g. decimal fractions 8/10, 1489/100, 24/100000, and 58900/10000 are expressed in decimal notation as 0.8, 14.89, 0.00024, 5.8900 respectively. In English-speaking, some Latin American and many Asian countries, a period (.) or raised period (·) is used as the decimal separator; in many other countries, particularly in Europe, a comma (,) is used.

The integer part, or integral part of a decimal number is the part to the left of the decimal separator. The part from the decimal separator to the right is the *fractional part*. It is usual for a decimal number that consists only of a fractional part (mathematically, a *proper fraction*) to have a leading zero in its notation (its *numeral*). This helps disambiguation between a decimal sign and other punctuation, and especially when the negative number sign is indicated, it helps visualize the sign of the numeral as a whole.

Trailing zeros after the decimal point are not necessary, although in science, engineering and statistics they can be retained to indicate a required precision or to show a level of confidence in the accuracy of the number: Although 0.080 and 0.08 are numerically equal, in engineering 0.080 suggests a measurement with an error of up to one part in two thousand (±0.0005), while 0.08 suggests a measurement with an error of up to one in two hundred.

Other Rational Numbers

Any rational number with a denominator whose only prime factors are 2 and/or 5 may be precisely expressed as a decimal fraction and has a finite decimal expansion.

$1/2 = 0.5$

$1/20 = 0.05$

$1/5 = 0.2$

$1/50 = 0.02$

$1/4 = 0.25$

$1/40 = 0.025$

$1/25 = 0.04$

$1/8 = 0.125$

$1/125 = 0.008$

$1/10 = 0.1$

If a fully reduced rational number's denominator has any prime factors other than 2 or 5, it cannot be expressed as a finite decimal fraction, and has a unique eventually repeating infinite decimal expansion.

$1/3 = 0.333333\ldots$ (with 3 repeating)

$1/9 = 0.111111\ldots$ (with 1 repeating)

$100 - 1 = 99 = 9 \times 11$:

$1/11 = 0.090909\ldots$

$1000 - 1 = 9 \times 111 = 27 \times 37$:

$1/27 = 0.037037037\ldots$

$1/37 = 0.027027027\ldots$

$1/111 = 0.009009009\ldots$

also:

$1/81 = 0.012345679012\ldots$ (with 012345679 repeating)

That a rational number must have a finite or recurring decimal expansion can be seen to be a consequence of the long division algorithm, in that there are at most $q - 1$ possible nonzero remainders on division by q, so that the recurring pattern will have a period less than q. For instance, to find 3/7 by long division:

```
     0.4285714...
  7 ) 3.00000000
      28               30 ÷ 7 = 4 with a remainder of 2
     ---------
       20
       14              20 ÷ 7 = 2 with a remainder of 6
      ---------
        60
        56             60 ÷ 7 = 8 with a remainder of 4
       ---------
         40
         35            40 ÷ 7 = 5 with a remainder of 5
        ---------
          50
          49           50 ÷ 7 = 7 with a remainder of 1
         ---------
           10
            7          10 ÷ 7 = 1 with a remainder of 3
          ---------
            30
            28         30 ÷ 7 = 4 with a remainder of 2
           ---------
            20
```

etc.

The converse to this observation is that every recurring decimal represents a rational number p/q. This is a consequence of the fact that the recurring part of a decimal representation is, in fact, an infinite geometric series which will sum to a rational number. For instance,

$$0.0123123123\cdots = \frac{123}{10000}\sum_{k=0}^{\infty} 0.001^k = \frac{123}{10000}\,\frac{1}{1-0.001} = \frac{123}{9990} = \frac{41}{3330}$$

Real Numbers

Every real number has a (possibly infinite) decimal representation; i.e., it can be written as

$$x = \text{sign} \sum_{i \in \mathbb{Z}} a_i 10^i$$

where

- sign $\in \{+,-\}$, which is related to the sign function,
- Z is the set of all integers (positive, negative, and zero), and
- $a_i \in \{ 0,1,\ldots,9 \}$ for all $i \in Z$ are its decimal digits, equal to zero for all i greater than some number (that number being the common logarithm of $|x|$).

Such a sum converges as more and more negative values of i are included, even if there are infinitely many non-zero a_i.

Rational numbers (e.g., p/q) with prime factors in the denominator other than 2 and 5 (when reduced to simplest terms) have a unique recurring decimal representation.

Non-uniqueness of Decimal Representation

Consider those rational numbers which have only the factors 2 and 5 in the denominator, i.e., which can be written as $p/2^a5^b$. In this case there is a terminating decimal representation. For instance, 1/1 = 1, 1/2 = 0.5, 3/5 = 0.6, 3/25 = 0.12 and 1306/1250 = 1.0448. Such numbers are the only real numbers which do not have a unique decimal representation, as they can also be written as a representation that has a recurring 9, for instance 1 = 0.99999..., 1/2 = 0.499999..., etc. The number 0 = 0/1 is special in that it has no representation with recurring 9.

This leaves the irrational numbers. They also have unique infinite decimal representations, and can be characterised as the numbers whose decimal representations neither terminate nor recur.

So in general the decimal representation is unique, if one excludes representations that end in a recurring 9.

The same trichotomy holds for other base-n positional numeral systems:

- Terminating representation: rational where the denominator divides some n^k
- Recurring representation: other rational
- Non-terminating, non-recurring representation: irrational

A version of this even holds for irrational-base numeration systems, such as golden mean base representation.

Decimal Computation

Decimal computation was carried out in ancient times in many ways, typically in rod calculus, with decimal multiplication table used in ancient China and with sand tables in India and Middle East or with a variety of abaci.

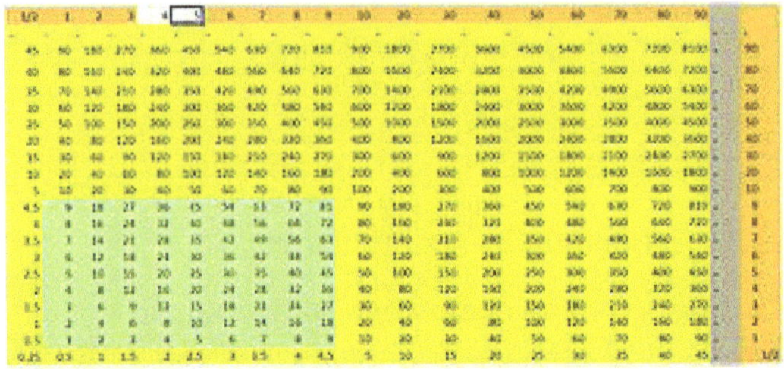

Diagram of the world's earliest decimal multiplication table (c. 305 BC) from the Warring States period

Modern computer hardware and software systems commonly use a binary representation internally (although many early computers, such as the ENIAC or the IBM 650, used decimal representation internally). For external use by computer specialists, this binary representation is sometimes presented in the related octal or hexadecimal systems.

For most purposes, however, binary values are converted to or from the equivalent decimal values for presentation to or input from humans; computer programs express literals in decimal by default. (123.1, for example, is written as such in a computer program, even though many computer languages are unable to encode that number precisely.)

Both computer hardware and software also use internal representations which are effectively decimal for storing decimal values and doing arithmetic. Often this arithmetic is done on data which are encoded using some variant of binary-coded decimal, especially in database implementations, but there are other decimal representations in use (such as in the new IEEE 754 Standard for Floating-Point Arithmetic).

Decimal arithmetic is used in computers so that decimal fractional results can be computed exactly, which is not possible using a binary fractional representation. This is often important for financial and other calculations.

History

Many ancient cultures calculated with numerals based on ten, as the human hands have 10 digits. Egyptian hieroglyphs, in evidence since around 3000 BC, used a purely decimal system, just as the Cretan hieroglyphs (ca. 1625–1500 BC) of the Minoans whose numerals are closely based on the Egyptian model. The decimal system was handed down to the consecutive Bronze Age cultures of Greece, including Linear A (ca. 18th century BC–1450 BC) and Linear B (ca. 1375–1200 BC) — the number system of classical Greece also used powers of ten, including, like the Roman numerals did, an intermediate base of 5. Notably, the polymath Archimedes (ca. 287–212 BC) invented a decimal positional system in his Sand Reckoner which was based on 10^8 and later led the German mathematician Carl Friedrich Gauss to lament what heights science would have already reached in his days if Archimedes had fully realized the potential of his ingenious discovery. The Hittites hieroglyphs (since 15th century BC), just like the Egyptian and early numerals in Greece, was strictly decimal.

The Egyptian hieratic numerals, the Greek alphabet numerals, the Hebrew alphabet numerals, the Roman numerals, the Chinese numerals and early Indian Brahmi numerals are all non-positional decimal systems, and required large numbers of symbols. For instance, Egyptian numerals used different symbols for 10, 20, to 90, 100, 200, to 900, 1000, 2000, 3000, 4000, to 10,000. The world's earliest positional decimal system was the Chinese rod calculus

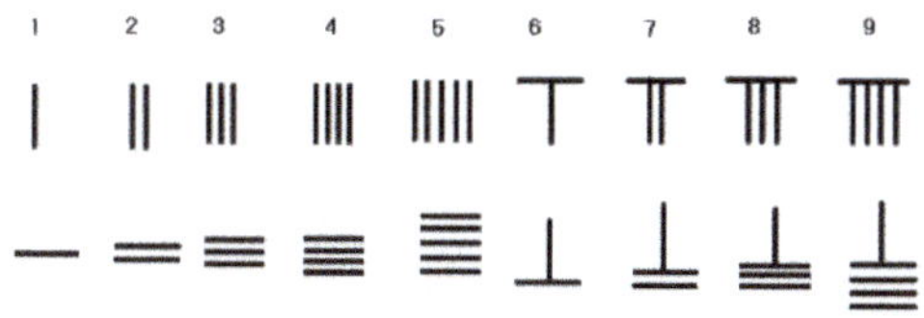

The world's earliest positional decimal system
Upper row vertical form
Lower row horizontal form

History of Decimal Fractions

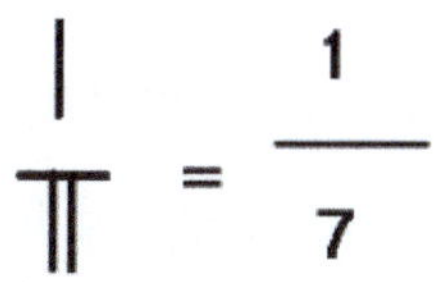

counting rod decimal fraction 1/7

According to Joseph Needham and Lam Lay Yong, decimal fractions were first developed and used by the Chinese in the 1st century BC, and then spread to the Middle East and from there to Europe. The written Chinese decimal fractions were non-positional. However, counting rod fractions were positional.

Qin Jiushao in his book Mathematical Treatise in Nine Sections (1247) denoted 0.96644 by

寸

, meaning

寸

096644

J. Lennart Berggren notes that positional decimal fractions appear for the first time in a book by the Arab mathematician Abu'l-Hasan al-Uqlidisi written in the 10th century. The Jewish mathematician Immanuel Bonfils used decimal fractions around 1350, anticipating Simon Stevin, but did not develop any notation to represent them. The Persian mathematician Jamshīd al-Kāshī claimed to have discovered decimal fractions himself in the 15th century.Al Khwarizmi introduced fraction to Islamic countries in the early 9th century, his fraction presentation was an exact copy of traditional Chinese mathematical fraction from The Mathematical Classic of Sunzi. This form of fraction with numerator on top and denominator at bottom without a horizontal bar was also used by 10th century Abu'l-Hasan al-Uqlidisi and 15th century Jamshīd al-Kāshī's work "Arithmetic Key".

Number: 184.54290

Simon Stevin's notation: 184⓪5①4②2③9④0

A forerunner of modern European decimal notation was introduced by Simon Stevin in the 16th century.

Natural Languages

The ingenious method of expressing every possible number using a set of ten symbols emerged in India. Several Indian languages show a straightforward decimal system. Many Indo-Aryan and Dravidian languages have numbers between 10 and 20 expressed in a regular pattern of addition to 10.

The Hungarian language also uses a straightforward decimal system. All numbers between 10 and 20 are formed regularly (e.g. 11 is expressed as "tizenegy" literally "one on ten"), as with those between 20 and 100 (23 as "huszonhárom" = "three on twenty").

A straightforward decimal rank system with a word for each order (10 十, 100 百, 1000 千, 10,000 万), and in which 11 is expressed as *ten-one* and 23 as *two-ten-three*, and 89,345 is expressed as 8 (ten thousands) 万 9 (thousand) 千 3 (hundred) 百 4 (tens) 十 5 is found in Chinese, and in Vietnamese with a few irregularities. Japanese, Korean, and Thai have imported the Chinese decimal system. Many other languages with a decimal system have special words for the numbers between 10 and 20, and decades. For example, in English 11 is "eleven" not "ten-one" or "one-teen".

Incan languages such as Quechua and Aymara have an almost straightforward decimal system, in which 11 is expressed as *ten with one* and 23 as *two-ten with three*.

Some psychologists suggest irregularities of the English names of numerals may hinder children's counting ability.

Other Bases

Some cultures do, or did, use other bases of numbers.

- Pre-Columbian Mesoamerican cultures such as the Maya used a base-20 system (presumably using all twenty fingers and toes).
- The Yuki language in California and the Pamean languages in Mexico have octal (base-8) systems because the speakers count using the spaces between their fingers rather than the fingers themselves.
- The existence of a non-decimal base in the earliest traces of the Germanic languages, is attested by the presence of words and glosses meaning that the count is in decimal (cognates to ten-count or tenty-wise), such would be expected if normal counting is not decimal, and unusual if it were. Where this counting system is known, it is based on the long hundred of

120 in number, and a long thousand of 1200 in number. The descriptions like 'long' only appear after the small hundred of 100 in number appeared with the Christians. Gordon's Introduction to Old Norse p 293, gives number names that belong to this system. An expression cognate to 'one hundred and eighty' is translated to 200, and the cognate to 'two hundred' is translated at 240. Goodare details the use of the long hundred in Scotland in the Middle Ages, giving examples, calculations where the carry implies i C (i.e. one hundred) as 120, etc. That the general population were not alarmed to encounter such numbers suggests common enough use. It is also possible to avoid hundred-like numbers by using intermediate units, such as stones and pounds, rather than a long count of pounds. Goodare gives examples of numbers like vii score, where one avoids the hundred by using extended scores. There is also a paper by W.H. Stevenson, on 'Long Hundred and its uses in England'.

- Many or all of the Chumashan languages originally used a base-4 counting system, in which the names for numbers were structured according to multiples of 4 and 16.
- Many languages use quinary (base-5) number systems, including Gumatj, Nunggubuyu, Kuurn Kopan Noot and Saraveca. Of these, Gumatj is the only true 5–25 language known, in which 25 is the higher group of 5.
- Some Nigerians use duodecimal systems. So did some small communities in India and Nepal, as indicated by their languages.
- The Huli language of Papua New Guinea is reported to have base-15 numbers. *Ngui* means 15, *ngui ki* means 15 × 2 = 30, and *ngui ngui* means 15 × 15 = 225.
- Umbu-Ungu, also known as Kakoli, is reported to have base-24 numbers. *Tokapu* means 24, *tokapu talu* means 24 × 2 = 48, and *tokapu tokapu* means 24 × 24 = 576.
- Ngiti is reported to have a base-32 number system with base-4 cycles.
- The Ndom language of Papua New Guinea is reported to have base-6 numerals. *Mer* means 6, *mer an thef* means 6 × 2 = 12, *nif* means 36, and *nif thef* means 36×2 = 72.

Binary Number

In mathematics and digital electronics, a binary number is a number expressed in the binary numeral system or base-2 numeral system which represents numeric values using two different symbols: typically 0 (zero) and 1 (one). The base-2 system is a positional notation with a radix of 2. Because of its straightforward implementation in digital electronic circuitry using logic gates, the binary system is used internally by almost all modern computers and computer-based devices. Each digit is referred to as a bit.

History

The modern binary number system was devised by Gottfried Leibniz in 1679 and appears in his article *Explication de l'Arithmétique Binaire* (published in 1703). Systems related to binary numbers have appeared earlier in multiple cultures including ancient Egypt, China, and India. Leibniz was specifically inspired by the Chinese I Ching.

Egypt

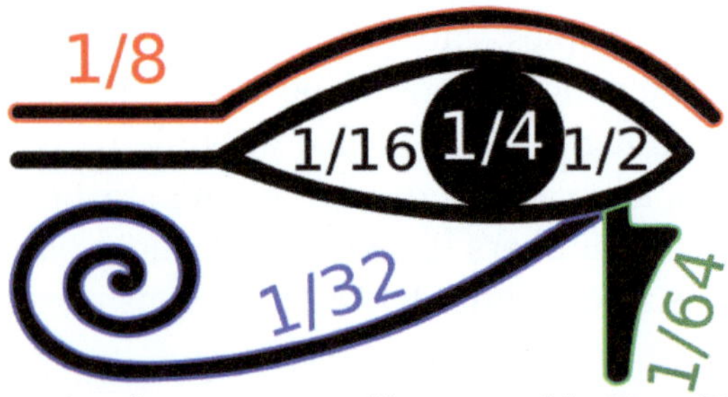

Arithmetic values represented by parts of the Eye of Horus

The scribes of ancient Egypt used two different systems for their fractions, Egyptian fractions (not related to the binary number system) and Horus-Eye fractions (so called because many historians of mathematics believe that the symbols used for this system could be arranged to form the eye of Horus, although this has been disputed). Horus-Eye fractions are a binary numbering system for fractional quantities of grain, liquids, or other measures, in which a fraction of a hekat is expressed as a sum of the binary fractions 1/2, 1/4, 1/8, 1/16, 1/32, and 1/64. Early forms of this system can be found in documents from the Fifth Dynasty of Egypt, approximately 2400 BC, and its fully developed hieroglyphic form dates to the Nineteenth Dynasty of Egypt, approximately 1200 BC.

The method used for ancient Egyptian multiplication is also closely related to binary numbers. In this method, multiplying one number by a second is performed by a sequence of steps in which a value (initially the first of the two numbers) is either doubled or has the first number added back into it; the order in which these steps are to be performed is given by the binary representation of the second number. This method can be seen in use, for instance, in the Rhind Mathematical Papyrus, which dates to around 1650 BC.

China

Daoist Bagua

The I Ching dates from the 9th century BC in China. The binary notation in the *I Ching* is used to interpret its quaternary divination technique.

It is based on taoistic duality of yin and yang. eight trigrams (Bagua) and a set of 64 hexagrams ("sixty-four" gua), analogous to the three-bit and six-bit binary numerals, were in use at least as early as the Zhou Dynasty of ancient China.

The contemporary scholar Shao Yong rearranged the hexagrams in a format that resembles modern binary numbers, although he did not intend his arrangement to be used mathematically. Viewing the least significant bit on top of single hexagrams in Shao Yong's square and reading along rows either from bottom right to top left with solid lines as 0 and broken lines as 1 or from top left to bottom right with solid lines as 1 and broken lines as 0 hexagrams can be interpreted as sequence from 0 to 63.

India

The Indian scholar Pingala (c. 2nd century BC) developed a binary system for describing prosody. He used binary numbers in the form of short and long syllables (the latter equal in length to two short syllables), making it similar to Morse code. Pingala's Hindu classic titled Chanda śāstra (8.23) describes the formation of a matrix in order to give a unique value to each meter. The binary representations in Pingala's system increases towards the right, and not to the left like in the binary numbers of the modern, Western positional notation.

Other Cultures

The residents of the island of Mangareva in French Polynesia were using a hybrid binary-decimal system before 1450. Slit drums with binary tones are used to encode messages across Africa and Asia. Sets of binary combinations similar to the I Ching have also been used in traditional African divination systems such as Ifá as well as in medieval Western geomancy. The base-2 system utilized in geomancy had long been widely applied in sub-Saharan Africa.

Western Predecessors to Leibniz

In 1605 Francis Bacon discussed a system whereby letters of the alphabet could be reduced to sequences of binary digits, which could then be encoded as scarcely visible variations in the font in any random text. Importantly for the general theory of binary encoding, he added that this method could be used with any objects at all: "provided those objects be capable of a twofold difference only; as by Bells, by Trumpets, by Lights and Torches, by the report of Muskets, and any instruments of like nature".

Leibniz and the I Ching

The full title of Leibniz's article is translated into English as the *"Explanation of the Binary Arithmetic, which uses only the characters 1 and 0, with some remarks on its usefulness, and on the light it throws on the ancient Chinese figures of Fu Xi"*. (1703). Leibniz's system uses 0 and 1, like the modern binary numeral system. An example of Leibniz's binary numeral system is as follows:

0 0 0 1 numerical value 2^0

0 0 1 0 numerical value 2^1

0 1 0 0 numerical value 2^2

1 0 0 0 numerical value 2^3

Gottfried Leibniz

Leibniz interpreted the hexagrams of the I Ching as evidence of binary calculus. As a Sinophile, Leibniz was aware of the I Ching, noted with fascination how its hexagrams correspond to the binary numbers from 0 to 111111, and concluded that this mapping was evidence of major Chinese accomplishments in the sort of philosophical mathematics he admired. Leibniz was first introduced to the *I Ching* through his contact with the French Jesuit Joachim Bouvet, who visited China in 1685 as a missionary. Leibniz saw the *I Ching* hexagrams as an affirmation of the universality of his own religious beliefs as a Christian. Binary numerals were central to Leibniz's theology. He believed that binary numbers were symbolic of the Christian idea of *creatio ex nihilo* or creation out of nothing.

[A concept that] is not easy to impart to the pagans, is the creation *ex nihilo* through God's almighty power. Now one can say that nothing in the world can better present and demonstrate this power than the origin of numbers, as it is presented here through the simple and unadorned presentation of One and Zero or Nothing.

—Leibniz's letter to the Duke of Brunswick attached with the I Ching hexagrams

Later Developments

In 1854, British mathematician George Boole published a landmark paper detailing an algebraic system of logic that would become known as Boolean algebra. His logical calculus was to become instrumental in the design of digital electronic circuitry.

In 1937, Claude Shannon produced his master's thesis at MIT that implemented Boolean algebra and binary arithmetic using electronic relays and switches for the first time in history. Entitled *A Symbolic Analysis of Relay and Switching Circuits*, Shannon's thesis essentially founded practical digital circuit design.

In November 1937, George Stibitz, then working at Bell Labs, completed a relay-based computer he dubbed the "Model K" (for "Kitchen", where he had assembled it), which calculated using binary addition. Bell Labs authorized a full research program in late 1938 with Stibitz at the helm. Their Complex Number Computer, completed 8 January 1940, was able to calculate complex numbers. In a demonstration to the American Mathematical Society conference at Dartmouth College on 11 September 1940, Stibitz was able to send the Complex Number Calculator remote commands over telephone lines by a teletype. It was the first computing machine ever used remotely over a phone line. Some participants of the conference who witnessed the demonstration were John von Neumann, John Mauchly and Norbert Wiener, who wrote about it in his memoirs.

George Boole

The Z1 computer, which was designed and built by Konrad Zuse between 1935 and 1938, used Boolean logic and binary floating point numbers.

Representation

Any number can be represented by any sequence of bits (binary digits), which in turn may be represented by any mechanism capable of being in two mutually exclusive states. Any of the following rows of symbols can be interpreted as the binary numeric value of 667.

1	0	1	0	0	1	1	0	1	1
¦	–	¦	–	–	¦	¦	–	¦	¦
x	o	x	o	o	x	x	o	x	x
y	n	y	n	n	y	y	n	y	y

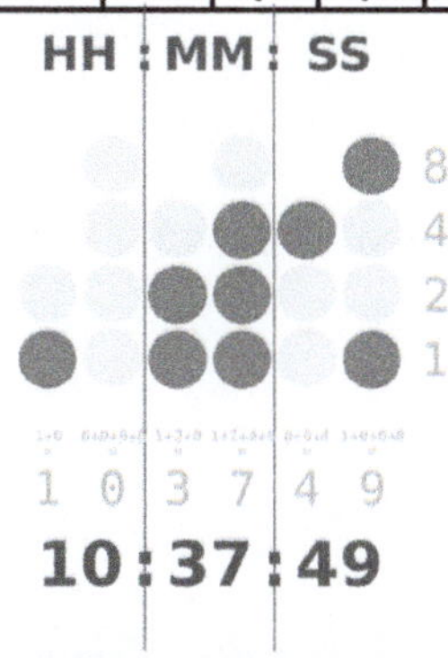

A binary clock might use LEDs to express binary values. In this clock, each column of LEDs shows a binary-coded decimal numeral of the traditional sexagesimal time.

The numeric value represented in each case is dependent upon the value assigned to each symbol.

In a computer, the numeric values may be represented by two different voltages; on a magnetic disk, magnetic polarities may be used. A "positive", "yes", or "on" state is not necessarily equivalent to the numerical value of one; it depends on the architecture in use.

In keeping with customary representation of numerals using Arabic numerals, binary numbers are commonly written using the symbols 0 and 1. When written, binary numerals are often subscripted, prefixed or suffixed in order to indicate their base, or radix. The following notations are equivalent:

- 100101 binary (explicit statement of format)
- 100101b (a suffix indicating binary format; also known as Intel convention)
- 100101B (a suffix indicating binary format)
- bin 100101 (a prefix indicating binary format)
- 100101_2 (a subscript indicating base-2 (binary) notation)
- %100101 (a prefix indicating binary format; also known as Motorola convention)
- 0b100101 (a prefix indicating binary format, common in programming languages)
- 6b100101 (a prefix indicating number of bits in binary format, common in programming languages)

When spoken, binary numerals are usually read digit-by-digit, in order to distinguish them from decimal numerals. For example, the binary numeral 100 is pronounced *one zero zero*, rather than *one hundred*, to make its binary nature explicit, and for purposes of correctness. Since the binary numeral 100 represents the value four, it would be confusing to refer to the numeral as *one hundred* (a word that represents a completely different value, or amount). Alternatively, the binary numeral 100 can be read out as "four" (the correct *value*), but this does not make its binary nature explicit.

Counting in Binary

Decimal pattern	Binary number
0	0
1	1
2	10
3	11
4	100
5	101
6	110
7	111
8	1000
9	1001
10	1010

11	1011
12	1100
13	1101
14	1110
15	1111

Counting in binary is similar to counting in any other number system. Beginning with a single digit, counting proceeds through each symbol, in increasing order. Before examining binary counting, it is useful to briefly discuss the more familiar decimal counting system as a frame of reference.

Decimal Counting

Decimal counting uses the ten symbols *0* through *9*. Counting begins with the incremental substitution of the least significant digit (rightmost digit) which is often called the *first digit*. When the available symbols for this position are exhausted, the least significant digit is reset to *0*, and the next digit of higher significance (one position to the left) is incremented (*overflow*), and incremental substitution of the low-order digit resumes. This method of reset and overflow is repeated for each digit of significance. Counting progresses as follows:

> 000, 001, 002, ... 007, 008, 009, (rightmost digit is reset to zero, and the digit to its left is incremented)
>
> 010, 011, 012, ...
>
> ...
>
> 090, 091, 092, ... 097, 098, 099, (rightmost two digits are reset to zeroes, and next digit is incremented)
>
> 100, 101, 102, ...

Binary Counting

2^4	2^3	2^2	2^1	2^0	
16	8	4	2	1	
0	0	0	0	0	00

This counter shows how to count in binary from numbers zero through thirty-one.

Binary counting follows the same procedure, except that only the two symbols *0* and *1* are available. Thus, after a digit reaches 1 in binary, an increment resets it to 0 but also causes an increment of the next digit to the left:

> 0000,
>
> 0001, (rightmost digit starts over, and next digit is incremented)
>
> 0010, 0011, (rightmost two digits start over, and next digit is incremented)

0100, 0101, 0110, 0111, (rightmost three digits start over, and the next digit is incremented)

1000, 1001, 1010, 1011, 1100, 1101, 1110, 1111 ...

In the binary system, each digit represents an increasing power of 2, with the rightmost digit representing 2^0, the next representing 2^1, then 2^2, and so on. The equivalent decimal representation of a binary number is sum of the powers of 2 which each digit represents. For example, the binary number 100101 is converted to decimal form as follows:

$$100101_2 = [(1) \times 2^5] + [(0) \times 2^4] + [(0) \times 2^3] + [(1) \times 2^2] + [(0) \times 2^1] + [(1) \times 2^0]$$

$$100101_2 = [1 \times 32] + [0 \times 16] + [0 \times 8] + [1 \times 4] + [0 \times 2] + [1 \times 1]$$

$$100101_2 = 37_{10}$$

Fractions

Fractions in binary only terminate if the denominator has 2 as the only prime factor. As a result, 1/10 does not have a finite binary representation, and this causes 10 × 0.1 not to be precisely equal to 1 in floating point arithmetic. As an example, to interpret the binary expression for 1/3 = .010101..., this means: $1/3 = 0 \times 2^{-1} + 1 \times 2^{-2} + 0 \times 2^{-3} + 1 \times 2^{-4} + \ldots = 0.3125 + \ldots$ An exact value cannot be found with a sum of a finite number of inverse powers of two, the zeros and ones in the binary representation of 1/3 alternate forever.

Fraction	Decimal	Binary	Fractional approximation
1/1	1 or 0.999...	1 or 0.111...	1/2 + 1/4 + 1/8...
1/2	0.5 or 0.4999...	0.1 or 0.0111...	1/4 + 1/8 + 1/16 . . .
1/3	0.333...	0.010101...	1/4 + 1/16 + 1/64 . . .
1/4	0.25 or 0.24999...	0.01 or 0.00111...	1/8 + 1/16 + 1/32 . . .
1/5	0.2 or 0.1999...	0.00110011...	1/8 + 1/16 + 1/128 . . .
1/6	0.1666...	0.0010101...	1/8 + 1/32 + 1/128 . . .
1/7	0.142857142857...	0.001001...	1/8 + 1/64 + 1/512 . . .
1/8	0.125 or 0.124999...	0.001 or 0.000111...	1/16 + 1/32 + 1/64 . . .
1/9	0.111...	0.000111000111...	1/16 + 1/32 + 1/64 . . .
1/10	0.1 or 0.0999...	0.000110011...	1/16 + 1/32 + 1/256 . . .
1/11	0.090909...	0.00010111010001011101...	1/16 + 1/64 + 1/128 . . .
1/12	0.08333...	0.00010101...	1/16 + 1/64 + 1/256 . . .
1/13	0.076923076923...	0.000100111011000100111011...	1/16 + 1/128 + 1/256 . . .
1/14	0.0714285714285...	0.0001001001...	1/16 + 1/128 + 1/1024 . . .
1/15	0.0666...	0.00010001...	1/16 + 1/256 . . .
1/16	0.0625 or 0.0624999...	0.0001 or 0.0000111...	1/32 + 1/64 + 1/128 . . .

Binary Arithmetic

Arithmetic in binary is much like arithmetic in other numeral systems. Addition, subtraction, multiplication, and division can be performed on binary numerals.

Addition

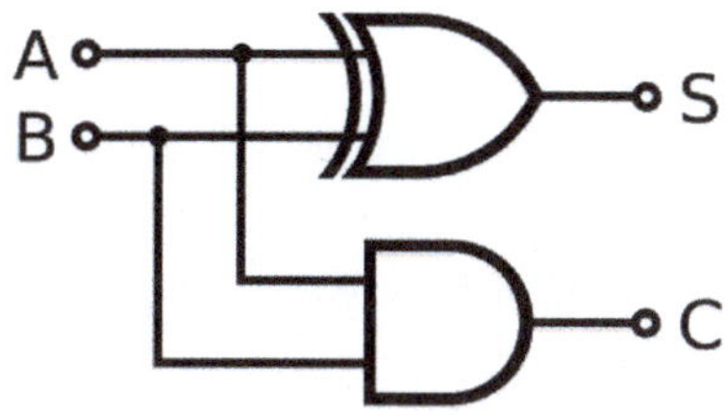

The circuit diagram for a binary half adder, which adds two bits together, producing sum and carry bits.

The simplest arithmetic operation in binary is addition. Adding two single-digit binary numbers is relatively simple, using a form of carrying:

$0 + 0 \rightarrow 0$

$0 + 1 \rightarrow 1$

$1 + 0 \rightarrow 1$

$1 + 1 \rightarrow 0$, carry 1 (since $1 + 1 = 2 = 0 + (1 \times 2^1)$)

Adding two "1" digits produces a digit "0", while 1 will have to be added to the next column. This is similar to what happens in decimal when certain single-digit numbers are added together; if the result equals or exceeds the value of the radix (10), the digit to the left is incremented:

$5 + 5 \rightarrow 0$, carry 1 (since $5 + 5 = 10 = 0 + (1 \times 10^1)$)

$7 + 9 \rightarrow 6$, carry 1 (since $7 + 9 = 16 = 6 + (1 \times 10^1)$)

This is known as *carrying*. When the result of an addition exceeds the value of a digit, the procedure is to "carry" the excess amount divided by the radix (that is, 10/10) to the left, adding it to the next positional value. This is correct since the next position has a weight that is higher by a factor equal to the radix. Carrying works the same way in binary:

```
   1 1 1 1 1    (carried digits)
     0 1 1 0 1
  +  1 0 1 1 1
  -------------
=  1 0 0 1 0 0 = 36
```

row shows the carry bits used. Starting in the rightmost column, $1 + 1 = 10_2$. The 1 is carried to the left, and the 0 is written at the bottom of the rightmost column. The second column from the right is added: $1 + 0 + 1 = 10_2$ again; the 1 is carried, and 0 is written at the bottom. The third column: $1 + 1 + 1 = 11_2$. This time, a 1 is carried, and a 1 is written in the bottom row. Proceeding like this gives the final answer 100100_2 (36 decimal).

When computers must add two numbers, the rule that: x xor y = (x + y) mod 2 for any two bits x and y allows for very fast calculation, as well.

Long Carry Method

A simplification for many binary addition problems is the Long Carry Method or Brookhouse Method of Binary Addition. This method is generally useful in any binary addition where one of the numbers contains a long "string" of ones. It is based on the simple premise that under the binary system, when given a "string" of digits composed entirely of n ones (*where: n* is any integer length), adding 1 will result in the number 1 followed by a string of n zeros. That concept follows, logically, just as in the decimal system, where adding 1 to a string of n 9s will result in the number 1 followed by a string of n 0s:

```
   Binary                       Decimal
    11111                        99999
 +      1       likewise     +       1
 ---------                   ---------
   100000                       100000
```

Such long strings are quite common in the binary system. From that one finds that large binary numbers can be added using two simple steps, without excessive carry operations. In the following example, two numerals are being added together: 1110111110_2 (958_{10}) and 1010110011_2 (691_{10}), using the traditional carry method on the left, and the long carry method on the right:

```
   Traditional Carry                          Long Carry Method
   Method

   111   11111            vs.               1  ←    1 ←
     11 10111110   (carried digits)          111 0 11 1 11 0   (struck through: 111, 11 1 11)
 +   1010110011                          +   1010110 011
 ----------------                        ----------------
 =  11001110001                          =  11001110001
```

carry the 1 until it is one digit past the "string" belowcross out the "string", and cross out the digit that was added to it

The top row shows the carry bits used. Instead of the standard carry from one column to the next, the lowest-ordered "1" with a "1" in the corresponding place value beneath it may be added and a "1" may be carried to one digit past the end of the series. The "used" numbers must be crossed off, since they are already added. Other long strings may likewise be cancelled using the same technique. Then, simply add together any remaining digits normally. Proceeding in this manner gives the final answer of 11001110001_2 (1649_{10}). In our simple example using small numbers, the traditional carry method required eight carry operations, yet the long carry method required only two, representing a substantial reduction of effort.

Addition Table

	0	1	10	11	100
0	0	1	10	11	100
1	1	10	11	100	101
10	10	11	100	101	110
11	11	100	101	110	111
100	100	101	110	111	1000

The binary addition table is similar, but not the same, as the truth table of the logical disjunction operation . The difference is that , while 1v1=1, while 1+1=10.

Subtraction

Further information: signed number representations and two's complement

Subtraction works in much the same way:

$0 - 0 \rightarrow 0$

$0 - 1 \rightarrow 1$, borrow 1

$1 - 0 \rightarrow 1$

$1 - 1 \rightarrow 0$

Subtracting a "1" digit from a "0" digit produces the digit "1", while 1 will have to be subtracted from the next column. This is known as *borrowing*. The principle is the same as for carrying. When the result of a subtraction is less than 0, the least possible value of a digit, the procedure is to "borrow" the deficit divided by the radix (that is, 10/10) from the left, subtracting it from the next positional value.

```
   *   ***        (starred columns are borrowed from)
  1101110
-   10111
----------
= 1010111
```

```
 *              (starred columns are borrowed from)
 1 01 1111
+  101011
----------
= 0110100
```

Subtracting a positive number is equivalent to *adding* a negative number of equal absolute value. Computers use signed number representations to handle negative numbers—most commonly the two's complement notation. Such representations eliminate the need for a separate "subtract" operation. Using two's complement notation subtraction can be summarized by the following formula:

A − B = A + not B + 1

Multiplication

Multiplication in binary is similar to its decimal counterpart. Two numbers *A* and *B* can be multiplied by partial products: for each digit in *B*, the product of that digit in *A* is calculated and written on a new line, shifted leftward so that its rightmost digit lines up with the digit in *B* that was used. The sum of all these partial products gives the final result.

Since there are only two digits in binary, there are only two possible outcomes of each partial multiplication:

- If the digit in *B* is 0, the partial product is also 0
- If the digit in *B* is 1, the partial product is equal to *A*

For example, the binary numbers 1011 and 1010 are multiplied as follows:

```
       1011      (A)
     x 1010       (B)
     -------
       0000    ← Corresponds to the rightmost 'zero' in B
  +   1011     ← Corresponds to the next 'one' in B
  +  0000
  + 1011
  ------------
  = 1101110
```

Binary numbers can also be multiplied with bits after a binary point:

```
          101.101       A (5.625 in decimal)
        x 110.01        B (6.25  in decimal)
      ---------------
            1.01101     ← Corresponds to a 'one' in B
  +        00.0000      ← Corresponds to a 'zero' in B
  +       000.000
  +      1011.01
  +     10110.1
  --------------------
  =    100011.00101     (35.15625 in decimal)
```

Booth's multiplication algorithm.

Multiplication Table

	0	1
0	0	0
1	0	1

The binary multiplication table is the same as the truth table of the logical conjunction operation

Division

Long division in binary is again similar to its decimal counterpart.

In the example below, the divisor is 101_2, or 5 decimal, while the dividend is 11011_2, or 27 decimal. The procedure is the same as that of decimal long division; here, the divisor 101_2 goes into the first three digits 110_2 of the dividend one time, so a "1" is written on the top line. This result is multiplied by the divisor, and subtracted from the first three digits of the dividend; the next digit (a "1") is included to obtain a new three-digit sequence:

```
          1
101 ) 11011
     - 101
     ---------
       001
```

The procedure is then repeated with the new sequence, continuing until the digits in the dividend have been exhausted:

```
          101
101 ) 11011
     - 101
     ---------
        111
      - 101
      ---------
         10
```

Thus, the quotient of 11011_2 divided by 101_2 is 101_2, as shown on the top line, while the remainder, shown on the bottom line, is 10_2. In decimal, 27 divided by 5 is 5, with a remainder of 2.

Square Root

The process of taking a binary square root digit by digit is the same as for a decimal square root, and is explained here. An example is:

```
            1001
         √ 1010001
           1
           ---------
101         01
             0
           ---------
1001         100
               0
             ------------
10001        10001
             10001
             -------------
                 0
```

Bitwise Operations

Though not directly related to the numerical interpretation of binary symbols, sequences of bits may be manipulated using Boolean logical operators. When a string of binary symbols is manipulated in this way, it is called a bitwise operation; the logical operators AND, OR, and XOR may be performed on corresponding bits in two binary numerals provided as input. The logical NOT operation may be performed on individual bits in a single binary numeral provided as input. Sometimes, such operations may be used as arithmetic short-cuts, and may have other computational benefits as well. For example, an arithmetic shift left of a binary number is the equivalent of multiplication by a (positive, integral) power of 2.

Conversion to and from other Numeral Systems

Decimal

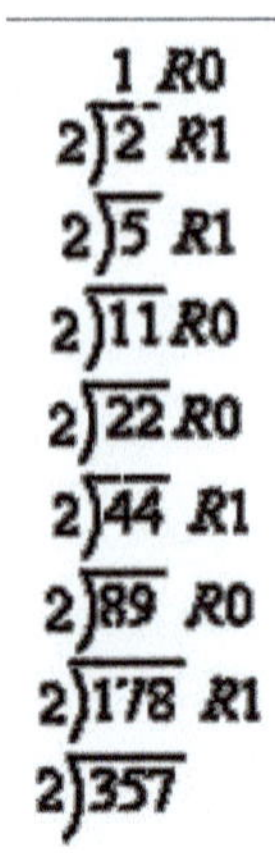

Conversion of $(357)_{10}$ to binary notation results in (101100101)

To convert from a base-10 integer to its base-2 (binary) equivalent, the number is divided by two. The remainder is the least-significant bit. The quotient is again divided by two; its remainder becomes the next least significant bit. This process repeats until a quotient of one is reached. The sequence of remainders (including the final quotient of one) forms the binary value, as each remainder must be either zero or one when dividing by two. For example, $(357)_{10}$ is expressed as $(101100101)_{2.}$

Conversion from base-2 to base-10 simply inverts the preceding algorithm. The bits of the binary number are used one by one, starting with the most significant (leftmost) bit. Beginning with the value 0, the prior value is doubled, and the next bit is then added to produce the next value. This can be organized in a multi-column table. For example, to convert 10010101101_2 to decimal:

Prior value	× 2 +	Next bit	Next value
0	× 2 +	1	= 1
1	× 2 +	0	= 2
2	× 2 +	0	= 4
4	× 2 +	1	= 9
9	× 2 +	0	= 18
18	× 2 +	1	= 37
37	× 2 +	0	= 74
74	× 2 +	1	= 149

149	× 2 +	1	= 299
299	× 2 +	0	= 598
598	× 2 +	1	= 1197

The result is 1197_{10}. Note that the first Prior Value of 0 is simply an initial decimal value. This method is an application of the Horner scheme.

Binary	1	0	0	1	0	1	0	1	1	0	1	
Decimal	1×2^{10} +	0×2^{9} +	0×2^{8} +	1×2^{7} +	0×2^{6} +	1×2^{5} +	0×2^{4} +	1×2^{3} +	1×2^{2} +	0×2^{1} +	1×2^{0} =	1197

The fractional parts of a number are converted with similar methods. They are again based on the equivalence of shifting with doubling or halving.

In a fractional binary number such as 0.11010110101_2, the first digit is $\frac{1}{2}$, the second $(\frac{1}{2})^2 = \frac{1}{4}$, etc.

So if there is a 1 in the first place after the decimal, then the number is at least $\frac{1}{2}$, and vice versa. Double that number is at least 1. This suggests the algorithm: Repeatedly double the number to be converted, record if the result is at least 1, and then throw away the integer part.

For example, $(\frac{1}{3})_{10}$, in binary, is:

Converting	Result
$\frac{1}{3}$	0.
$\frac{1}{3}\times2=\frac{2}{3}<1$	0.0
$\frac{2}{3}\times2=1\frac{1}{3}\geq1$	0.01
$\frac{1}{3}\times2=\frac{2}{3}<1$	0.010
$\frac{2}{3}\times2=1\frac{1}{3}\geq1$	0.0101

Thus the repeating decimal fraction 0.3... is equivalent to the repeating binary fraction 0.01... .

Or for example, 0.1_{10}, in binary, is:

Converting	Result
0.1	0.
$0.1 \times 2 = \mathbf{0.2} < 1$	0.0
$0.2 \times 2 = \mathbf{0.4} < 1$	0.00
$0.4 \times 2 = \mathbf{0.8} < 1$	0.000
$0.8 \times 2 = \mathbf{1.6} \geq 1$	0.0001
$0.6 \times 2 = \mathbf{1.2} \geq 1$	0.00011
$0.2 \times 2 = \mathbf{0.4} < 1$	0.000110
$0.4 \times 2 = \mathbf{0.8} < 1$	0.0001100
$0.8 \times 2 = \mathbf{1.6} \geq 1$	0.00011001
$0.6 \times 2 = \mathbf{1.2} \geq 1$	0.000110011
$0.2 \times 2 = \mathbf{0.4} < 1$	0.0001100110

This is also a repeating binary fraction $0.0\overline{0011}\ldots$. It may come as a surprise that terminating decimal fractions can have repeating expansions in binary. It is for this reason that many are surprised to discover that 0.1 + ... + 0.1, (10 additions) differs from 1 in floating point arithmetic. In fact, the only binary fractions with terminating expansions are of the form of an integer divided by a power of 2, which 1/10 is not.

The final conversion is from binary to decimal fractions. The only difficulty arises with repeating fractions, but otherwise the method is to shift the fraction to an integer, convert it as above, and then divide by the appropriate power of two in the decimal base. For example:

$$x = 1100.1\overline{01110}\ldots$$

$$x \times 2^6 = 1100101110.\overline{01110}\ldots$$

$$x \times 2 = 11001.01110\ldots$$

$$x \times (2^6 - 2) = 1100010101$$

$$x = 1100010101/111110$$

$$x = (789/62)_{10}$$

Another way of converting from binary to decimal, often quicker for a person familiar with hexadecimal, is to do so indirectly—first converting (x in binary) into (x in hexadecimal) and then converting (x in hexadecimal) into (x in decimal).

For very large numbers, these simple methods are inefficient because they perform a large number of multiplications or divisions where one operand is very large. A simple divide-and-conquer algorithm is more effective asymptotically: given a binary number, it is divided by 10^k, where k is chosen so that the quotient roughly equals the remainder; then each of these pieces is converted to decimal and the two are concatenated. Given a decimal number, it can be split into two pieces of about the same size, each of which is converted to binary, whereupon the first converted piece is multiplied by 10^k and added to the second converted piece, where k is the number of decimal digits in the second, least-significant piece before conversion.

Hexadecimal

Binary may be converted to and from hexadecimal somewhat more easily. This is because the radix of the hexadecimal system (16) is a power of the radix of the binary system (2). More specifically, $16 = 2^4$, so it takes four digits of binary to represent one digit of hexadecimal.

To convert a hexadecimal number into its binary equivalent, simply substitute the corresponding binary digits:

$3A_{16} = 0011\ 1010_2$

$E7_{16} = 1110\ 0111_2$

To convert a binary number into its hexadecimal equivalent, divide it into groups of four bits. If the number of bits isn't a multiple of four, simply insert extra 0 bits at the left (called padding). For example:

1010010_2 = 0101 0010 grouped with padding = 52_{16}

11011101_2 = 1101 1101 grouped = DD_{16}

To convert a hexadecimal number into its decimal equivalent, multiply the decimal equivalent of each hexadecimal digit by the corresponding power of 16 and add the resulting values:

$C0E7_{16} = (12 \times 16^3) + (0 \times 16^2) + (14 \times 16^1) + (7 \times 16^0) = (12 \times 4096) + (0 \times 256) + (14 \times 16) + (7 \times 1) = 49{,}383_{10}$

Octal

Binary is also easily converted to the octal numeral system, since octal uses a radix of 8, which is a power of two (namely, 2^3, so it takes exactly three binary digits to represent an octal digit). The correspondence between octal and binary numerals is the same as for the first eight digits of hexadecimal in the table above. Binary 000 is equivalent to the octal digit 0, binary 111 is equivalent to octal 7, and so forth.

Octal	0	1	2	3	4	5	6	7
Binary	000	001	010	011	100	101	110	111

Converting from octal to binary proceeds in the same fashion as it does for hexadecimal:

$65_8 = 110\ 101_2$

$17_8 = 001\ 111_2$

And from binary to octal:

$101100_2 = 101\ 100_2$ grouped $= 54_8$

$10011_2 = 010\ 011_2$ grouped with padding $= 23_8$

And from octal to decimal:

$65_8 = (6 \times 8^1) + (5 \times 8^0) = (6 \times 8) + (5 \times 1) = 53_{10}$

$127_8 = (1 \times 8^2) + (2 \times 8^1) + (7 \times 8^0) = (1 \times 64) + (2 \times 8) + (7 \times 1) = 87_{10}$

Representing Real Numbers

Non-integers can be represented by using negative powers, which are set off from the other digits by means of a radix point (called a decimal point in the decimal system). For example, the binary number 11.01_2 thus means:

1×2^1	$(1 \times 2 = 2)$	plus
1×2^0	$(1 \times 1 = 1)$	plus
0×2^{-1}	$(0 \times 1/2 = 0)$	plus
1×2^{-2}	$(1 \times 1/4 = 0.25)$	

For a total of 3.25 decimal.

All dyadic rational numbers $\frac{p}{2^a}$ have a *terminating* binary numeral—the binary representation has a finite number of terms after the radix point. Other rational numbers have binary representation, but instead of terminating, they *recur*, with a finite sequence of digits repeating indefinitely. For instance

$$\frac{1_{10}}{3_{10}} = \frac{1_2}{11_2} = 0.0101010\overline{101}\ldots_2$$

$$\frac{12_{10}}{17_{10}} = \frac{1100_2}{10001_2} = 0.10110100\ 10110100\ \overline{10110100}\ldots_2$$

The phenomenon that the binary representation of any rational is either terminating or recurring also occurs in other radix-based numeral systems. See, for instance, the explanation in decimal. Another similarity is the existence of alternative representations for any terminating representation, relying on the fact that 0.111111... is the sum of the geometric series $2^{-1} + 2^{-2} + 2^{-3} + \ldots$ which is 1.

Binary numerals which neither terminate nor recur represent irrational numbers. For instance,

- 0.10100100010000100000100... does have a pattern, but it is not a fixed-length recurring pattern, so the number is irrational
- 1.0110101000001001111001100110011111110... is the binary representation of , the square root of 2, another irrational. It has no discernible pattern. See irrational number.

Indeterminate Form

In calculus and other branches of mathematical analysis, limits involving an algebraic combination of functions in an independent variable may often be evaluated by replacing these functions by their limits; if the expression obtained after this substitution does not give enough information to determine the original limit, it is said to take on an indeterminate form. The term was originally introduced by Cauchy's student Moigno in the middle of the 19th century.

The indeterminate forms typically considered in the literature are denoted $0/0$, ∞/∞, $0 \times \infty$, $\infty - \infty$, 0^0, 1^∞ and ∞^0.

Discussion

The most common example of an indeterminate form occurs as the ratio of two functions, in which both of these functions tend to zero in the limit, and is referred to as "the indeterminate form 0/0". As x approaches 0, the ratios x/x^3, x/x, and x^2/x go to ∞, 1, and 0 respectively. In each case, if the limits of the numerator and denominator are substituted, the resulting expression is $0/0$, which is undefined. So, in a manner of speaking, $0/0$ can take on the values 0, 1, or ∞, and it is possible to construct similar examples for which the limit is any particular value.

More formally, the fact that the functions $f(x)$ and $g(x)$ both approach 0 as x approaches some limit point c is not enough information to evaluate the limit

$$\lim_{x \to c} \frac{f(x)}{g(x)}.$$

Not every undefined algebraic expression corresponds to an indeterminate form. For example, the expression $1/0$ is undefined as a real number but does not correspond to an indeterminate form, because any limit that gives rise to this form will diverge to infinity.

An indeterminate form expression may have a value in some contexts. For example, if κ is an infinite cardinal number, then expressions 0^κ, 0^0, 1^κ and κ^0 are well defined in the context of cardinal arithmetic. Note that zero to the power infinity is not an indeterminate form.

Some Examples and Non-examples

Indeterminate Form 0/0

- 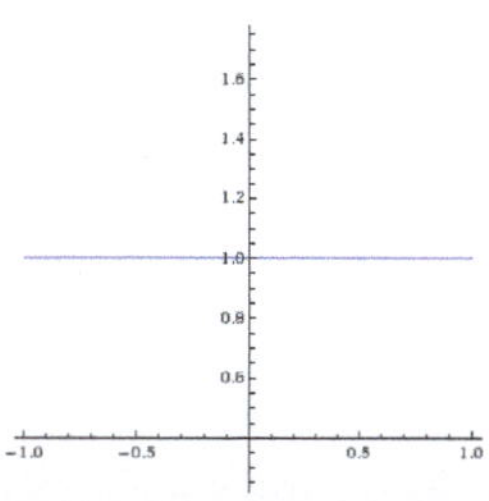

(1)

- 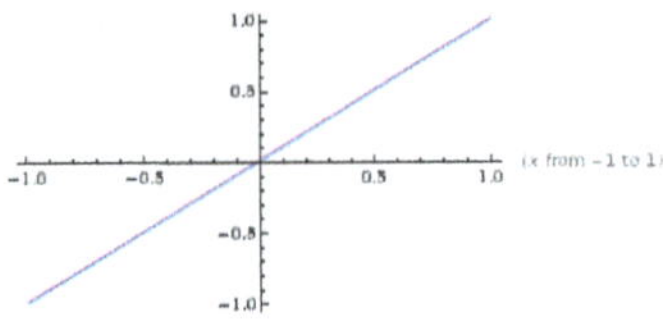

(2)

- 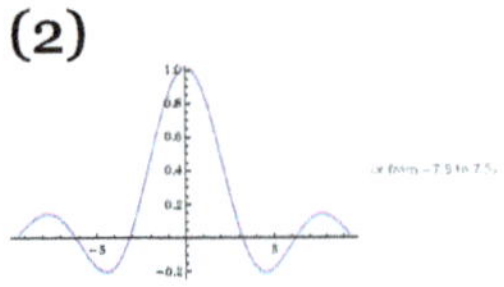

(3)

- 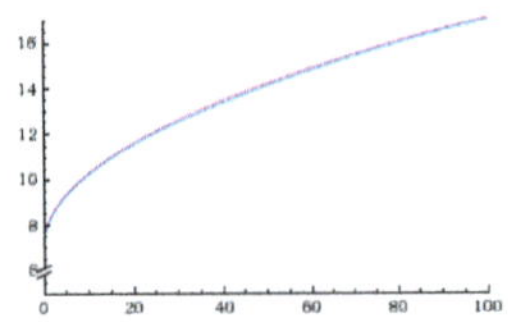

(4)

- 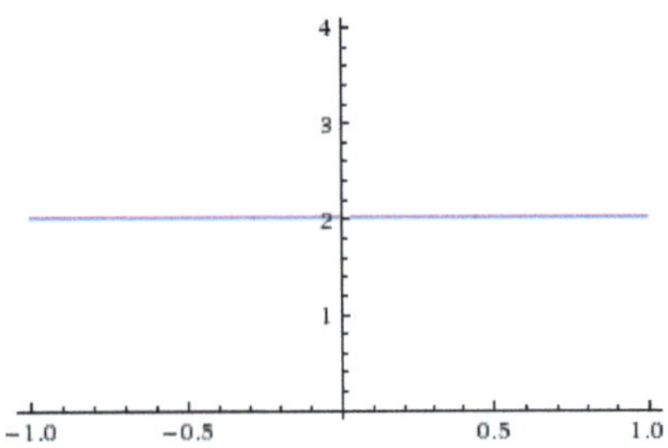

(5)

- 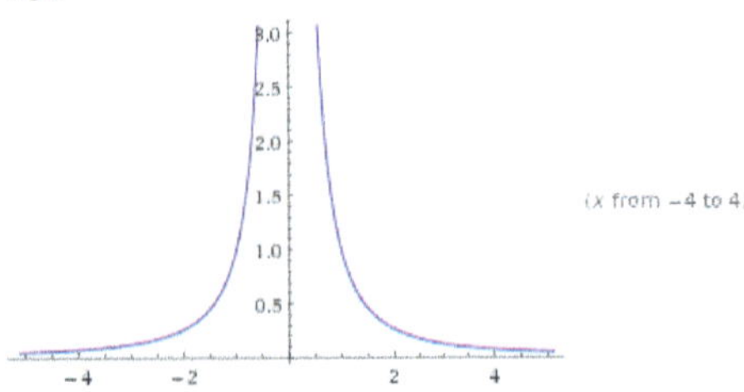

(6)

The indeterminate form 0/0 is particularly common in calculus because it often arises in the evaluation of derivatives using their limit definition.

As mentioned above,

$$\lim_{x\to 0}\frac{x}{x}=1, \quad (1)$$

while

$$\lim_{x\to 0}\frac{x^2}{x}=0,\ (2)$$

This is enough to show that 0/0 is an indeterminate form. Other examples with this indeterminate form include

$$\lim_{x\to 0}\frac{\sin(x)}{x}=1,\ (3)$$

and

$$\lim_{x\to 49}\frac{x-49}{\sqrt{x}-7}=14,\ (4)$$

Direct substitution of the number that x approaches into any of these expressions shows that these are examples of the indeterminate form 0/0, but these limits take many different values. Any desired value a can be obtained for this indeterminate form as follows:

$$\lim_{x\to 0}\frac{ax}{x}=a.\ (5)$$

The value infinity can also be obtained (in the sense of divergence to infinity):

$$\lim_{x\to 0}\frac{x}{x^3}=\infty.\ (6)$$

Indeterminate form 0^0

- 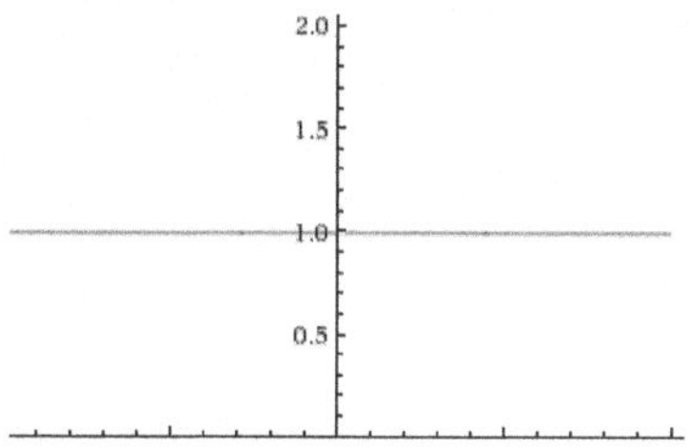

(7)

- 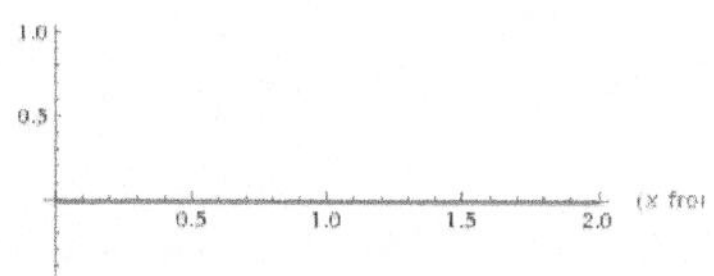

(8)

The following limits illustrate that the expression 0^0 is an indeterminate form:

$$\lim_{x\to 0}x^0=1,\ (7)$$

$$\lim_{x\to 0^+}0^x=0.\ (8)$$

Thus, in general, knowing that $\lim_{x\to c} f(x)=0$ and $\lim_{x\to c} g(x)=0$ is not sufficient to calculate the limit

$$\lim_{x\to c} f(x)^{g(x)}.$$

If the functions f and g are analytic at c, and f is positive for x sufficiently close (but not equal) to c, then the limit of $f(x)^{g(x)}$ will be 1. Otherwise, use the transformation in the table below to evaluate the limit.

Expressions that are not Indeterminate Forms

The expression 1/0 is not commonly regarded as an indeterminate form because there is not an infinite range of values that f/g could approach. Specifically, if f approaches 1 and g approaches 0, then f and g may be chosen so that (1) f/g approaches $+\infty$, (2) f/g approaches $-\infty$, or (3) the limit fails to exist. In each case the absolute value $|f/g|$ approaches $+\infty$, and so the quotient f/g must diverge, in the sense of the extended real numbers. (In the framework of the projectively extended real line, the limit is the unsigned infinity ∞ in all three cases.) Similarly, any expression of the form $a/0$, with $a \neq 0$ (including $a = +\infty$ and $a = -\infty$), is not an indeterminate form since a quotient giving rise to such an expression will always diverge.

The expression 0^{∞} is not an indeterminate form. The expression $0^{+\infty}$ has the limiting value 0 for the given individual limits, and the expression $0^{-\infty}$ is equivalent to 1/0.

Evaluating Indeterminate Forms

The adjective *indeterminate* does *not* imply that the limit does not exist, as many of the examples above show. In many cases, algebraic elimination, L'Hôpital's rule, or other methods can be used to manipulate the expression so that the limit can be evaluated.

For example, the expression x^2/x can be simplified to x at any point other than $x = 0$. Thus, the limit of this expression as x approaches 0 (which depends only on points *near* 0, not at $x = 0$ itself) is the limit of x, which is 0. Most of the other examples above can also be evaluated using algebraic simplification.

Equivalent Infinitesimal

When two variables α and β converge to zero at the same point and $\lim\frac{\beta}{\alpha}=1$, they are called *equivalent infinitesimal.*

For the evaluation of the indeterminate form 0/0, we can use the following equivalent infinitesimals:

$x \sim \sin x$,

$x \sim \arcsin x$,

$x \sim \tan x$,

$x \sim \arctan x$,

$x \sim \ln(1+x),$

$1-\cos x \sim \frac{1}{2}x^2,$

$a^x - 1 \sim x \ln a,$

$(1+x)^a - 1 \sim ax.$

For example:

$$\lim_{x\to 0}\frac{1}{x^3}[(\frac{2+\cos x}{3})^x - 1] = \lim_{x\to 0}\frac{e^{x\ln\frac{2+\cos x}{3}} - 1}{x^3} = \lim_{x\to 0}\frac{1}{x^2}\ln\frac{2+\cos x}{3} = \lim_{x\to 0}\frac{1}{x^2}\ln(\frac{\cos x - 1}{3} + 1) = \lim_{x\to 0}\frac{\cos x - 1}{3x^2} = -\frac{1}{6}$$

Here is a brief proof:

Suppose there are two equivalent infinitesimals $\alpha \sim \alpha'$ and $\beta \sim \beta'$.

$$\lim\frac{\beta}{\alpha} = \lim\frac{\beta\beta'\alpha'}{\beta'\alpha'\alpha} = \lim\frac{\beta}{\beta'}\lim\frac{\alpha'}{\alpha}\lim\frac{\beta'}{\alpha'} = \lim\frac{\beta'}{\alpha'}$$

L'Hôpital's Rule

L'Hôpital's rule is a general method for evaluating the indeterminate forms 0/0 and ∞/∞. This rule states that (under appropriate conditions)

$$\lim_{x\to c}\frac{f(x)}{g(x)} = \lim_{x\to c}\frac{f'(x)}{g'(x)},$$

where *f'* and *g'* are the derivatives of *f* and *g*. (Note that this rule does *not* apply to expressions ∞/0, 1/0, and so on; these expressions are not indeterminate forms.) These derivatives will allow one to perform algebraic simplification and eventually evaluate the limit.

L'Hôpital's rule can also be applied to other indeterminate forms, using first an appropriate algebraic transformation. For example, to evaluate the form 0^0:

$$\ln\lim_{x\to c} f(x)^{g(x)} = \lim_{x\to c}\frac{\ln f(x)}{1/g(x)}.$$

The right-hand side is of the form ∞/∞, so L'Hôpital's rule applies to it. Note that this equation is valid (as long as the right-hand side is defined) because the natural logarithm (ln) is a continuous function; it is irrelevant how well-behaved *f* and *g* may (or may not) be as long as *f* is asymptotically positive.

Although L'Hôpital's rule applies both to 0/0 and to ∞/∞, one of these forms may be more useful than the other in a particular case (because of the possibility of algebraic simplification afterwards). One can change between these forms, if necessary, by transforming f/g to $(1/g)/(1/f)$.

List of Indeterminate Forms

Indeterminate form	Conditions	Transformation to 0/0	Transformation to ∞/∞
$0/0$	$\lim_{x\to c} f(x)=0,\ \lim_{x\to c} g(x)=0$	–	$\lim_{x\to c}\frac{f(x)}{g(x)}=\lim_{x\to c}\frac{1/g(x)}{1/f(x)}$
∞/∞	$\lim_{x\to c} f(x)=\infty,\ \lim_{x\to c} g(x)=\infty$	$\lim_{x\to c}\frac{f(x)}{g(x)}=\lim_{x\to c}\frac{1/g(x)}{1/f(x)}$	–
$0\times\infty$	$\lim_{x\to c} f(x)=0,\ \lim_{x\to c} g(x)=\infty$	$\lim_{x\to c} f(x)g(x)=\lim_{x\to c}\frac{f(x)}{1/g(x)}$	$\lim_{x\to c} f(x)g(x)=\lim_{x\to c}\frac{g(x)}{1/f(x)}$
$\infty-\infty$	$\lim_{x\to c} f(x)=\infty,\ \lim_{x\to c} g(x)=\infty$	$\lim_{x\to c}(f(x)-g(x))=\lim_{x\to c}\frac{1/g(x)-1/f(x)}{1/(f(x)g(x))}$	$\lim_{x\to c}(f(x)-g(x))=\ln\lim_{x\to c}\frac{e^{f(x)}}{e^{g(x)}}$
0^0	$\lim_{x\to c} f(x)=0^+,\ \lim_{x\to c} g(x)=0$	$\lim_{x\to c} f(x)^{g(x)}=\exp\lim_{x\to c}\frac{g(x)}{1/\ln f(x)}$	$\lim_{x\to c} f(x)^{g(x)}=\exp\lim_{x\to c}\frac{\ln f(x)}{1/g(x)}$
1^∞	$\lim_{x\to c} f(x)=1,\ \lim_{x\to c} g(x)=\infty$	$\lim_{x\to c} f(x)^{g(x)}=\exp\lim_{x\to c}\frac{\ln f(x)}{1/g(x)}$	$\lim_{x\to c} f(x)^{g(x)}=\exp\lim_{x\to c}\frac{g(x)}{1/\ln f(x)}$
∞^0	$\lim_{x\to c} f(x)=\infty,\ \lim_{x\to c} g(x)=0$	$\lim_{x\to c} f(x)^{g(x)}=\exp\lim_{x\to c}\frac{g(x)}{1/\ln f(x)}$	$\lim_{x\to c} f(x)^{g(x)}=\exp\lim_{x\to c}\frac{\ln f(x)}{1/g(x)}$

Fundamental Theorem of Arithmetic

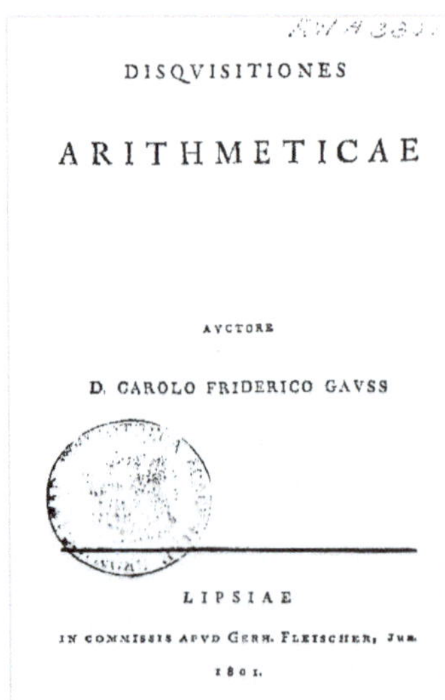
DISQVISITIONES

ARITHMETICAE

AVCTORE

D. CAROLO FRIDERICO GAVSS

LIPSIAE

IN COMMISSIS APVD GERH. FLEISCHER, IUN.

1801.

The unique factorization theorem was proved by Gauss with his 1801 book *Disquisitiones Arithmeticae*. In this book, Gauss used the fundamental theorem for proving the law of quadratic reciprocity.

In number theory, the fundamental theorem of arithmetic, also called the unique factorization theorem or the unique-prime-factorization theorem, states that every integer greater than 1 either is prime itself or is the product of prime numbers, and that this product is unique, up to the order of the factors. For example,

$1200 = 2^4 \times 3^1 \times 5^2 = 3 \times 2 \times 2 \times 2 \times 2 \times 5 \times 5 = 5 \times 2 \times 3 \times 2 \times 5 \times 2 \times 2 =$ etc.

The theorem is stating two things: first, that 1200 *can* be represented as a product of primes, and second, no matter how this is done, there will always be four 2s, one 3, two 5s, and no other primes in the product.

The requirement that the factors be prime is necessary: factorizations containing composite numbers may not be unique (e.g. $12 = 2 \times 6 = 3 \times 4$).

This theorem is one of the main reasons why 1 is not considered a prime number: if 1 were prime, the factorization would not be unique, as, for example, $2 = 2\times1 = 2\times1\times1 = \ldots$

History

Book VII, propositions 30, 31 and 32, and Book IX, proposition 14 of Euclid's Elements are essentially the statement and proof of the fundamental theorem.

If two numbers by multiplying one another make some number, and any prime number measure the product, it will also measure one of the original numbers.

—*Euclid, Elements Book VII, Proposition 30*

Proposition 30 is referred to as Euclid's lemma. And it is the key in the proof of the fundamental theorem of arithmetic.

Any composite number is measured by some prime number.

—*Euclid, Elements Book VII, Proposition 31*

Proposition 31 is proved directly by infinite descent.

Any number either is prime or is measured by some prime number.

—*Euclid, Elements Book VII, Proposition 32*

Proposition 32 is derived from proposition 31, and prove that the decomposition is possible.

If a number be the least that is measured by prime numbers, it will not be measured by any other prime number except those originally measuring it.

—*Euclid, Elements Book IX, Proposition 14*

Book IX, proposition 14 is derived from Book VII, proposition 30, and prove partially that the decomposition is unique – a point critically noted by André Weil. Indeed, in this proposition the exponents are all equal to one, so nothing is said for the general case.

Article 16 of Gauss' *Disquisitiones Arithmeticae* is an early modern statement and proof employing modular arithmetic.

Applications

Canonical Representation of a Positive Integer

Every positive integer $n > 1$ can be represented in exactly one way as a product of prime powers:

$$n = p_1^{\alpha_1} p_2^{\alpha_2} \cdots p_k^{\alpha_k} = \prod_{i=1}^{k} p_i^{\alpha_i}$$

where $p_1 < p_2 < \ldots < p_k$ are primes and the α_i are positive integers. This representation is commonly extended to all positive integers, including one, by the convention that the empty product is equal to 1 (the empty product corresponds to $k = 0$).

This representation is called the canonical representation of n, or the standard form of n.

For example $999 = 3^3 \times 37$, $1000 = 2^3 \times 5^3$, $1001 = 7 \times 11 \times 13$

Note that factors $p^0 = 1$ may be inserted without changing the value of n (e.g. $1000 = 2^3 \times 3^0 \times 5^3$). In fact, any positive integer can be uniquely represented as an infinite product taken over all the positive prime numbers,

$$n = 2^{n_1} 3^{n_2} 5^{n_3} 7^{n_4} \cdots = \prod p_i^{n_i},$$

where a finite number of the n_i are positive integers, and the rest are zero. Allowing negative exponents provides a canonical form for positive rational numbers.

Arithmetic Operations

The canonical representation, when it is known, is convenient for easily computing products, gcd, and lcm:

$$a \cdot b = 2^{a_2+b_2} 3^{a_3+b_3} 5^{a_5+b_5} 7^{a_7+b_7} \cdots = \prod p_i^{a_{p_i}+b_{p_i}},$$

$$\gcd(a,b) = 2^{\min(a_2,b_2)} 3^{\min(a_3,b_3)} 5^{\min(a_5,b_5)} 7^{\min(a_7,b_7)} \cdots = \prod p_i^{\min(a_{p_i},b_{p_i})},$$

$$\operatorname{lcm}(a,b) = 2^{\max(a_2,b_2)} 3^{\max(a_3,b_3)} 5^{\max(a_5,b_5)} 7^{\max(a_7,b_7)} \cdots = \prod p_i^{\max(a_{p_i},b_{p_i})}.$$

However, as Integer factorization of large integers is much harder than computing their product, gcd or lcm, these formulas have, in practice, a limited usage.

Arithmetical Functions

Many arithmetical functions are defined using the canonical representation. In particular, the values of additive and multiplicative functions are determined by their values on the powers of prime numbers.

Proof

The proof uses Euclid's lemma (*Elements* VII, 30): if a prime p divides the product of two natural numbers a and b, then p divides a or p divides b.

Existence

We need to show that every integer greater than 1 is either prime or a product of primes. For the base case, note that 2 is prime. By induction: assume true for all numbers between 1 and n. If n is prime, there is nothing more to prove. Otherwise, there are integers a and b, where $n = ab$ and $1 < a \le b < n$. By the induction hypothesis, $a = p_1p_2...p_j$ and $b = q_1q_2...q_k$ are products of primes. But then $n = ab = p_1p_2...p_jq_1q_2...q_k$ is a product of primes.

Uniqueness

Assume that $s > 1$ is the product of prime numbers in two different ways:

$$\begin{aligned} s &= p_1 p_2 \cdots p_m \\ &= q_1 q_2 \cdots q_n. \end{aligned}$$

We must show $m = n$ and that the q_j are a rearrangement of the p_i.

By Euclid's lemma, p_1 must divide one of the q_j; relabeling the q_j if necessary, say that p_1 divides q_1. But q_1 is prime, so its only divisors are itself and 1. Therefore, $p_1 = q_1$, so that

$$\begin{aligned} \frac{s}{p_1} &= p_2 \cdots p_m \\ &= q_2 \cdots q_n. \end{aligned}$$

Reasoning the same way, p_2 must equal one of the remaining q_j. Relabeling again if necessary, say $p_2 = q_2$. Then

$$\begin{aligned} \frac{s}{p_1 p_2} &= p_3 \cdots p_m \\ &= q_3 \cdots q_n. \end{aligned}$$

This can be done for each of the m p_i's, showing that $m \le n$ and every p_i is a q_j. Applying the same argument with the p's and q's reversed shows $n \le m$ (hence $m = n$) and every q_j is a p_i.

Elementary Proof of Uniqueness

The fundamental theorem of arithmetic can also be proved without using Euclid's lemma, as follows:

Assume that $s > 1$ is the smallest positive integer which is the product of prime numbers in two different ways. If s were prime then it would factor uniquely as itself, so there must be at least two primes in each factorization of s:

$$\begin{aligned} s &= p_1 p_2 \cdots p_m \\ &= q_1 q_2 \cdots q_n. \end{aligned}$$

If any $p_i = q_j$ then, by cancellation, $s/p_i = s/q_j$ would be another positive integer, different from s, which is greater than 1 and also has two distinct factorizations. But s/p_i is smaller than s, meaning s would not actually be the smallest such integer. Therefore every p_i must be distinct from every q_j.

Without loss of generality, take $p_1 < q_1$ (if this is not already the case, switch the p and q designations.) Consider

$$t = (q_1 - p_1)(q_2 \cdots q_n),$$

and note that $1 < q_2 \leq t < s$. Therefore t must have a unique prime factorization. By rearrangement we see,

$$\begin{aligned} t &= q_1(q_2 \cdots q_n) - p_1(q_2 \cdots q_n) \\ &= s - p_1(q_2 \cdots q_n) \\ &= p_1((p_2 \cdots p_m) - (q_2 \cdots q_n)). \end{aligned}$$

Here $u = ((p_2 \ldots p_m) - (q_2 \ldots q_n))$ is positive, for if it were negative or zero then so would be its product with p_1, but that product equals t which is positive. So u is either 1 or factors into primes. In either case, $t = p_1 u$ yields a prime factorization of t, which we know to be unique, so p_1 appears in the prime factorization of t.

If $(q_1 - p_1)$ equaled 1 then the prime factorization of t would be all q's, which would preclude p_1 from appearing. Thus $(q_1 - p_1)$ is not 1, but is positive, so it factors into primes: $(q_1 - p_1) = (r_1 \ldots r_h)$. This yields a prime factorization of

$$t = (r_1 \cdots r_h)(q_2 \cdots q_n),$$

which we know is unique. Now, p_1 appears in the prime factorization of t, and it is not equal to any q, so it must be one of the r's. That means p_1 is a factor of $(q_1 - p_1)$, so there exists a positive integer k such that $p_1 k = (q_1 - p_1)$, and therefore

$$p_1(k+1) = q_1.$$

But that means q_1 has a proper factorization, so it is not a prime number. This contradiction shows that s does not actually have two different prime factorizations. As a result, there is no smallest positive integer with multiple prime factorizations, hence all positive integers greater than 1 factor uniquely into primes.

Generalizations

The first generalization of the theorem is found in Gauss's second monograph (1832) on biquadratic reciprocity. This paper introduced what is now called the ring of Gaussian integers, the set of all complex numbers $a + bi$ where a and b are integers. It is now denoted by $\mathbb{Z}[i]$. He showed that this ring has the four units ± 1 and $\pm i$, that the non-zero, non-unit numbers fall into two classes, primes and composites, and that (except for order), the composites have unique factorization as a product of primes.

Similarly, in 1844 while working on cubic reciprocity, Eisenstein introduced the ring $\mathbb{Z}[\omega]$, where $\omega = \frac{-1+\sqrt{-3}}{2}$, $\omega^3 = 1$ is a cube root of unity. This is the ring of Eisenstein integers, and he proved it has the six units $\pm 1, \pm\omega, \pm\omega^2$ and that it has unique factorization.

However, it was also discovered that unique factorization does not always hold. An example is given by $\mathbb{Z}[\sqrt{-5}]$. In this ring one has

$$6 = 2 \cdot 3 = (1+\sqrt{-5})(1-\sqrt{-5}).$$

Examples like this caused the notion of "prime" to be modified. In $\mathbb{Z}[\sqrt{-5}]$ it can be proven that if any of the factors above can be represented as a product, e.g. 2 = *ab*, then one of *a* or *b* must be a unit. This is the traditional definition of "prime". It can also be proven that none of these factors obeys Euclid's lemma; e.g. 2 divides neither $(1 + \sqrt{-5})$ nor $(1 - \sqrt{-5})$ even though it divides their product 6. In algebraic number theory 2 is called irreducible in $\mathbb{Z}[\sqrt{-5}]$ (only divisible by itself or a unit) but not prime in $\mathbb{Z}[\sqrt{-5}]$ (if it divides a product it must divide one of the factors). The mention of $\mathbb{Z}[\sqrt{-5}]$ is required because 2 is prime and irreducible in $\mathbb{Z}$. Using these definitions it can be proven that in any ring a prime must be irreducible. Euclid's classical lemma can be rephrased as "in the ring of integers $\mathbb{Z}$ every irreducible is prime". This is also true in $\mathbb{Z}[i]$ and $\mathbb{Z}[\omega]$, but not in $\mathbb{Z}[\sqrt{-5}]$.

The rings in which factorization into irreducibles is essentially unique are called unique factorization domains. Important examples are polynomial rings over the integers or over a field, Euclidean domains and principal ideal domains.

In 1843 Kummer introduced the concept of ideal number, which was developed further by Dedekind (1876) into the modern theory of ideals, special subsets of rings. Multiplication is defined for ideals, and the rings in which they have unique factorization are called Dedekind domains.

There is a version of unique factorization for ordinals, though it requires some additional conditions to ensure uniqueness.

Greatest Common Divisor

In mathematics, the greatest common divisor (gcd) of two or more integers, when at least one of them is not zero, is the largest positive integer that divides the numbers without a remainder. For example, the GCD of 8 and 12 is 4.

The greatest common divisor is also known as the greatest common factor (gcf), highest common factor (hcf), greatest common measure (gcm), or highest common divisor.

This notion can be extended to polynomials and other commutative rings.

Overview

Notation

In this article we will denote the greatest common divisor of two integers a and b as gcd(a,b). Some textbooks use (a,b).

Example

What is the greatest common divisor of 54 and 24?

The number 54 can be expressed as a product of two integers in several different ways:

$$54 \times 1 = 27 \times 2 = 18 \times 3 = 9 \times 6.$$

Thus the divisors of 54 are: $1, 2, 3, 6, 9, 18, 27, 54$.

Similarly, the divisors of 24 are: $1, 2, 3, 4, 6, 8, 12, 24$.

The numbers that these two lists share in common are the common divisors of 54 and 24:

$$1, 2, 3, 6.$$

The greatest of these is 6. That is, the greatest common divisor of 54 and 24. One writes: $\gcd(54, 24) = 6$.

Reducing Fractions

The greatest common divisor is useful for reducing fractions to be in lowest terms. For example, gcd(42, 56) = 14, therefore,

$$\frac{42}{56} = \frac{3 \cdot 14}{4 \cdot 14} = \frac{3}{4}.$$

Coprime Numbers

Two numbers are called *relatively prime*, or *coprime*, if their greatest common divisor equals 1. For example, 9 and 28 are relatively prime.

A Geometric View

For example, a 24-by-60 rectangular area can be divided into a grid of: 1-by-1 squares, 2-by-2 squares, 3-by-3 squares, 4-by-4 squares, 6-by-6 squares or 12-by-12 squares. Therefore, 12 is the greatest common divisor of 24 and 60. A 24-by-60 rectangular area can be divided into a grid of 12-by-12 squares, with two squares along one edge (24/12 = 2) and five squares along the other (60/12 = 5).

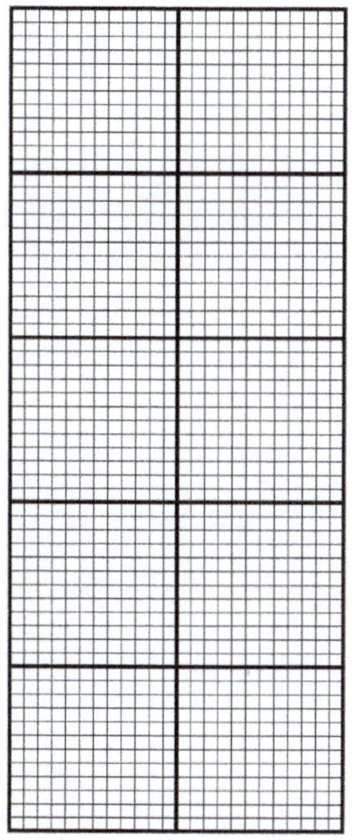

A 24-by-60 rectangle is covered with ten 12-by-12 square tiles, where 12 is the GCD of 24 and 60. More generally, an a-by-b rectangle can be covered with square tiles of side length c only if c is a common divisor of a and b.

Calculation

Using Prime Factorizations

Greatest common divisors can in principle be computed by determining the prime factorizations of the two numbers and comparing factors, as in the following example: to compute gcd(18, 84), we find the prime factorizations $18 = 2 \cdot 3^2$ and $84 = 2^2 \cdot 3 \cdot 7$ and notice that the "overlap" of the two expressions is $2 \cdot 3$; so gcd(18, 84) = 6. In practice, this method is only feasible for small numbers; computing prime factorizations in general takes far too long.

Here is another concrete example, illustrated by a Venn diagram. Suppose it is desired to find the greatest common divisor of 48 and 180. First, find the prime factorizations of the two numbers:

$$48 = 2 \times 2 \times 2 \times 2 \times 3,$$

$$180 = 2 \times 2 \times 3 \times 3 \times 5.$$

What they share in common is two "2"s and a "3":

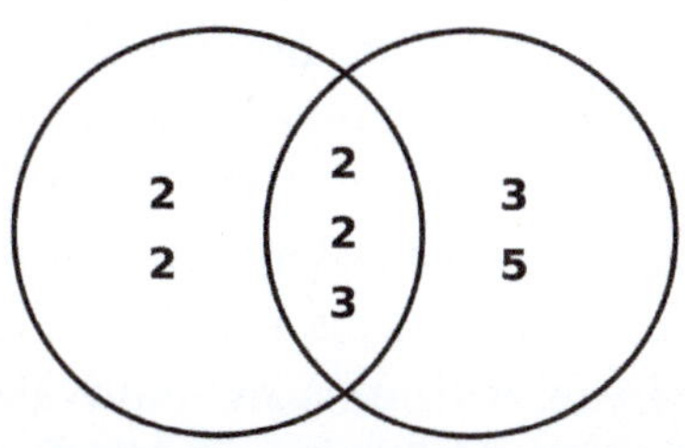

Least common multiple = $2 \times 2 \times (2 \times 2 \times 3) \times 3 \times 5 = 720$

Greatest common divisor = $2 \times 2 \times 3 = 12$.

Using Euclid's Algorithm

A much more efficient method is the Euclidean algorithm, which uses a division algorithm such as long division in combination with the observation that the gcd of two numbers also divides their difference. To compute gcd(48,18), divide 48 by 18 to get a quotient of 2 and a remainder of 12. Then divide 18 by 12 to get a quotient of 1 and a remainder of 6. Then divide 12 by 6 to get a remainder of 0, which means that 6 is the gcd. Note that we ignored the quotient in each step except to notice when the remainder reached 0, signalling that we had arrived at the answer. Formally the algorithm can be described as:

$$\gcd(a,0) = a$$

$$\gcd(a,b) = \gcd(b, a \bmod b),$$

where

$$a \bmod b = a - b\left\lfloor \frac{a}{b} \right\rfloor.$$

If the arguments are both greater than zero then the algorithm can be written in more elementary terms as follows:

$$\gcd(a,a) = a,$$

$$\gcd(a,b) = \gcd(a-b,b) \quad ,\text{if } a > b$$

$$\gcd(a,b) = \gcd(a,b-a) \quad ,\text{if } b > a$$

Animation showing an application of the Euclidean Algorithm to find the Great Common Divisor of 62 and 36 which is 2.

Complexity of Euclidean Method

The existence of the Euclidean algorithm places (the decision problem version of) the greatest common divisor problem in P, the class of problems solvable in polynomial time. The GCD problem is not known to be in NC, and so there is no known way to parallelize its computation across many processors; nor is it known to be P-complete, which would imply that it is unlikely to be possible to parallelize GCD computation. In this sense the GCD problem is analogous to e.g. the integer factorization problem, which has no known polynomial-time algorithm, but is not known to be NP-complete. Shallcross et al. showed that a related problem (EUGCD, determining the remainder sequence arising during the Euclidean algorithm) is NC-equivalent to the problem of integer linear programming with two variables; if either problem is in NC or is P-complete, the other is as well. Since NC contains NL, it is also unknown whether a space-efficient algorithm for computing the GCD exists, even for nondeterministic Turing machines.

Although the problem is not known to be in **NC**, parallel algorithms asymptotically faster than the Euclidean algorithm exist; the best known deterministic algorithm is by Chor and Goldreich, which (in the CRCW-PRAM model) can solve the problem in $O(n/\log n)$ time with $n^{1+\varepsilon}$ processors.

Randomized algorithms can solve the problem in $O((\log n)^2)$ time on $\exp\left[O\left(\sqrt{n \log n}\right)\right]$ processors (note this is superpolynomial).

Binary Method

An alternative method of computing the gcd is the binary gcd method which uses only subtraction and division by 2. In outline the method is as follows: Let a and b be the two non negative integers. Also set the integer d to 0. There are five possibilities:

- $a = b$.

As $\gcd(a, a) = a$, the desired gcd is $a \times 2^d$ (as a and b are changed in the other cases, and d records the number of times that a and b have been both divided by 2 in the next step, the gcd of the initial pair is the product of a by 2^d).

- Both a and b are even.

In this case 2 is a common divisor. Divide both a and b by 2, increment d by 1 to record the number of times 2 is a common divisor and continue.

- a is even and b is odd.

In this case 2 is not a common divisor. Divide a by 2 and continue.

- a is odd and b is even.

As in the previous case 2 is not a common divisor. Divide b by 2 and continue.

- Both a and b are odd.

As $\gcd(a,b) = \gcd(b,a)$ and we have already considered the case $a = b$, we may assume that $a > b$. The number $c = a - b$ is smaller than a yet still positive. Any number that divides a and b must also divide c so every common divisor of a and b is also a common divisor of b and c. Similarly, $a = b + c$ and every common divisor of b and c is also a common divisor of a and b. So the two pairs (a, b) and (b, c) have the same common divisors, and thus $\gcd(a,b) = \gcd(b,c)$. Moreover, as a and b are both odd, c is even, and one may replace c by $c/2$ without changing the gcd. Thus the process can be continued with the pair (a, b) replaced by the smaller numbers $(c/2, b)$.

Each of the above steps reduces at least one of a and b towards 0 and so can only be repeated a finite number of times. Thus one must eventually reach the case $a = b$, which is the only stopping case. Then, as quoted above, the gcd is $a \times 2^d$.

This algorithm may easily programmed as follows:

Input: a, b positive integers

Output: g and d such that g is odd and $\gcd(a, b) = g \times 2^d$

$d := 0$

while a and b are both even do

$a := a/2$

$b := b/2$

$d := d + 1$

while $a \neq b$ do

if a is even then $a := a/2$

else if b is even then $b := b/2$

else if $a > b$ then $a := (a - b)/2$

else $b := (b - a)/2$

$g := a$

output g, d

Example: $(a, b, d) = (48, 18, 0) \rightarrow (24, 9, 1) \rightarrow (12, 9, 1) \rightarrow (6, 9, 1) \rightarrow (3, 9, 1) \rightarrow (3, 6, 1) \rightarrow (3, 3, 1)$; the original gcd is thus $2^d = 2^1$ times $a = b = 3$, that is 6.

The Binary GCD algorithm is particularly easy to implement on binary computers. The test for whether a number is divisible by two can be performed by testing the lowest bit in the number. Division by two can be achieved by shifting the input number by one bit. Each step of the algorithm makes at least one such shift. Subtracting two numbers smaller than a and b costs $O(\log a + \log b)$ bit operations. Each step makes at most one such subtraction. The total number of steps is at most the sum of the numbers of bits of a and b, hence the computational complexity is

$$O((\log a + \log b)^2).$$

Other Methods

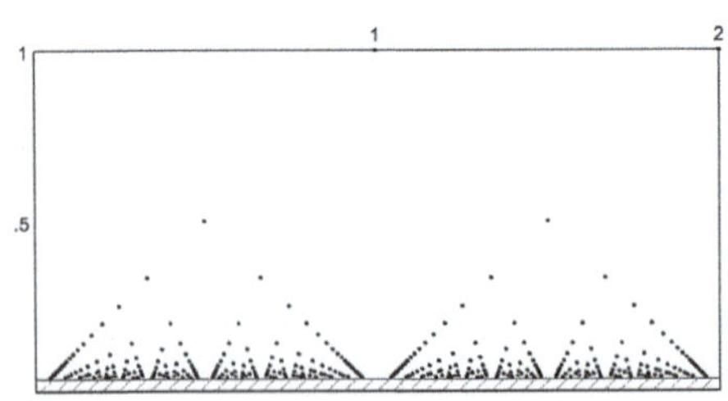

$\gcd(1, x) = y$, or Thomae's function. Hatching at bottom indicates ellipses.

If a and b are both nonzero, the greatest common divisor of a and b can be computed by using least common multiple (lcm) of a and b:

$$\gcd(a, b) = \frac{a \cdot b}{\operatorname{lcm}(a, b)},$$

but more commonly the lcm is computed from the gcd.

Using Thomae's function *f*,

$$\gcd(a,b) = af\left(\frac{b}{a}\right),$$

which generalizes to *a* and *b* rational numbers or commensurable real numbers.

Keith Slavin has shown that for odd $a \geq 1$:

$$\gcd(a,b) = \log_2 \prod_{k=0}^{a-1} (1 + e^{-2i\pi kb/a})$$

which is a function that can be evaluated for complex *b*. Wolfgang Schramm has shown that

$$\gcd(a,b) = \sum_{k=1}^{a} \exp(2\pi ikb/a) \cdot \sum_{d|a} \frac{c_d(k)}{d}$$

is an entire function in the variable *b* for all positive integers *a* where $c_d(k)$ is Ramanujan's sum. Donald Knuth proved the following reduction:

$$\gcd(2^a - 1, 2^b - 1) = 2^{\gcd(a,b)} - 1$$

for non-negative integers *a* and *b*, where *a* and *b* are not both zero. More generally

$$\gcd(n^a - 1, n^b - 1) = n^{\gcd(a,b)} - 1$$

which can be proven by considering the Euclidean algorithm in base *n*. Another useful identity relates $\gcd(a,b)$ to the Euler's totient function:

$$\gcd(a,b) = \sum_{k|a \text{ and } k|b} \varphi(k).$$

Properties

- Every common divisor of *a* and *b* is a divisor of gcd(*a*, *b*).
- gcd(*a*, *b*), where *a* and *b* are not both zero, may be defined alternatively and equivalently as the smallest positive integer *d* which can be written in the form $d = a \cdot p + b \cdot q$, where *p* and *q* are integers. This expression is called Bézout's identity. Numbers *p* and *q* like this can be computed with the extended Euclidean algorithm.

- gcd(a, 0) = $|a|$, for $a \neq 0$, since any number is a divisor of 0, and the greatest divisor of a is $|a|$. This is usually used as the base case in the Euclidean algorithm.
- If a divides the product $b \cdot c$, and gcd(a, b) = d, then a/d divides c.
- If m is a non-negative integer, then gcd($m \cdot a$, $m \cdot b$) = $m \cdot$gcd(a, b).
- If m is any integer, then gcd($a + m \cdot b$, b) = gcd(a, b).
- If m is a nonzero common divisor of a and b, then gcd(a/m, b/m) = gcd(a, b)/m.
- The gcd is a multiplicative function in the following sense: if a_1 and a_2 are relatively prime, then gcd($a_1 \cdot a_2$, b) = gcd(a_1, b)·gcd(a_2, b). In particular, recalling that gcd is a positive integer valued function (i.e., gets natural values only) we obtain that gcd(a, $b \cdot c$) = 1 if and only if gcd(a, b) = 1 and gcd(a, c) = 1.
- The gcd is a commutative function: gcd(a, b) = gcd(b, a).
- The gcd is an associative function: gcd(a, gcd(b, c)) = gcd(gcd(a, b), c).
- The gcd of three numbers can be computed as gcd(a, b, c) = gcd(gcd(a, b), c), or in some different way by applying commutativity and associativity. This can be extended to any number of numbers.
- gcd(a, b) is closely related to the least common multiple lcm(a, b): we have

  gcd(a, b)·lcm(a, b) = $a \cdot b$.

  This formula is often used to compute least common multiples: one first computes the gcd with Euclid's algorithm and then divides the product of the given numbers by their gcd.
- The following versions of distributivity hold true:

  gcd(a, lcm(b, c)) = lcm(gcd(a, b), gcd(a, c))

  lcm(a, gcd(b, c)) = gcd(lcm(a, b), lcm(a, c)).
- If we have the unique prime factorizations of $a = p_1^{e_1} p_2^{e_2} \ldots p_m^{e_m}$ and $b = p_1^{f_1} p_2^{f_2} \ldots p_m^{f_m}$ where $e_i \geq 0$ and $f_i \geq 0$, then the gcd of a and b is

  $$\text{gcd}(a,b) = p_1^{\min(e_1,f_1)} p_2^{\min(e_2,f_2)} \ldots p_m^{\min(e_m,f_m)}$$
- It is sometimes useful to define gcd(0, 0) = 0 and lcm(0, 0) = 0 because then the natural numbers become a complete distributive lattice with gcd as meet and lcm as join operation. This extension of the definition is also compatible with the generalization for commutative rings given below.
- In a Cartesian coordinate system, gcd(a, b) can be interpreted as the number of segments between points with integral coordinates on the straight line segment joining the points (0, 0) and (a, b).

Probabilities and Expected Value

In 1972, James E. Nymann showed that k integers, chosen independently and uniformly from $\{1,\ldots,n\}$, are coprime with probability $1/\zeta(k)$ as n goes to infinity, where ζ refers to the Riemann zeta function. This result was extended in 1987 to show that the probability that k random integers have greatest common divisor d is $d^{-k}/\zeta(k)$.

Using this information, the expected value of the greatest common divisor function can be seen (informally) to not exist when $k = 2$. In this case the probability that the gcd equals d is $d^{-2}/\zeta(2)$, and since $\zeta(2) = \pi^2/6$ we have

$$\mathrm{E}(2) = \sum_{d=1}^{\infty} d \frac{6}{\pi^2 d^2} = \frac{6}{\pi^2} \sum_{d=1}^{\infty} \frac{1}{d}.$$

This last summation is the harmonic series, which diverges. However, when $k \geq 3$, the expected value is well-defined, and by the above argument, it is

$$\mathrm{E}(k) = \sum_{d=1}^{\infty} d^{1-k} \zeta(k)^{-1} = \frac{\zeta(k-1)}{\zeta(k)}.$$

For $k = 3$, this is approximately equal to 1.3684. For $k = 4$, it is approximately 1.1106.

The gcd in Commutative Rings

The notion of greatest common divisor can more generally be defined for elements of an arbitrary commutative ring, although in general there need not exist one for every pair of elements.

If R is a commutative ring, and a and b are in R, then an element d of R is called a *common divisor* of a and b if it divides both a and b (that is, if there are elements x and y in R such that $d{\cdot}x = a$ and $d{\cdot}y = b$). If d is a common divisor of a and b, and every common divisor of a and b divides d, then d is called a *greatest common divisor* of a and b.

Note that with this definition, two elements a and b may very well have several greatest common divisors, or none at all. If R is an integral domain then any two gcd's of a and b must be associate elements, since by definition either one must divide the other; indeed if a gcd exists, any one of its associates is a gcd as well. Existence of a gcd is not assured in arbitrary integral domains. However if R is a unique factorization domain, then any two elements have a gcd, and more generally this is true in gcd domains. If R is a Euclidean domain in which euclidean division is given algorithmically (as is the case for instance when $R = F[X]$ where F is a field, or when R is the ring of Gaussian integers), then greatest common divisors can be computed using a form of the Euclidean algorithm based on the division procedure.

The following is an example of an integral domain with two elements that do not have a gcd:

$$R = \mathbb{Z}\left[\sqrt{-3}\right], \quad a = 4 = 2{\cdot}2 = \left(1+\sqrt{-3}\right)\left(1-\sqrt{-3}\right), \quad b = \left(1+\sqrt{-3}\right){\cdot}2.$$

The elements 2 and $1+\sqrt{-3}$ are two "maximal common divisors" (i.e. any common divisor which is a multiple of 2 is associated to 2, the same holds for $1+\sqrt{-3}$, but they are not associated, so there is no greatest common divisor of a and b.

Corresponding to the Bézout property we may, in any commutative ring, consider the collection of elements of the form $pa + qb$, where p and q range over the ring. This is the ideal generated by a and b, and is denoted simply (a, b). In a ring all of whose ideals are principal (a principal ideal domain or PID), this ideal will be identical with the set of multiples of some ring element d; then this d is a greatest common divisor of a and b. But the ideal (a, b) can be useful even when there is no greatest common divisor of a and b. (Indeed, Ernst Kummer used this ideal as a replacement for a gcd in his treatment of Fermat's Last Theorem, although he envisioned it as the set of multiples of some hypothetical, or *ideal*, ring element d, whence the ring-theoretic term.)

Least Common Multiple

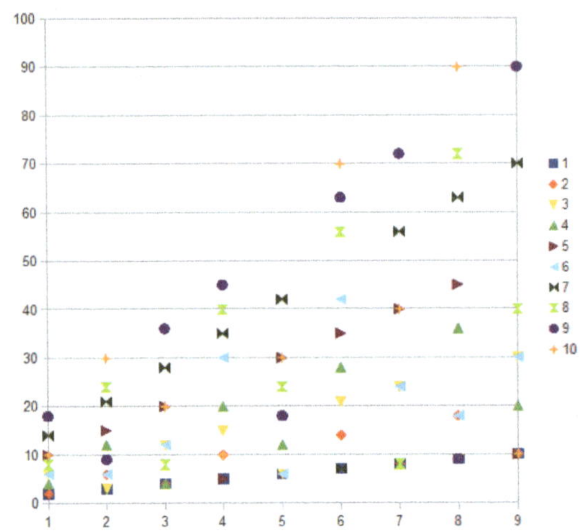

LCMs of numbers 1 through 10 with numbers 2 through 10. Line labels = first number. X axis = second number minus 1. Y axis = LCM of the two numbers.

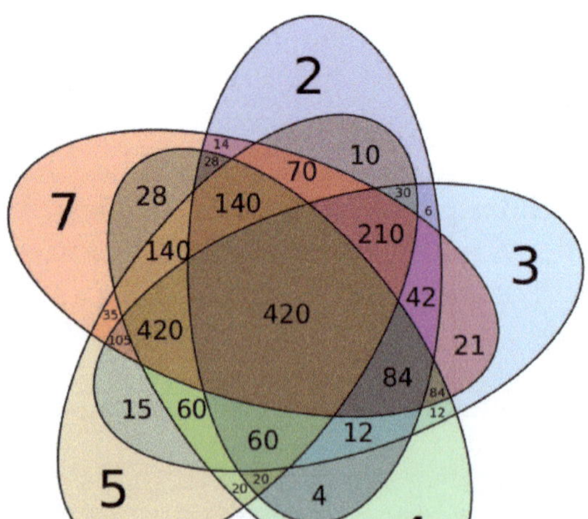

A Venn Diagram showing the least common multiples of combinations of 2, 3, 4, 5 and 7 (6 is skipped as it is 2 × 3, both of which are already represented).

In arithmetic and number theory, the least common multiple, lowest common multiple, or smallest common multiple of two integers *a* and *b*, usually denoted by LCM(***a***, ***b***), is the smallest positive integer that is divisible by both *a* and *b*. Since division of integers by zero is undefined, this definition has meaning only if *a* and *b* are both different from zero. However, some authors define lcm(*a*,0) as 0 for all *a*, which is the result of taking the lcm to be the least upper bound in the lattice of divisibility.

The LCM is familiar from elementary-school arithmetic as the "lowest common denominator" (LCD) that must be determined before fractions can be added, subtracted or compared. The LCM of more than two integers is also well-defined: it is the smallest positive integer that is divisible by each of them.

Overview

A multiple of a number is the product of that number and an integer. For example, 10 is a multiple of 5 because $5 \times 2 = 10$, so 10 is divisible by 5 and 2. Because 10 is the smallest positive integer that is divisible by both 5 and 2, it is the least common multiple of 5 and 2. By the same principle, 10 is the least common multiple of −5 and −2 as well.

Notation

In this article we will denote the least common multiple of two integers *a* and *b* as lcm(*a*, *b*).

Some older textbooks use [*a*, *b*].

The programming language J uses a*.b

Example

What is the LCM of 4 and 6?

Multiples of 4 are:

4, 8, 12, 16, 20, 24, 28, 32, 36, 40, 44, 48, 52, 56, 60, 64, 68, 72, 76, ...

and the multiples of 6 are:

6, 12, 18, 24, 30, 36, 42, 48, 54, 60, 66, 72, ...

Common multiples of 4 and 6 are simply the numbers that are in both lists:

12, 24, 36, 48, 60, 72, ...

So, from this list of the first few common multiples of the numbers 4 and 6, their least common multiple is 12.

Applications

When adding, subtracting, or comparing vulgar fractions, it is useful to find the least common multiple of the denominators, often called the lowest common denominator, because each of the fractions can be expressed as a fraction with this denominator. For instance,

$$\frac{2}{21}+\frac{1}{6}=\frac{4}{42}+\frac{7}{42}=\frac{11}{42}$$

where the denominator 42 was used because it is the least common multiple of 21 and 6.

Computing the Least Common Multiple

Reduction by the Greatest Common Divisor

The following formula reduces the problem of computing the least common multiple to the problem of computing the greatest common divisor (GCD), also known as the greatest common factor:

$$\operatorname{lcm}(a,b)=\frac{|a\cdot b|}{\gcd(a,b)}.$$

This formula is also valid when exactly one of a and b is 0, since $\gcd(a, 0) = |a|$. (However, if both a and b are 0, this formula would cause division by zero; $\operatorname{lcm}(0, 0) = 0$ is a special case.

There are fast algorithms for computing the GCD that do not require the numbers to be factored, such as the Euclidean algorithm. To return to the example above,

$$\operatorname{lcm}(21,6)=\frac{21\cdot 6}{\gcd(21,6)}=\frac{21\cdot 6}{\gcd(3,6)}=\frac{21\cdot 6}{3}=\frac{126}{3}=42.$$

Because $\gcd(a, b)$ is a divisor of both a and b, it is more efficient to compute the LCM by dividing *before* multiplying:

$$\operatorname{lcm}(a,b)=\left(\frac{|a|}{\gcd(a,b)}\right)\cdot|b|=\left(\frac{|b|}{\gcd(a,b)}\right)\cdot|a|.$$

This reduces the size of one input for both the division and the multiplication, and reduces the required storage needed for intermediate results (overflow in the $a\times b$ computation). Because $\gcd(a, b)$ is a divisor of both a and b, the division is guaranteed to yield an integer, so the intermediate result can be stored in an integer. Done this way, the previous example becomes:

$$\operatorname{lcm}(21,6)=\frac{21}{\gcd(21,6)}\cdot 6=\frac{21}{\gcd(3,6)}\cdot 6=\frac{21}{3}\cdot 6=7\cdot 6=42.$$

Finding Least Common Multiples by Prime Factorization

The unique factorization theorem says that every positive integer greater than 1 can be written in only one way as a product of prime numbers. The prime numbers can be considered as the atomic elements which, when combined together, make up a composite number.

For example:

$$90=2^1\cdot 3^2\cdot 5^1=2\cdot 3\cdot 3\cdot 5.$$

Here we have the composite number 90 made up of one atom of the prime number 2, two atoms of the prime number 3 and one atom of the prime number 5.

This knowledge can be used to find the LCM of a set of numbers.

Example: Find the value of lcm(8,9,21).

First, factor each number and express it as a product of prime number powers.

$$8 = 2^3$$
$$9 = 3^2$$
$$21 = 3^1 \cdot 7^1$$

The lcm will be the product of multiplying the highest power of each prime number together. The highest power of the three prime numbers 2, 3, and 7 is 2^3, 3^2, and 7^1, respectively. Thus,

$$\operatorname{lcm}(8,9,21) = 2^3 \cdot 3^2 \cdot 7^1 = 8 \cdot 9 \cdot 7 = 504.$$

This method is not as efficient as reducing to the greatest common divisor, since there is no known general efficient algorithm for integer factorization, but is useful for illustrating concepts.

This method can be illustrated using a Venn diagram as follows. Find the prime factorization of each of the two numbers. Put the prime factors into a Venn diagram with one circle for each of the two numbers, and *all* factors they share in common in the intersection. To find the LCM, just multiply all of the prime numbers in the diagram.

Here is an example:

$48 = 2 \times 2 \times 2 \times 2 \times 3,$

$180 = 2 \times 2 \times 3 \times 3 \times 5,$

and what they share in common is two "2"s and a "3":

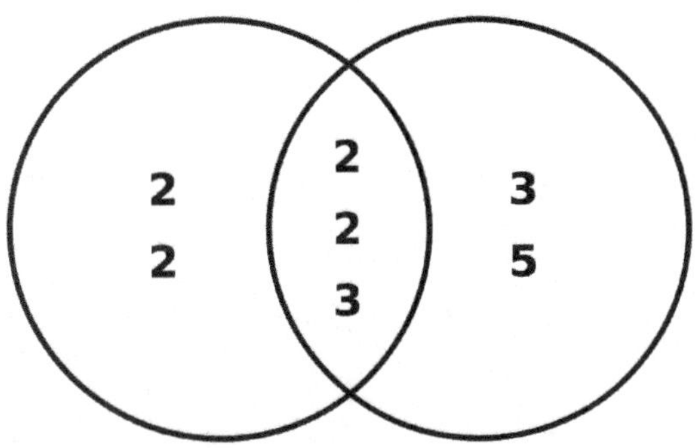

Least common multiple $= 2 \times 2 \times 2 \times 2 \times 3 \times 3 \times 5 = 720$

Greatest common divisor $= 2 \times 2 \times 3 = 12$

This also works for the greatest common divisor (GCD), except that instead of multiplying all of the numbers in the Venn diagram, one multiplies only the prime factors that are in the intersection. Thus the GCD of 48 and 180 is $2 \times 2 \times 3 = 12$.

A Simple Algorithm

This method works as easily for finding the LCM of several integers.

Let there be a finite sequence of positive integers $X = (x_1, x_2, ..., x_n)$, $n > 1$. The algorithm proceeds in steps as follows: on each step m it examines and updates the sequence $X^{(m)} = (x_1^{(m)}, x_2^{(m)}, ..., x_n^{(m)})$, $X^{(1)} = X$, where $X^{(m)}$ is the mth iteration of X, i.e. X at step m of the algorithm, etc. The purpose of the examination is to pick the least (perhaps, one of many) element of the sequence $X^{(m)}$. Assuming $x_{ko}^{(m)}$ is the selected element, the sequence $X^{(m+1)}$ is defined as

$$x_k^{(m+1)} = x_k^{(m)},\ k \neq k_o$$

$$x_{ko}^{(m+1)} = x_{ko}^{(m)} + x_{ko}^{(1)}.$$

In other words, the least element is increased by the corresponding x whereas the rest of the elements pass from $X^{(m)}$ to $X^{(m+1)}$ unchanged.

The algorithm stops when all elements in sequence $X^{(m)}$ are equal. Their common value L is exactly LCM(X).

A Method using a Table

This method works for any number of factors. One begins by listing all of the numbers vertically in a table (in this example 4, 7, 12, 21, and 42):

4

7

12

21

42

The process begins by dividing all of the factors by 2. If any of them divides evenly, write 2 at the top of the table and the result of division by 2 of each factor in the space to the right of each factor and below the 2. If a number does not divide evenly, just rewrite the number again. If 2 does not divide evenly into any of the numbers, try 3.

x	**2**
4	2
7	7
12	6
21	21
42	21

Now, check if 2 divides again:

x	2	2
4	2	1
7	7	7
12	6	3
21	21	21
42	21	21

Once 2 no longer divides, divide by 3. If 3 no longer divides, try 5 and 7. Keep going until all of the numbers have been reduced to 1.

x	2	2	3	7
4	2	1	1	1
7	7	7	7	1
12	6	3	1	1
21	21	21	7	1
42	21	21	7	1

Now, multiply the numbers on the top and you have the LCM. In this case, it is $2 \times 2 \times 3 \times 7 = 84$. You will get to the LCM the quickest if you use prime numbers and start from the lowest prime, 2.

As a general computational algorithm, the above is quite inefficient. One would never want to implement it in software: it takes too many steps, and requires too much storage space. A far more efficient numerical algorithm can be obtained simply by using Euclid's algorithm to compute the gcd first, and then obtaining the lcm by division.

Formulas

Fundamental Theorem of Arithmetic

According to the fundamental theorem of arithmetic a positive integer is the product of prime numbers, and, except for their order, this representation is unique:

$$n = 2^{n_2} 3^{n_3} 5^{n_5} 7^{n_7} \cdots = \prod_p p^{n_p},$$

where the exponents n_2, n_3, ... are non-negative integers; for example, $84 = 2^2\, 3^1\, 5^0\, 7^1\, 11^0\, 13^0 \ldots$

Given two positive integers $a = \prod_p p^{a_p}$ and $b = \prod_p p^{b_p}$ their least common multiple and greatest common divisor are given by the formulas

$$\gcd(a,b) = \prod_p p^{\min(a_p, b_p)}$$

and

$$\operatorname{lcm}(a,b) = \prod_p p^{\max(a_p, b_p)}.$$

Since

$$\min(x,y) + \max(x,y) = x + y,$$

this gives

$$\gcd(a,b)\operatorname{lcm}(a,b) = ab.$$

In fact, any rational number can be written uniquely as the product of primes if negative exponents are allowed. When this is done, the above formulas remain valid. For example:

$4 = 2^2 3^0,$	$6 = 2^1 3^1,$	$\gcd(4,6) = 2^1 3^0,$	$\operatorname{lcm}(4,6) = 2^2 3^1.$
$\frac{1}{3} = 2^0 3^{-1} 5^0,$	$\frac{2}{5} = 2^1 3^0 5^{-1},$	$\gcd(\frac{1}{3}, \frac{2}{5}) = 2^0 3^{-1} 5^{-1},$	$\operatorname{lcm}(\frac{1}{3}, \frac{2}{5}) = 2^1 3^0 5^0,$
$\frac{1}{6} = 2^{-1} 3^{-1},$	$\frac{3}{4} = 2^{-2} 3^1,$	$\gcd(\frac{1}{6}, \frac{3}{4}) = 2^{-2} 3^{-1},$	$\operatorname{lcm}(\frac{1}{6}, \frac{3}{4}) = 2^{-1} 3^1.$

Lattice-theoretic

The positive integers may be partially ordered by divisibility: if a divides b (i.e. if b is an integer multiple of a) write $a \le b$ (or equivalently, $b \ge a$).

Under this ordering, the positive integers become a lattice with meet given by the gcd and join given by the lcm. The proof is straightforward, if a bit tedious; it amounts to checking that lcm and gcd satisfy the axioms for meet and join. Putting the lcm and gcd into this more general context establishes a duality between them:

If a formula involving integer variables, gcd, lcm, ≤ and ≥ is true, then the formula obtained by switching gcd with lcm and switching ≥ with ≤ is also true. (Remember ≤ is defined as divides).

The following pairs of dual formulas are special cases of general lattice-theoretic identities.

Commutative laws	Associative laws	Absorption laws
$\operatorname{lcm}(a,b) = \operatorname{lcm}(b,a),$	$\operatorname{lcm}(a, \operatorname{lcm}(b,c)) = \operatorname{lcm}(\operatorname{lcm}(a,b),c),$	$\operatorname{lcm}(a, \gcd(a,b)) = a,$
$\gcd(a,b) = \gcd(b,a).$	$\gcd(a, \gcd(b,c)) = \gcd(\gcd(a,b),c).$	$\gcd(a, \operatorname{lcm}(a,b)) = a.$

Idempotent laws	Define divides in terms of lcm and gcd
$\operatorname{lcm}(a,a) = a,$	$a \ge b \Leftrightarrow a = \operatorname{lcm}(a,b),$
$\gcd(a,a) = a.$	$a \le b \Leftrightarrow a = \gcd(a,b).$

It can also be shown that this lattice is distributive, i.e. that lcm distributes over gcd and, dually, that gcd distributes over lcm:

$$\text{lcm}(a, \gcd(b,c)) = \gcd(\text{lcm}(a,b), \text{lcm}(a,c)),$$

$$\gcd(a, \text{lcm}(b,c)) = \text{lcm}(\gcd(a,b), \gcd(a,c)).$$

This identity is self-dual:

Other

Let D be the product of $\omega(D)$ distinct prime numbers (i.e. D is squarefree).

Then

$$\gcd(\text{lcm}(a,b), \text{lcm}(b,c), \text{lcm}(a,c)) = \text{lcm}(\gcd(a,b), \gcd(b,c), \gcd(a,c)).$$

where the absolute bars || denote the cardinality of a set.

The LCM in Commutative Rings

The least common multiple can be defined generally over commutative rings as follows: Let a and b be elements of a commutative ring R. A common multiple of a and b is an element m of R such that both a and b divide m (i.e. there exist elements x and y of R such that $ax = m$ and $by = m$). A least common multiple of a and b is a common multiple that is minimal in the sense that for any other common multiple n of a and b, m divides n.

In general, two elements in a commutative ring can have no least common multiple or more than one. However, any two least common multiples of the same pair of elements are associates. In a unique factorization domain, any two elements have a least common multiple. In a principal ideal domain, the least common multiple of a and b can be characterised as a generator of the intersection of the ideals generated by a and b (the intersection of a collection of ideals is always an ideal).

References

- Schmid, Hermann (1983) [1974]. Decimal Computation (1 (reprint) ed.). Malabar, Florida, USA: Robert E. Krieger Publishing Company. ISBN 0-89874-318-4.
- Decimal Floating-Point: Algorism for Computers, Cowlishaw, M. F., Proceedings 16th IEEE Symposium on Computer Arithmetic, ISBN 0-7695-1894-X, pp104-111, IEEE Comp. Soc., June 2003
- Menninger, Karl: Zahlwort und Ziffer. Eine Kulturgeschichte der Zahl, Vandenhoeck und Ruprecht, 3rd. ed., 1979, ISBN 3-525-40725-4, pp. 150-153
- Mazaudon, Martine (2002). "Les principes de construction du nombre dans les langues tibéto-birmanes". In François, Jacques. La Pluralité (PDF). Leuven: Peeters. pp. 91–119. ISBN 90-429-1295-2
- Gauss, Carl Friedrich; Clarke, Arthur A. (translator into English) (1986), Disquisitiones Arithemeticae (Second, corrected edition), New York: Springer, ISBN 978-0-387-96254-2
- Baker, Alan (1984), A Concise Introduction to the Theory of Numbers, Cambridge, UK: Cambridge University Press, ISBN 978-0-521-28654-1

3

Probability

Probability is the possibility of an event happening. It can be measured between 0 to 1, where 0 is uncertainty and 1 signifies certainty. Sample space, event, random variable, expected value and conditional probability are some of the topics elucidated in the text. This chapter elucidates the crucial theories and principles of probability.

Probability

Probability is the measure of the likelihood that an event will occur. Probability is quantified as a number between 0 and 1 (where 0 indicates impossibility and 1 indicates certainty). The higher the probability of an event, the more certain that the event will occur. A simple example is the tossing of a fair (unbiased) coin. Since the coin is unbiased, the two outcomes ("head" and "tail") are both equally probable; the probability of "head" equals the probability of "tail." Since no other outcomes are possible, the probability is 1/2 (or 50%), of either "head" or "tail". In other words, the probability of "head" is 1 out of 2 outcomes and the probability of "tail" is also 1 out of 2 outcomes, expressed as 0.5 when converted to decimal, with the above-mentioned quantification system. This type of probability is also called a priori probability.

These concepts have been given an axiomatic mathematical formalization in probability theory, which is used widely in such areas of study as mathematics, statistics, finance, gambling, science (in particular physics), artificial intelligence/machine learning, computer science, game theory, and philosophy to, for example, draw inferences about the expected frequency of events. Probability theory is also used to describe the underlying mechanics and regularities of complex systems.

Interpretations

When dealing with experiments that are random and well-defined in a purely theoretical setting (like tossing a fair coin), probabilities can be numerically described by the number of desired outcomes divided by the total number of all outcomes. For example, tossing a fair coin twice will yield "head-head", "head-tail", "tail-head", and "tail-tail" outcomes. The probability of getting an outcome of "head-head" is 1 out of 4 outcomes or 1/4 or 0.25 (or 25%). When it comes to practical application however, there are two major competing categories of probability interpretations, whose adherents possess different views about the fundamental nature of probability:

- Objectivists assign numbers to describe some objective or physical state of affairs. The most popular version of objective probability is frequentist probability, which claims that the probability of a random event denotes the *relative frequency of occurrence* of an experiment's outcome, when repeating the experiment. This interpretation considers probability to be the relative frequency "in the long run" of outcomes. A modification of this is propensity probability, which interprets probability as the tendency of some experiment to yield a certain outcome, even if it is performed only once.

- Subjectivists assign numbers per subjective probability, i.e., as a degree of belief. The degree of belief has been interpreted as, "the price at which you would buy or sell a bet that pays 1 unit of utility if E, 0 if not E." The most popular version of subjective probability is Bayesian probability, which includes expert knowledge as well as experimental data to produce probabilities. The expert knowledge is represented by some (subjective) prior probability distribution. These data are incorporated in a likelihood function. The product of the prior and the likelihood, normalized, results in a posterior probability distribution that incorporates all the information known to date. Starting from arbitrary, subjective probabilities for a group of agents, some Bayesians claim that all agents will eventually have sufficiently similar assessments of probabilities, given enough evidence.

Etymology

The word *probability* derives from the Latin *probabilitas*, which can also mean "probity", a measure of the authority of a witness in a legal case in Europe, and often correlated with the witness's nobility. In a sense, this differs much from the modern meaning of *probability*, which, in contrast, is a measure of the weight of empirical evidence, and is arrived at from inductive reasoning and statistical inference.

History

The scientific study of probability is a modern development of mathematics. Gambling shows that there has been an interest in quantifying the ideas of probability for millennia, but exact mathematical descriptions arose much later. There are reasons of course, for the slow development of the mathematics of probability. Whereas games of chance provided the impetus for the mathematical study of probability, fundamental issues are still obscured by the superstitions of gamblers.

Christiaan Huygens probably published the first book on probability

According to Richard Jeffrey, "Before the middle of the seventeenth century, the term 'probable' (Latin *probabilis*) meant *approvable*, and was applied in that sense, univocally, to opinion and to action. A probable action or opinion was one such as sensible people would undertake or hold, in the circumstances." However, in legal contexts especially, 'probable' could also apply to propositions for which there was good evidence.

Gerolamo Cardano

The sixteenth century Italian polymath Gerolamo Cardano demonstrated the efficacy of defining odds as the ratio of favourable to unfavourable outcomes (which implies that the probability of an event is given by the ratio of favourable outcomes to the total number of possible outcomes). Aside from the elementary work by Cardano, the doctrine of probabilities dates to the correspondence of Pierre de Fermat and Blaise Pascal (1654). Christiaan Huygens (1657) gave the earliest known scientific treatment of the subject. Jakob Bernoulli's *Ars Conjectandi* (posthumous, 1713) and Abraham de Moivre's *Doctrine of Chances* (1718) treated the subject as a branch of mathematics.

The theory of errors may be traced back to Roger Cotes's *Opera Miscellanea* (posthumous, 1722), but a memoir prepared by Thomas Simpson in 1755 (printed 1756) first applied the theory to the discussion of errors of observation. The reprint (1757) of this memoir lays down the axioms that positive and negative errors are equally probable, and that certain assignable limits define the range of all errors. Simpson also discusses continuous errors and describes a probability curve.

The first two laws of error that were proposed both originated with Pierre-Simon Laplace. The first law was published in 1774 and stated that the frequency of an error could be expressed as an exponential function of the numerical magnitude of the error, disregarding sign. The second law of error was proposed in 1778 by Laplace and stated that the frequency of the error is an exponential function of the square of the error. The second law of error is called the normal distribution or the Gauss law. "It is difficult historically to attribute that law to Gauss, who in spite of his well-known precocity had probably not made this discovery before he was two years old."

Daniel Bernoulli (1778) introduced the principle of the maximum product of the probabilities of a system of concurrent errors.

Carl Friedrich Gauss

Adrien-Marie Legendre (1805) developed the method of least squares, and introduced it in his *Nouvelles méthodes pour la détermination des orbites des comètes* (*New Methods for Determining the Orbits of Comets*). In ignorance of Legendre's contribution, an Irish-American writer, Robert Adrain, editor of "The Analyst" (1808), first deduced the law of facility of error,

$$\phi(x) = ce^{-h^2x^2},$$

where h is a constant depending on precision of observation, and c is a scale factor ensuring that the area under the curve equals 1. He gave two proofs, the second being essentially the same as John Herschel's (1850). Gauss gave the first proof that seems to have been known in Europe (the third after Adrain's) in 1809. Further proofs were given by Laplace (1810, 1812), Gauss (1823), James Ivory (1825, 1826), Hagen (1837), Friedrich Bessel (1838), W. F. Donkin (1844, 1856), and Morgan Crofton (1870). Other contributors were Ellis (1844), De Morgan (1864), Glaisher (1872), and Giovanni Schiaparelli (1875). Peters's (1856) formula for r, the probable error of a single observation, is well known.

In the nineteenth century authors on the general theory included Laplace, Sylvestre Lacroix (1816), Littrow (1833), Adolphe Quetelet (1853), Richard Dedekind (1860), Helmert (1872), Hermann Laurent (1873), Liagre, Didion, and Karl Pearson. Augustus De Morgan and George Boole improved the exposition of the theory.

Andrey Markov introduced the notion of Markov chains (1906), which played an important role in stochastic processes theory and its applications. The modern theory of probability based on the measure theory was developed by Andrey Kolmogorov (1931).

On the geometric side contributors to *The Educational Times* were influential (Miller, Crofton, McColl, Wolstenholme, Watson, and Artemas Martin).

Theory

Like other theories, the theory of probability is a representation of probabilistic concepts in formal terms—that is, in terms that can be considered separately from their meaning. These formal terms are manipulated by the rules of mathematics and logic, and any results are interpreted or translated back into the problem domain.

There have been at least two successful attempts to formalize probability, namely the Kolmogorov formulation and the Cox formulation. In Kolmogorov's formulation, sets are interpreted as events and probability itself as a measure on a class of sets. In Cox's theorem, probability is taken as a primitive (that is, not further analyzed) and the emphasis is on constructing a consistent assignment of probability values to propositions. In both cases, the laws of probability are the same, except for technical details.

There are other methods for quantifying uncertainty, such as the Dempster–Shafer theory or possibility theory, but those are essentially different and not compatible with the laws of probability as usually understood.

Applications

Probability theory is applied in everyday life in risk assessment and modeling. The insurance industry and markets use actuarial science to determine pricing and make trading decisions. Governments apply probabilistic methods in environmental regulation, entitlement analysis (Reliability theory of aging and longevity), and financial regulation.

A good example of the use of probibility theory in equity trading is the effect of the perceived probability of any widespread Middle East conflict on oil prices, which have ripple effects in the economy as a whole. An assessment by a commodity trader that a war is more likely can send that commodity's prices up or down, and signals other traders of that opinion. Accordingly, the probabilities are neither assessed independently nor necessarily very rationally. The theory of behavioral finance emerged to describe the effect of such groupthink on pricing, on policy, and on peace and conflict.

In addition to financial assessment, probability can be used to analyze trends in biology (e.g. disease spread) as well as ecology (e.g. biological Punnett squares). As with finance, risk assessment can be used as a statistical tool to calculate the likelihood of undesirable events occurring and can assist with implementing protocols to avoid encountering such circumstances. Probability is used to design games of chance so that casinos can make a guaranteed profit, yet provide payouts to players that are frequent enough to encourage continued play.

The discovery of rigorous methods to assess and combine probability assessments has changed society. It is important for most citizens to understand how probability assessments are made, and how they contribute to decisions.

Another significant application of probability theory in everyday life is reliability. Many consumer products, such as automobiles and consumer electronics, use reliability theory in product design to reduce the probability of failure. Failure probability may influence a manufacturer's decisions on a product's warranty.

The cache language model and other statistical language models that are used in natural language processing are also examples of applications of probability theory.

Mathematical Treatment

Consider an experiment that can produce a number of results. The collection of all possible results is called the sample space of the experiment. The power set of the sample space is formed by considering all different collections of possible results. For example, rolling a dice can produce six possible results. One collection of possible results gives an odd number on the dice. Thus, the subset {1,3,5} is an element of the power set of the sample space of dice rolls. These collections are called "events." In this case, {1,3,5} is the event that the dice falls on some odd number. If the results that actually occur fall in a given event, the event is said to have occurred.

A probability is a way of assigning every event a value between zero and one, with the requirement that the event made up of all possible results (in our example, the event {1,2,3,4,5,6}) is assigned a value of one. To qualify as a probability, the assignment of values must satisfy the requirement that if you look at a collection of mutually exclusive events (events with no common results, e.g., the events {1,6}, {3}, and {2,4} are all mutually exclusive), the probability that at least one of the events will occur is given by the sum of the probabilities of all the individual events.

The probability of an event A is written as $P(A)$, $p(A)$, or $\Pr(A)$. This mathematical definition of probability can extend to infinite sample spaces, and even uncountable sample spaces, using the concept of a measure.

The *opposite* or *complement* of an event A is the event [not A] (that is, the event of A not occurring), often denoted as $\overline{A}, A^C, \neg A$, or $\sim A$; its probability is given by $P(\text{not } A) = 1 - P(A)$. As an example, the chance of not rolling a six on a six-sided die is 1 – (chance of rolling a six) $= 1 - \frac{1}{6} = \frac{5}{6}$.

If two events A and B occur on a single performance of an experiment, this is called the intersection or joint probability of A and B, denoted as $P(A \cap B)$.

Independent Events

If two events, A and B are independent then the joint probability is

$$P(A \text{ and } B) = P(A \cap B) = P(A)P(B),$$

for example, if two coins are flipped the chance of both being heads is $\frac{1}{2} \times \frac{1}{2} = \frac{1}{4}$.

Mutually Exclusive Events

If either event A or event B occurs on a single performance of an experiment this is called the union of the events A and B denoted as $P(A \cup B)$. If two events are mutually exclusive then the probability of either occurring is

$$P(A \text{ or } B) = P(A \cup B) = P(A) + P(B).$$

For example, the chance of rolling a 1 or 2 on a six-sided die is $P(1 \text{ or } 2) = P(1) + P(2) = \frac{1}{6} + \frac{1}{6} = \frac{1}{3}$.

Not Mutually Exclusive Events

If the events are not mutually exclusive then

$$P(A \text{ or } B) = P(A) + P(B) - P(A \text{ and } B).$$

For example, when drawing a single card at random from a regular deck of cards, the chance of getting a heart or a face card (J,Q,K) (or one that is both) is $\frac{13}{52}+\frac{12}{52}-\frac{3}{52}=\frac{11}{26}$, because of the 52 cards of a deck 13 are hearts, 12 are face cards, and 3 are both: here the possibilities included in the "3 that are both" are included in each of the "13 hearts" and the "12 face cards" but should only be counted once.

Conditional Probability

Conditional probability is the probability of some event *A*, given the occurrence of some other event *B*. Conditional probability is written $P(A \mid B)$, and is read "the probability of *A*, given *B*". It is defined by

$$P(A \mid B)=\frac{P(A \cap B)}{P(B)}.$$

If $P(B)=0$ then $P(A \mid B)$ is formally undefined by this expression. However, it is possible to define a conditional probability for some zero-probability events using a σ-algebra of such events (such as those arising from a continuous random variable).

For example, in a bag of 2 red balls and 2 blue balls (4 balls in total), the probability of taking a red ball is $1/2$; however, when taking a second ball, the probability of it being either a red ball or a blue ball depends on the ball previously taken, such as, if a red ball was taken, the probability of picking a red ball again would be $1/3$ since only 1 red and 2 blue balls would have been remaining.

Inverse Probability

In probability theory and applications, Bayes' rule relates the odds of event A_1 to event A_2, before (prior to) and after (posterior to) conditioning on another event B. The odds on A_1 to event is simply the ratio of the probabilities of the two events. When arbitrarily many events A are of interest, not just two, the rule can be rephrased as posterior is proportional to prior times likelihood, $P(A \mid B) \propto P(A)\ P(B \mid A)$ where the proportionality symbol means that the left hand side is proportional to (i.e., equals a constant times) the right hand side as *A* varies, for fixed or given *B* (Lee, 2012; Bertsch McGrayne, 2012). In this form it goes back to Laplace (1774) and to Cournot (1843).

Summary of Probabilities

Summary of probabilities

Event	Probability
A	$P(A) \in [0,1]$
not A	$P(A^c)=1-P(A)$
A or B	$P(A \cup B)=P(A)+P(B)-P(A \cap B)$, $P(A \cup B)=P(A)+P(B)$ if A and B are mutually exclusive
A and B	$P(A \cap B)=P(A \mid B)P(B)=P(B \mid A)P(A)P(A \cap B)=P(A)P(B)$ if A and B are independent

A given B	$P(A\mid B)=\dfrac{P(A\cap B)}{P(B)}=\dfrac{P(B\mid A)P(A)}{P(B)}$

Relation to Randomness

In a deterministic universe, based on Newtonian concepts, there would be no probability if all conditions were known (Laplace's demon), (but there are situations in which sensitivity to initial conditions exceeds our ability to measure them, i.e. know them). In the case of a roulette wheel, if the force of the hand and the period of that force are known, the number on which the ball will stop would be a certainty (though as a practical matter, this would likely be true only of a roulette wheel that had not been exactly levelled — as Thomas A. Bass' Newtonian Casino revealed). Of course, this also assumes knowledge of inertia and friction of the wheel, weight, smoothness and roundness of the ball, variations in hand speed during the turning and so forth. A probabilistic description can thus be more useful than Newtonian mechanics for analyzing the pattern of outcomes of repeated rolls of a roulette wheel. Physicists face the same situation in kinetic theory of gases, where the system, while deterministic *in principle*, is so complex (with the number of molecules typically the order of magnitude of Avogadro constant $6.02\cdot10^{23}$) that only a statistical description of its properties is feasible.

Probability theory is required to describe quantum phenomena. A revolutionary discovery of early 20th century physics was the random character of all physical processes that occur at sub-atomic scales and are governed by the laws of quantum mechanics. The objective wave function evolves deterministically but, according to the Copenhagen interpretation, it deals with probabilities of observing, the outcome being explained by a wave function collapse when an observation is made. However, the loss of determinism for the sake of instrumentalism did not meet with universal approval. Albert Einstein famously remarked in a letter to Max Born: "I am convinced that God does not play dice". Like Einstein, Erwin Schrödinger, who discovered the wave function, believed quantum mechanics is a statistical approximation of an underlying deterministic reality. In some modern interpretations of the statistical mechanics of measurement, quantum decoherence is invoked to account for the appearance of subjectively probabilistic experimental outcomes.

Sample Space

In probability theory, the sample space of an experiment or random trial is the set of all possible outcomes or results of that experiment. A sample space is usually denoted using set notation, and the possible outcomes are listed as elements in the set. It is common to refer to a sample space by the labels S, Ω, or U (for "universal set").

For example, if the experiment is tossing a coin, the sample space is typically the set {head, tail}. For tossing two coins, the corresponding sample space would be {(head,head), (head,tail), (tail,-head), (tail,tail)}. For tossing a single six-sided die, the typical sample space is {1, 2, 3, 4, 5, 6} (in which the result of interest is the number of pips facing up).

A well-defined sample space is one of three basic elements in a probabilistic model (a probability space); the other two are a well-defined set of possible events (a sigma-algebra) and a probability assigned to each event (a probability measure function).

Multiple Sample Spaces

For many experiments, there may be more than one plausible sample space available, depending on what result is of interest to the experimenter. For example, when drawing a card from a standard deck of fifty-two playing cards, one possibility for the sample space could be the various ranks (Ace through King), while another could be the suits (clubs, diamonds, hearts, or spades). A more complete description of outcomes, however, could specify both the denomination and the suit, and a sample space describing each individual card can be constructed as the Cartesian product of the two sample spaces noted above (this space would contain fifty-two equally likely outcomes). Still other sample spaces are possible, such as {right-side up, up-side down} if some cards have been flipped when shuffling.

Equally Likely Outcomes

Flipping a coin leads to a **sample space** composed of two outcomes that are almost equally likely.

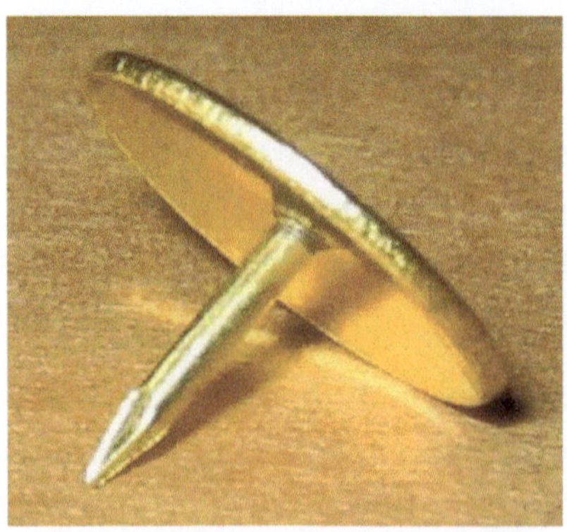

Up or down? Flipping a brass tack leads to a **sample space** composed of two outcomes that are not equally likely.

Some treatments of probability assume that the various outcomes of an experiment are always defined so as to be equally likely. However, there are experiments that are not easily described by a sample space of equally likely outcomes—for example, if one were to toss a thumb tack many times and observe whether it landed with its point upward or downward, there is no symmetry to suggest that the two outcomes should be equally likely.

Though most random phenomena do not have equally likely outcomes, it can be helpful to define a sample space in such a way that outcomes are at least approximately equally likely, since this condition significantly simplifies the computation of probabilities for events within the sample space. If each individual outcome occurs with the same probability, then the probability of any event becomes simply:

$$P(event) = \frac{\text{number of outcomes in event}}{\text{number of outcomes in sample space}}$$

Simple Random Sample

In statistics, inferences are made about characteristics of a population by studying a sample of that population's individuals. In order to arrive at a sample that presents an unbiased estimate of the true characteristics of the population, statisticians often seek to study a simple random sample—that is, a sample in which every individual in the population is equally likely to be included. The result of this is that every possible combination of individuals who could be chosen for the sample is also equally likely (that is, the space of simple random samples of a given size from a given population is composed of equally likely outcomes).

Infinitely Large Sample Spaces

In an elementary approach to probability, any subset of the sample space is usually called an event. However, this gives rise to problems when the sample space is infinite, so that a more precise definition of an event is necessary. Under this definition only measurable subsets of the sample space, constituting a σ-algebra over the sample space itself, are considered events.

Event (Probability Theory)

In probability theory, an event is a set of outcomes of an experiment (a subset of the sample space) to which a probability is assigned. A single outcome may be an element of many different events, and different events in an experiment are usually not equally likely, since they may include very different groups of outcomes. An event defines a complementary event, namely the complementary set (the event *not* occurring), and together these define a Bernoulli trial: did the event occur or not?

Typically, when the sample space is finite, any subset of the sample space is an event (*i.e.* all elements of the power set of the sample space are defined as events). However, this approach does not work well in cases where the sample space is uncountably infinite, most notably when the outcome is a real number. So, when defining a probability space it is possible, and often necessary, to exclude certain subsets of the sample space from being events.

A Simple Example

If we assemble a deck of 52 playing cards with no jokers, and draw a single card from the deck, then the sample space is a 52-element set, as each card is a possible outcome. An event, however, is any subset of the sample space, including any singleton set (an elementary event), the empty set (an impossible event, with probability zero) and the sample space itself (a certain event, with probability one). Other events are proper subsets of the sample space that contain multiple elements. So, for example, potential events include:

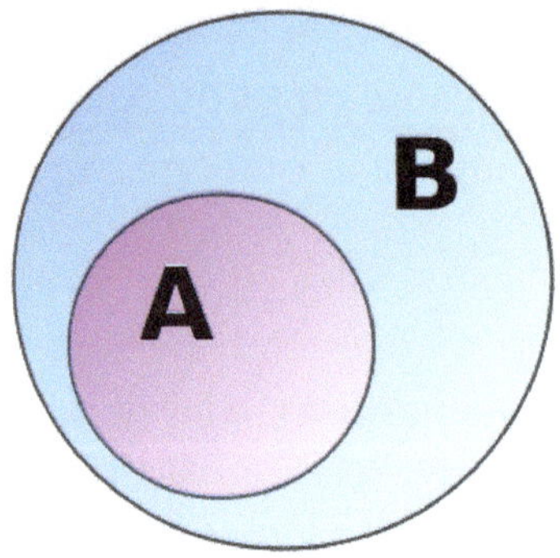

A Venn diagram of an event. *B* is the sample space and *A* is an event.

By the ratio of their areas, the probability of *A* is approximately 0.4.

- "Red and black at the same time without being a joker" (0 elements),
- "The 5 of Hearts" (1 element),
- "A King" (4 elements),
- "A Face card" (12 elements),
- "A Spade" (13 elements),
- "A Face card or a red suit" (32 elements),
- "A card" (52 elements).

Since all events are sets, they are usually written as sets (e.g. {1, 2, 3}), and represented graphically using Venn diagrams. Given that each outcome in the sample space Ω is equally likely, the probability *P* of an event *A* is the following formula:

$$P(A) = \frac{|A|}{|\Omega|} \left(\text{alternatively: } \Pr(A) = \frac{|A|}{|\Omega|} \right)$$

This rule can readily be applied to each of the example events above.

Events in Probability Spaces

Defining all subsets of the sample space as events works well when there are only finitely many outcomes, but gives rise to problems when the sample space is infinite. For many standard probability distributions, such as the normal distribution, the sample space is the set of real numbers or some subset of the real numbers. Attempts to define probabilities for all subsets of the real numbers run into difficulties when one considers 'badly behaved' sets, such as those that are nonmeasurable. Hence, it is necessary to restrict attention to a more limited family of subsets. For the standard tools of probability theory, such as joint and conditional probabilities, to work, it is necessary to use a σ-algebra, that is, a family closed under complementation and countable unions of its members. The most natural choice is the Borel measurable set derived from unions and intersections of intervals. However, the larger class of Lebesgue measurable sets proves more useful in practice.

In the general measure-theoretic description of probability spaces, an event may be defined as an element of a selected σ-algebra of subsets of the sample space. Under this definition, any subset of the sample space that is not an element of the σ-algebra is not an event, and does not have a probability. With a reasonable specification of the probability space, however, all *events of interest* are elements of the σ-algebra.

A Note on Notation

Even though events are subsets of some sample space Ω, they are often written as propositional formulas involving random variables. For example, if X is a real-valued random variable defined on the sample space Ω, the event

$$\{\omega \in \Omega \mid u < X(\omega) \leq v\}$$

can be written more conveniently as, simply,

$$u < X \leq v.$$

This is especially common in formulas for a probability, such as

$$\Pr(u < X \leq v) = F(v) - F(u).$$

The set $u < X \leq v$ is an example of an inverse image under the mapping X because $\omega \in X^{-1}((u,v])$ if and only if $u < X(\omega) \leq v$.

Random Variable

In probability and statistics, a random variable, random quantity, aleatory variable or stochastic variable is a quantity whose value depends in some clearly-defined way on a set of possible random events. A random variable can take on a set of possible different values (similarly to other mathematical variables), each with an associated probability. Like a traditional mathematical variable, its value is thus unknown a priori (before the outcome of the events is known). However, in a more abstract sense, it is defined as a function that maps from an outcome of the events (that is, from a point in a probability space) to a mathematically convenient outcome label, usually a real number. In this sense, it is a procedure for assigning a number to an outcome, and paradoxically this procedure itself is neither random nor variable. The function which characterizes a random variable must also be measurable, which rules out certain pathological cases such as those in which the random variable's quantity is infinitely sensitive to any small change in the outcome.

A random variable's possible values might represent the possible outcomes of a yet-to-be-performed experiment, or the possible outcomes of a past experiment whose already-existing value is uncertain (for example, due to imprecise measurements or quantum uncertainty). They may also conceptually represent either the results of an "objectively" random process (such as rolling a die) or the "subjective" randomness that results from incomplete knowledge of a quantity. The meaning of the probabilities assigned to the potential values of a random variable is not part of probability theory itself but is instead related to philosophical arguments over the interpretation of probability. The mathematics works the same regardless of the particular interpretation in use.

A random variable has a probability distribution, which specifies the probability that its value falls in any given interval. Random variables can be discrete, that is, taking any of a specified finite or countable list of values, endowed with a probability mass function characteristic of the random variable's probability distribution; or continuous, taking any numerical value in an interval or collection of intervals, via a probability density function that is characteristic of the random variable's probability distribution; or a mixture of both types. Two random variables with the same probability distribution can still differ in terms of their associations with, or independence from, other random variables. The realizations of a random variable, that is, the results of randomly choosing values according to the variable's probability distribution function, are called random variates.

The formal mathematical treatment of random variables is a topic in probability theory. In that context, a random variable is understood as a function defined on a sample space whose outputs are numerical values.

Definition

A *random variable* $X : \Omega \to E$ is a measurable function from the set of possible outcomes Ω to some set E. The technical axiomatic definition requires Ω to be a probability space and E to be a measurable space.

Note that although X is usually a real-valued function ($E = \mathbb{R}$), it does *not* return a probability. The probabilities of different outcomes or sets of outcomes (events) are already given by the probability measure P with which Ω is equipped. Rather, X describes some numerical property that outcomes in Ω may have — e.g., the number of heads in a random collection of coin flips, or the height of a random person. The probability that X takes value ≤ 3 is the measure of the set of outcomes $\{\omega \in \Omega : X(\omega) \leq 3\}$, denoted $P(X \leq 3)$.

Standard Case

In many cases, $E = \mathbb{R}$. In some contexts, the term *random element* is used to denote a random variable not of this form.

When the image (or range) of X is finite or countably infinite, the random variable is called a discrete random variable and its distribution can be described by a probability mass function which assigns a probability to each value in the image of X. If the image is uncountably infinite then X

is called a continuous random variable. In the special case that it is absolutely continuous, its distribution can be described by a probability density function, which assigns probabilities to intervals; in particular, each individual point must necessarily have probability zero for an absolutely continuous random variable. Not all continuous random variables are absolutely continuous, for example a mixture distribution. Such random variables cannot be described by a probability density or a probability mass function.

Any random variable can be described by its cumulative distribution function, which describes the probability that the random variable will be less than or equal to a certain value.

Extensions

The term "random variable" in statistics is traditionally limited to the real-valued case ($E = \mathbb{R}$). This ensures that it is possible to define quantities such as the expected value and variance of a random variable, its cumulative distribution function, and the moments of its distribution.

However, the definition above is valid for any measurable space E of values. Thus one can consider random elements of other sets E, such as random boolean values, categorical values, complex numbers, vectors, matrices, sequences, trees, sets, shapes, manifolds, and functions. One may then specifically refer to a *random variable of type* E, or an E*-valued random variable*.

This more general concept of a random element is particularly useful in disciplines such as graph theory, machine learning, natural language processing, and other fields in discrete mathematics and computer science, where one is often interested in modeling the random variation of non-numerical data structures. In some cases, it is nonetheless convenient to represent each element of E using one or more real numbers. In this case, a random element may optionally be represented as a vector of real-valued random variables (all defined on the same underlying probability space Ω, which allows the different random variables to covary). For example:

- A random word may be represented as a random integer that serves as an index into the vocabulary of possible words. Alternatively, it can be represented as a random indicator vector whose length equals the size of the vocabulary, where the only values of positive probability are $(1\,0\,0\,0\cdots)$, $(0\,1\,0\,0\cdots)$, $(0\,0\,1\,0\cdots)$ and the position of the 1 indicates the word.
- A random sentence of given length N may be represented as a vector of N random words.
- A random graph on N given vertices may be represented as a $N \times N$ matrix of random variables, whose values specify the adjacency matrix of the random graph.
- A random function F may be represented as a collection of random variables $F(x)$, giving the function's values at the various points x in the function's domain. The $F(x)$ are ordinary real-valued random variables provided that the function is real-valued. For example, a stochastic process is a random function of time, a random vector is a random function of some index set such as $1, 2, \ldots n$, and random field is a random function on any set (typically time, space, or a discrete set).

Examples

Discrete Random Variable

In an experiment a person may be chosen at random, and one random variable may be the person's height. Mathematically, the random variable is interpreted as a function which maps the person to the person's height. Associated with the random variable is a probability distribution that allows the computation of the probability that the height is in any subset of possible values, such as the probability that the height is between 180 and 190 cm, or the probability that the height is either less than 150 or more than 200 cm.

Another random variable may be the person's number of children; this is a discrete random variable with non-negative integer values. It allows the computation of probabilities for individual integer values – the probability mass function (PMF) – or for sets of values, including infinite sets. For example, the event of interest may be "an even number of children". For both finite and infinite event sets, their probabilities can be found by adding up the PMFs of the elements; that is, the probability of an even number of children is the infinite sum $\mathrm{PMF}(0)+\mathrm{PMF}(2)+\mathrm{PMF}(4)+\cdots$.

In examples such as these, the sample space (the set of all possible persons) is often suppressed, since it is mathematically hard to describe, and the possible values of the random variables are then treated as a sample space. But when two random variables are measured on the same sample space of outcomes, such as the height and number of children being computed on the same random persons, it is easier to track their relationship if it is acknowledged that both height and number of children come from the same random person, for example so that questions of whether such random variables are correlated or not can be posed.

Coin Toss

The possible outcomes for one coin toss can be described by the sample space $\Omega = \{\text{heads}, \text{tails}\}$. We can introduce a real-valued random variable Y that models a \$1 payoff for a successful bet on heads as follows:

$$Y(\omega) = \begin{cases} 1, & \text{if } \omega = \text{heads}, \\ 0, & \text{if } \omega = \text{tails}. \end{cases}$$

If the coin is a fair coin, Y has a probability mass function f_Y given by:

$$f_Y(y) = \begin{cases} \frac{1}{2}, & \text{if } y = 1, \\ \frac{1}{2}, & \text{if } y = 0, \end{cases}$$

Dice Roll

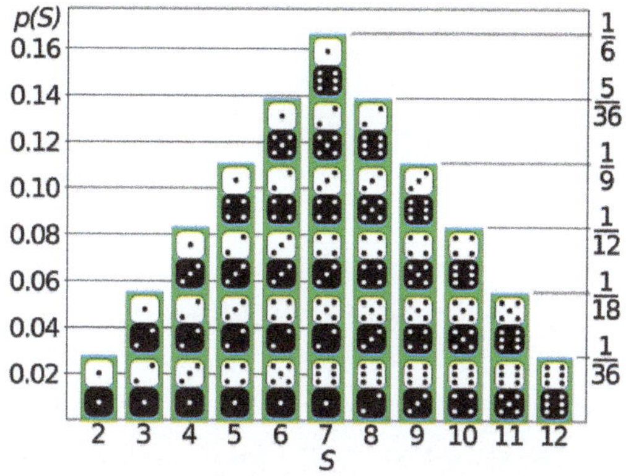

If the sample space is the set of possible numbers rolled on two dice, and the random variable of interest is the sum S of the numbers on the two dice, then S is a discrete random variable whose distribution is described by the probability mass function plotted as the height of picture columns here.

A random variable can also be used to describe the process of rolling dice and the possible outcomes. The most obvious representation for the two-dice case is to take the set of pairs of numbers n_1 and n_2 from {1, 2, 3, 4, 5, 6} (representing the numbers on the two dice) as the sample space. The total number rolled (the sum of the numbers in each pair) is then a random variable X given by the function that maps the pair to the sum:

$$X((n_1, n_2)) = n_1 + n_2$$

and (if the dice are fair) has a probability mass function f_X given by:

$$f_X(S) = \frac{\min(S-1, 13-S)}{36}, \text{for } S \in \{2,3,4,5,6,7,8,9,10,11,12\}$$

Continuous Random Variable

An example of a continuous random variable would be one based on a spinner that can choose a horizontal direction. Then the values taken by the random variable are directions. We could represent these directions by North, West, East, South, Southeast, etc. However, it is commonly more convenient to map the sample space to a random variable which takes values which are real numbers. This can be done, for example, by mapping a direction to a bearing in degrees clockwise from North. The random variable then takes values which are real numbers from the interval [0, 360), with all parts of the range being "equally likely". In this case, $\boldsymbol{X}$ = the angle spun. Any real number has probability zero of being selected, but a positive probability can be assigned to any *range* of values. For example, the probability of choosing a number in [0, 180] is $\frac{1}{2}$. Instead of speaking of a probability mass function, we say that the probability *density* of $\boldsymbol{X}$ is 1/360. The probability of a subset of [0, 360) can be calculated by multiplying the measure of the set by 1/360. In general, the probability of a set for a given continuous random variable can be calculated by integrating the density over the given set.

Mixed Type

An example of a random variable of mixed type would be based on an experiment where a coin is flipped and the spinner is spun only if the result of the coin toss is heads. If the result is tails, $X = -1$; otherwise X = the value of the spinner as in the preceding example. There is a probability of $\frac{1}{2}$ that this random variable will have the value −1. Other ranges of values would have half the probabilities of the last example.

Measure-theoretic Definition

The most formal, axiomatic definition of a random variable involves measure theory. Continuous random variables are defined in terms of sets of numbers, along with functions that map such sets to probabilities. Because of various difficulties (e.g. the Banach–Tarski paradox) that arise if such sets are insufficiently constrained, it is necessary to introduce what is termed a sigma-algebra to constrain the possible sets over which probabilities can be defined. Normally, a particular such sigma-algebra is used, the Borel σ-algebra, which allows for probabilities to be defined over any sets that can be derived either directly from continuous intervals of numbers or by a finite or countably infinite number of unions and/or intersections of such intervals.

The measure-theoretic definition is as follows.

Let $(\Omega, \mathcal{F}, P)$ be a probability space and $(E, \mathcal{E})$ a measurable space. Then an $(E, \mathcal{E})$-valued random variable is a function $X : \Omega \to E$ which is $(\mathcal{F}, \mathcal{E})$-measurable. The latter means that, for every subset $B \in \mathcal{E}$, its preimage $X^{-1}(B) \in \mathcal{F}$ where $X^{-1}(B) = \{\omega : X(\omega) \in B\}$. This definition enables us to measure any subset $B \in \mathcal{E}$ in the target space by looking at its preimage, which by assumption is measurable.

In more intuitive terms, Ω represents the "outcome", $\mathcal{F}$ represents the measurable subsets of possible outcomes, P represents the function that gives the probability of any such subset, E represents the kind of quantity the random value should take (such as real numbers), and $\mathcal{E}$ represents all the "well-behaved" (measurable) subsets of E (those for which you might want to find the probability). The random variable is then a function from any outcome to a quantity, such that the outcomes leading to any useful subset of quantities for the random variable have a well-defined probability.

When E is a topological space, then the most common choice for the σ-algebra $\mathcal{E}$ is the Borel σ-algebra $\mathcal{B}(E)$, which is the σ-algebra generated by the collection of all open sets in E. In such case the $(E, \mathcal{E})$-valued random variable is called the E-valued random variable. Moreover, when space E is the real line $\mathbb{R}$, then such a real-valued random variable is called simply the random variable.

Real-valued Random Variables

In this case the observation space is the set of real numbers. Recall, $(\Omega, \mathcal{F}, P)$ is the probability space. For real observation space, the function $X : \Omega \to \mathbb{R}$ is a real-valued random variable if

$$\{\omega : X(\omega) \le r\} \in \mathcal{F} \qquad \forall r \in \mathbb{R}.$$

This definition is a special case of the above because the set $\{(-\infty, r] : r \in \mathbb{R}\}$ generates the Borel σ-algebra on the set of real numbers, and it suffices to check measurability on any generating set. Here we can prove measurability on this generating set by using the fact that $\{\omega : X(\omega) \le r\} = X^{-1}((-\infty, r])$.

Distribution Functions of Random Variables

If a random variable $X:\Omega\rightarrow\mathbb{R}$ defined on the probability space $(\Omega,\mathcal{F},P)$ is given, we can ask questions like "How likely is it that the value of X is equal to 2?". This is the same as the probability of the event $\{\omega: X(\omega)=2\}$ which is often written as $P(X=2)$ or $p_X(2)$ for short.

Recording all these probabilities of output ranges of a real-valued random variable X yields the probability distribution of X. The probability distribution "forgets" about the particular probability space used to define X and only records the probabilities of various values of X. Such a probability distribution can always be captured by its cumulative distribution function

$$F_X(x)=\mathrm{P}(X\leq x)$$

and sometimes also using a probability density function, p_X. In measure-theoretic terms, we use the random variable X to "push-forward" the measure P on Ω to a measure p_X on $\mathbb{R}$. The underlying probability space Ω is a technical device used to guarantee the existence of random variables, sometimes to construct them, and to define notions such as correlation and dependence or independence based on a joint distribution of two or more random variables on the same probability space. In practice, one often disposes of the space Ω altogether and just puts a measure on $\mathbb{R}$ that assigns measure 1 to the whole real line, i.e., one works with probability distributions instead of random variables.

Moments

The probability distribution of a random variable is often characterised by a small number of parameters, which also have a practical interpretation. For example, it is often enough to know what its "average value" is. This is captured by the mathematical concept of expected value of a random variable, denoted v, and also called the first moment. In general, $\mathrm{E}[f(X)]$ is not equal to $f(\mathrm{E}[X])$. Once the "average value" is known, one could then ask how far from this average value the values of X typically are, a question that is answered by the variance and standard deviation of a random variable. $\mathrm{E}[X]$ can be viewed intuitively as an average obtained from an infinite population, the members of which are particular evaluations of X.

Mathematically, this is known as the (generalised) problem of moments: for a given class of random variables X, find a collection $\{f_i\}$ of functions such that the expectation values fully characterise the distribution of the random variable X.

Moments can only be defined for real-valued functions of random variables (or complex-valued, etc.). If the random variable is itself real-valued, then moments of the variable itself can be taken, which are equivalent to moments of the identity function $f(X)=X$ of the random variable. However, even for non-real-valued random variables, moments can be taken of real-valued functions of those variables. For example, for a categorical random variable X that can take on the nominal values "red", "blue" or "green", the real-valued function $[X=\text{green}]$ can be constructed; this uses the Iverson bracket, and has the value 1 if X has the value "green", 0 otherwise. Then, the expected value and other moments of this function can be determined.

Functions of Random Variables

A new random variable Y can be defined by applying a real Borel measurable function $g:\mathbb{R}\to\mathbb{R}$ to the outcomes of a real-valued random variable X. The cumulative distribution function of Y is

$$F_Y(y) = \mathrm{P}(g(X) \le y).$$

If function g is invertible, i.e., g^{-1} exists, and is either increasing or decreasing, then the previous relation can be extended to obtain

$$F_Y(y) = \mathrm{P}(g(X) \le y) = \begin{cases} \mathrm{P}(X \le g^{-1}(y)) = F_X(g^{-1}(y)), & \text{if } g^{-1} \text{ increasing,} \\ \mathrm{P}(X \ge g^{-1}(y)) = 1 - F_X(g^{-1}(y)), & \text{if } g^{-1} \text{ decreasing.} \end{cases}$$

and, again with the same hypotheses of invertibility of g, assuming also differentiability, we can find the relation between the probability density functions by differentiating both sides with respect to y, in order to obtain

$$f_Y(y) = f_X(g^{-1}(y))\left|\frac{dg^{-1}(y)}{dy}\right|.$$

If there is no invertibility of g but each y admits at most a countable number of roots (i.e., a finite, or countably infinite, number of x_i such that $y = g(x_i)$ then the previous relation between the probability density functions can be generalized with

$$f_Y(y) = \sum_i f_X(g_i^{-1}(y))\left|\frac{dg_i^{-1}(y)}{dy}\right|$$

where $x_i = g_i^{-1}(y)$. The formulas for densities do not demand g to be increasing.

In the measure-theoretic, axiomatic approach to probability, if we have a random variable X on Ω and a Borel measurable function $g:\mathbb{R}\to\mathbb{R}$, then $Y = g(X)$ will also be a random variable on Ω, since the composition of measurable functions is also measurable. (However, this is not true if g is Lebesgue measurable.) The same procedure that allowed one to go from a probability space (Ω, P) to $(\mathbb{R}, dF_X)$ can be used to obtain the distribution of Y.

Example 1

Let X be a real-valued, continuous random variable and let $Y = X^2$.

$$F_Y(y) = \mathrm{P}(X^2 \le y).$$

If $y < 0$, then $P(X^2 \le y) = 0$, so

$$F_Y(y) = 0 \quad \text{if} \quad y < 0.$$

If $y \geq 0$, then

$$P(X^2 \leq y) = P(|X| \leq \sqrt{y}) = P(-\sqrt{y} \leq X \leq \sqrt{y}),$$

so

$$F_Y(y) = F_X(\sqrt{y}) - F_X(-\sqrt{y}) \quad \text{if} \quad y \geq 0.$$

Example 2

Suppose X is a random variable with a cumulative distribution

$$F_X(x) = P(X \leq x) = \frac{1}{(1+e^{-x})^{\theta}}$$

where $\theta > 0$ is a fixed parameter. Consider the random variable $Y = \log(1+e^{-X})$. Then,

$$F_Y(y) = P(Y \leq y) = P(\log(1+e^{-X}) \leq y) = P(X > -\log(e^y - 1)).$$

The last expression can be calculated in terms of the cumulative distribution of so

$$F_Y(y) = 1 - F_X(-\log(e^y - 1))$$

$$= 1 - \frac{1}{(1+e^{\log(e^y-1)})^{\theta}}$$

$$= 1 - \frac{1}{(1+e^y-1)^{\theta}}$$
$$= 1 - e^{-y\theta}.$$

which is the cdf of an exponential distribution.

Example 3

Suppose X is a random variable with a standard normal distribution, whose density is

$$f_X(x) = \frac{1}{\sqrt{2\pi}} e^{-x^2/2}.$$

Consider the random variable $Y = X^2$. We can find the density using the above formula for a change of variables:

$$f_Y(y) = \sum_i f_X(g_i^{-1}(y)) \left| \frac{dg_i^{-1}(y)}{dy} \right|.$$

In this case the change is not monotonic, because every value of Y has two corresponding values of X (one positive and negative). However, because of symmetry, both halves will transform identically, i.e.,

$$f_Y(y) = 2 f_X(g^{-1}(y)) \left| \frac{dg^{-1}(y)}{dy} \right|.$$

The inverse transformation is

$$x = g^{-1}(y) = \sqrt{y}$$

and its derivative is

$$\frac{dg^{-1}(y)}{dy} = \frac{1}{2\sqrt{y}}.$$

Then,

$$f_Y(y) = 2 \frac{1}{\sqrt{2\pi}} e^{-y/2} \frac{1}{2\sqrt{y}} = \frac{1}{\sqrt{2\pi y}} e^{-y/2}.$$

This is a chi-squared distribution with one degree of freedom.

Equivalence of Random Variables

There are several different senses in which random variables can be considered to be equivalent. Two random variables can be equal, equal almost surely, or equal in distribution.

In increasing order of strength, the precise definition of these notions of equivalence is given below.

Equality in Distribution

If the sample space is a subset of the real line, random variables X and Y are *equal in distribution* (denoted $X \stackrel{d}{=} Y$) if they have the same distribution functions:

$$P(X \le x) = P(Y \le x) \quad \text{for all} \quad x.$$

Two random variables having equal moment generating functions have the same distribution. This provides, for example, a useful method of checking equality of certain functions of i.i.d. random variables. However, the moment generating function exists only for distributions that have a defined Laplace transform.

Almost Sure Equality

Two random variables X and Y are *equal almost surely* if, and only if, the probability that they are different is zero:

$$P(X \ne Y) = 0.$$

For all practical purposes in probability theory, this notion of equivalence is as strong as actual equality. It is associated to the following distance:

$$d_\infty(X,Y) = \operatorname*{ess\,sup}_\omega |X(\omega) - Y(\omega)|,$$

where "ess sup" represents the essential supremum in the sense of measure theory.

Equality

Finally, the two random variables X and Y are *equal* if they are equal as functions on their measurable space:

$$X(\omega) = Y(\omega) \qquad \text{for all } \omega.$$

Convergence

A significant theme in mathematical statistics consists of obtaining convergence results for certain sequences of random variables; for instance the law of large numbers and the central limit theorem.

There are various senses in which a sequence (X_n) of random variables can converge to a random variable X. These are explained in the article on convergence of random variables.

Expected Value

In probability theory, the expected value of a random variable, intuitively, is the long-run average value of repetitions of the experiment it represents. For example, the expected value in rolling a six-sided die is 3.5 because, roughly speaking, the average of all the numbers that come up in an extremely large number of rolls is very nearly always quite close to three and a half. Less roughly, the law of large numbers states that the arithmetic mean of the values almost surely converges to the expected value as the number of repetitions approaches infinity. The expected value is also known as the expectation, mathematical expectation, EV, average, mean value, mean, or first moment.

More practically, the expected value of a discrete random variable is the probability-weighted average of all possible values. In other words, each possible value the random variable can assume is multiplied by its probability of occurring, and the resulting products are summed to produce the expected value. The same principle applies to a continuous random variable, except that an integral of the variable with respect to its probability density replaces the sum. The formal definition subsumes both of these and also works for distributions which are neither discrete nor continuous: the expected value of a random variable is the integral of the random variable with respect to its probability measure.

The expected value does not exist for random variables having some distributions with large "tails", such as the Cauchy distribution. For random variables such as these, the long-tails of the distribution prevent the sum/integral from converging.

The expected value is a key aspect of how one characterizes a probability distribution; it is one type of location parameter. By contrast, the variance is a measure of dispersion of the possible values of the random variable around the expected value. The variance itself is defined in terms of two expectations: it is the expected value of the squared deviation of the variable's value from the variable's expected value.

The expected value plays important roles in a variety of contexts. In regression analysis, one desires a formula in terms of observed data that will give a "good" estimate of the parameter giving the effect of some explanatory variable upon a dependent variable. The formula will give different estimates using different samples of data, so the estimate it gives is itself a random variable. A formula is typically considered good in this context if it is an unbiased estimator—that is, if the expected value of the estimate (the average value it would give over an arbitrarily large number of separate samples) can be shown to equal the true value of the desired parameter.

In decision theory, and in particular in choice under uncertainty, an agent is described as making an optimal choice in the context of incomplete information. For risk neutral agents, the choice involves using the expected values of uncertain quantities, while for risk averse agents it involves maximizing the expected value of some objective function such as a von Neumann–Morgenstern utility function. One example of using expected value in reaching optimal decisions is the Gordon–Loeb model of information security investment. According to the model, one can conclude that the amount a firm spends to protect information should generally be only a small fraction of the expected loss (i.e., the expected value of the loss resulting from a cyber/information security breach).

Definition

Univariate Discrete Random Variable, Finite Case

Suppose random variable X can take value x_1 with probability p_1, value x_2 with probability p_2, and so on, up to value x_k with probability p_k. Then the expectation of this random variable X is defined as

$$\operatorname{E}[X] = x_1 p_1 + x_2 p_2 + \cdots + x_k p_k .$$

Since all probabilities p_i add up to one ($p_1 + p_2 + \ldots + p_k = 1$), the expected value can be viewed as the weighted average, with p_i's being the weights:

$$\operatorname{E}[X] = \frac{x_1 p_1 + x_2 p_2 + \cdots + x_k p_k}{1} = \frac{x_1 p_1 + x_2 p_2 + \cdots + x_k p_k}{p_1 + p_2 + \cdots + p_k} .$$

If all outcomes x_i are equally likely (that is, $p_1 = p_2 = \ldots = p_k$), then the weighted average turns into the simple average. This is intuitive: the expected value of a random variable is the average of all values it can take; thus the expected value is what one expects to happen *on average*. If the outcomes x_i are not equally probable, then the simple average must be replaced with the weighted average, which takes into account the fact that some outcomes are more likely than the others. The intuition however remains the same: the expected value of X is what one expects to happen *on average*.

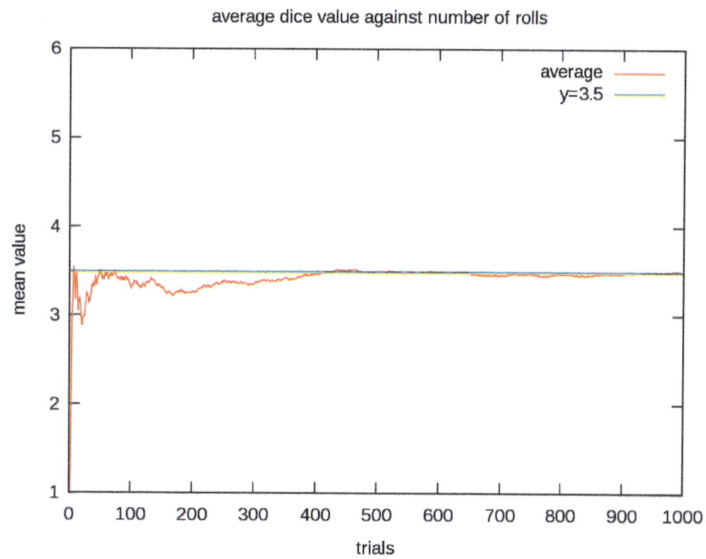

An illustration of the convergence of sequence averages of rolls of a die to the expected value of 3.5 as the number of rolls (trials) grows.

Example 1. Let X represent the outcome of a roll of a fair six-sided die. More specifically, X will be the number of pips showing on the top face of the die after the toss. The possible values for X are 1, 2, 3, 4, 5, and 6, all equally likely (each having the probability of 1/6). The expectation of X is

$$\mathrm{E}[X] = 1\cdot\frac{1}{6} + 2\cdot\frac{1}{6} + 3\cdot\frac{1}{6} + 4\cdot\frac{1}{6} + 5\cdot\frac{1}{6} + 6\cdot\frac{1}{6} = 3.5.$$

If one rolls the die n times and computes the average (arithmetic mean) of the results, then as n grows, the average will almost surely converge to the expected value, a fact known as the strong law of large numbers. One example sequence of ten rolls of the die is 2, 3, 1, 2, 5, 6, 2, 2, 2, 6, which has the average of 3.1, with the distance of 0.4 from the expected value of 3.5. The convergence is relatively slow: the probability that the average falls within the range 3.5 ± 0.1 is 21.6% for ten rolls, 46.1% for a hundred rolls and 93.7% for a thousand rolls. See the figure for an illustration of the averages of longer sequences of rolls of the die and how they converge to the expected value of 3.5. More generally, the rate of convergence can be roughly quantified by e.g. Chebyshev's inequality and the Berry–Esseen theorem.

Example 2. The roulette game consists of a small ball and a wheel with 38 numbered pockets around the edge. As the wheel is spun, the ball bounces around randomly until it settles down in one of the pockets. Suppose random variable X represents the (monetary) outcome of a \$1 bet on a single number ("straight up" bet). If the bet wins (which happens with probability 1/38 in American roulette), the payoff is \$35; otherwise the player loses the bet. The expected profit from such a bet will be

$$\mathrm{E}[\text{gain from \$1 bet}] = -\$1\cdot\frac{37}{38} + \$35\cdot\frac{1}{38} = -\$0.0526.$$

i.e. the bet of \$1 stands to lose \$0.0526, so its expected value is -\$0.0526.

Univariate Discrete Random Variable, Countable Case

Let X be a discrete random variable taking values x 1, x 2, ... with probabilities p 1, p 2, ... respectively. Then the expected value of this random variable is the infinite sum

$$\mathrm{E}[X]=\sum_{i=1}^{\infty}x_i\,p_i,$$

providedthatthisseriesconvergesabsolutely(thatis,thesummustremainfiniteifweweretoreplaceall$x$ i's with their absolute values). If this series does not converge absolutely, we say that the expected value of X does not exist.

For example, suppose random variable X takes values 1, −2, 3, −4, ..., with respective probabilities $c/1^2, c/2^2, c/3^2, c/4^2, \ldots$, where $c = 6/\pi^2$ is a normalizing constant that ensures the probabilities sum up to one. Then the infinite sum

$$\sum_{i=1}^{\infty}x_i\,p_i=c\left(1-\frac{1}{2}+\frac{1}{3}-\frac{1}{4}+\cdots\right)$$

converges and its sum is equal to $\ln(2)6/\pi^2 \simeq 0.421383$. However it would be incorrect to claim that the expected value of X is equal to this number—in fact E[X] does not exist, as this series does not converge absolutely.

Univariate Continuous Random Variable

If the probability distribution of X admits a probability density function $f(x)$, then the expected value can be computed as

$$\mathrm{E}[X]=\int_{-\infty}^{\infty}xf(x)\mathrm{d}x.$$

General Definition

In general, if X is a random variable defined on a probability space (Ω, Σ, P), then the expected value of X, denoted by E[X], $\langle X\rangle$, X or E[X], is defined as the Lebesgue integral

$$\mathrm{E}[X]=\int_{\Omega}X\,\mathrm{d}P=\int_{\Omega}X(\omega)P(\mathrm{d}\omega)$$

When this integral exists, it is defined as the expectation of X. Not all random variables have a finite expected value, since the integral may not converge absolutely; furthermore, for some it is not defined at all (e.g., Cauchy distribution). Two variables with the same probability distribution will have the same expected value, if it is defined.

It follows directly from the discrete case definition that if X is a constant random variable, i.e. $X = b$ for some fixed real number b, then the expected value of X is also b.

The expected value of a measurable function of X, $g(X)$, given that X has a probability density function $f(x)$, is given by the inner product of f and g:

$$\mathrm{E}[g(X)] = \int_{-\infty}^{\infty} g(x) f(x) \mathrm{d}x.$$

This is sometimes called the law of the unconscious statistician. Using representations as Riemann–Stieltjes integral and integration by parts the formula can be restated as

$$\mathrm{E}[g(X)] = \int_{-\infty}^{\infty} g(x) \mathrm{dP}(X \leq x) = \begin{cases} g(a) + \int_a^{\infty} g'(x) \mathrm{P}(X > x) \mathrm{d}x & \text{if } \mathrm{P}(X \geq a) = 1 \\ g(b) - \int_{-\infty}^{b} g'(x) \mathrm{P}(X \leq x) \mathrm{d}x & \text{if } \mathrm{P}(X \leq b) = 1. \end{cases}$$

As a special case let α denote a positive real number. Then

$$\mathrm{E}\left[|X|^{\alpha}\right] = \alpha \int_0^{\infty} t^{\alpha-1} \mathrm{P}(|X| > t) \mathrm{d}t.$$

In particular, if $\alpha = 1$ and $\Pr[X \geq 0] = 1$, then this reduces to

$$\mathrm{E}[|X|] = \mathrm{E}[X] = \int_0^{\infty} \{1 - F(t)\} \mathrm{d}t,$$

where F is the cumulative distribution function of X. This last identity is an instance of what, in a non-probabilistic setting, has been called the layer cake representation.

The law of the unconscious statistician applies also to a measurable function g of several random variables $X_1, \ldots X_n$ having a joint density f:

$$\mathrm{E}[g(X_1, \ldots, X_n)] = \int_{-\infty}^{\infty} \cdots \int_{-\infty}^{\infty} g(x_1, \ldots, x_n) f(x_1, \ldots, x_n) \mathrm{d}x_1 \cdots \mathrm{d}x_n.$$

Properties

Constants

The expected value of a constant is equal to the constant itself; i.e., if is a constant, then $\mathrm{E}[c] = c$.

Monotonicity

If X and Y are random variables such that $X \leq Y$ almost surely, then $\mathrm{E}[X] \leq \mathrm{E}[Y]$.

Linearity

The expected value operator (or expectation operator) $\mathrm{E}[\cdot]$ is linear in the sense that

$$\mathrm{E}[X+c]=\mathrm{E}[X]+c$$
$$\mathrm{E}[X+Y]=\mathrm{E}[X]+\mathrm{E}[Y]$$
$$\mathrm{E}[aX]=a\,\mathrm{E}[X]$$

The second result is valid even if X is not statistically independent of Y. Combining the results from the previous three equations, we can see that

$$\mathrm{E}[aX+bY+c]=a\,\mathrm{E}[X]+b\,\mathrm{E}[Y]+c$$

for any two random variables X and Y (which need to be defined on the same probability space) and any real numbers a, b and c.

Iterated Expectation

Iterated Expectation for Discrete Random Variables

For any two discrete random variables X, Y one may define the conditional expectation:

$$\mathrm{E}[X \mid Y=y]=\sum_x x\cdot \mathrm{P}(X=x \mid Y=y),$$

which means that $\mathrm{E}[X \mid Y=y]$ is a function of y. Let $g(y)$ be that function of y; then the notation $\mathrm{E}[X \mid Y]$ is a random variable in its own right, equal to $g(Y)$.

Lemma. Then the expectation of X satisfies:

$$\mathrm{E}[X]=\mathrm{E}\left[\mathrm{E}[X \mid Y]\right].$$

Proof.

$$\begin{aligned}
\mathrm{E}\left[\mathrm{E}[X \mid Y]\right] &= \sum_y \mathrm{E}[X \mid Y=y]\cdot \mathrm{P}(Y=y) \\
&= \sum_y \left(\sum_x x\cdot \mathrm{P}(X=x \mid Y=y) \right)\cdot \mathrm{P}(Y=y) \\
&= \sum_y \sum_x x\cdot \mathrm{P}(X=x \mid Y=y)\cdot \mathrm{P}(Y=y) \\
&= \sum_y \sum_x x\cdot \mathrm{P}(Y=y \mid X=x)\cdot \mathrm{P}(X=x) \\
&= \sum_x x\cdot \mathrm{P}(X=x)\cdot \left(\sum_y \mathrm{P}(Y=y \mid X=x) \right) \\
&= \sum_x x\cdot \mathrm{P}(X=x) \\
&= \mathrm{E}[X]
\end{aligned}$$

The left-hand side of this equation is referred to as the *iterated expectation*. The equation is sometimes called the *tower rule* or the *tower property*; it is treated under law of total expectation.

Iterated Expectation for Continuous Random Variables

In the continuous case, the results are completely analogous. The definition of conditional expectation would use inequalities, density functions, and integrals to replace equalities, mass functions, and summations, respectively. However, the main result still holds:

$$\mathrm{E}[X] = \mathrm{E}[\mathrm{E}[X \mid Y]]$$

Inequality

If a random variable X is always less than or equal to another random variable Y, the expectation of X is less than or equal to that of Y:

If $X \leq Y$, then $\mathrm{E}[X] \leq \mathrm{E}[Y]$.

In particular, if we set Y to $|X|$ we know $X \leq Y$ and $-X \leq Y$. Therefore, we know $\mathrm{E}[X] \leq \mathrm{E}[Y]$ and $\mathrm{E}[-X] \leq \mathrm{E}[Y]$. From the linearity of expectation we know $-\mathrm{E}[X] \leq \mathrm{E}[Y]$. Therefore, the absolute value of expectation of a random variable is less than or equal to the expectation of its absolute value:

$$|\mathrm{E}[X]| \leq \mathrm{E}[|X|]$$

This is a special case of Jensen's inequality.

Non-multiplicativity

If one considers the joint probability density function of X and Y, say $j(x,y)$, then the expectation of XY is

$$\mathrm{E}[XY] = \iint xy\, j(x,y)\,\mathrm{d}x\mathrm{d}y.$$

In general, the expected value operator is not multiplicative, i.e. $\mathrm{E}[XY]$ is not necessarily equal to $\mathrm{E}[X]\cdot\mathrm{E}[Y]$. In fact, the amount by which multiplicativity fails is called the covariance:

$$\mathrm{Cov}(X,Y) = \mathrm{E}[XY] - \mathrm{E}[X]\mathrm{E}[Y].$$

Thus multiplicativity holds precisely when $\mathrm{Cov}(X, Y) = 0$, in which case X and Y are said to be uncorrelated (independent variables are a notable case of uncorrelated variables).

Now if X and Y are independent, then by definition $j(x,y) = f(x)g(y)$ where f and g are the marginal PDFs for X and Y. Then

$$\mathrm{E}[XY] = \iint xy\, j(x,y)\,\mathrm{d}x\mathrm{d}y = \iint xyf(x)g(y)\,\mathrm{d}y\mathrm{d}x$$
$$= \left[\int xf(x)\,\mathrm{d}x\right]\left[\int yg(y)\,\mathrm{d}y\right] = \mathrm{E}[X]\mathrm{E}[Y]$$

and $\mathrm{Cov}(X, Y) = 0$.

Observe that independence of X and Y is required only to write $j(x, y) = f(x)g(y)$, and this is required to establish the second equality above. The third equality follows from a basic application of the Fubini–Tonelli theorem.

Functional Non-invariance

In general, with the exception of linear functions, the expectation operator and functions of random variables do not commute; that is

$$\mathrm{E}[g(X)] = \int_{\Omega} g(X)\mathrm{dP} \neq g(\mathrm{E}[X]).$$

A notable inequality concerning this topic is Jensen's inequality, involving expected values of convex (or concave) functions.

Relation to Characteristic Function

The probability density function of a scalar random variable X is related to its characteristic function φ_X by the inversion formula:

$$f_X(x) = \frac{1}{2\pi}\int_{\mathbf{R}} e^{-itx}\varphi_X(t)dt.$$

We can use this inversion formula in expected value of a function $g(X)$ to obtain

$$\mathrm{E}[g(X)] = \frac{1}{2\pi}\int\int g(x)e^{-itx}\varphi_X(t)dt\,dx$$

Changing the order of integration, we get

$$\mathrm{E}[g(X)] = \frac{1}{2\pi}\int G(t)\varphi_X(t)dt,$$

where

$$G(t) = \int g(x)e^{-itx}\,dx$$

is the Fourier transform of $g(x)$. This can also be seen as a direct consequence of Plancherel theorem.

Uses and Applications

It is possible to construct an expected value equal to the probability of an event by taking the expectation of an indicator function that is one if the event has occurred and zero otherwise. This relationship can be used to translate properties of expected values into properties of probabilities, e.g. using the law of large numbers to justify estimating probabilities by frequencies.

The expected values of the powers of X are called the moments of X; the moments about the mean of X are expected values of powers of X – E[X]. The moments of some random variables can be used to specify their distributions, via their moment generating functions.

To empirically estimate the expected value of a random variable, one repeatedly measures observations of the variable and computes the arithmetic mean of the results. If the expected value exists, this procedure estimates the true expected value in an unbiased manner and has the property of minimizing the sum of the squares of the residuals (the sum of the squared differences between the observations and the estimate). The law of large numbers demonstrates (under fairly mild conditions) that, as the size of the sample gets larger, the variance of this estimate gets smaller.

This property is often exploited in a wide variety of applications, including general problems of statistical estimation and machine learning, to estimate (probabilistic) quantities of interest via Monte Carlo methods, since most quantities of interest can be written in terms of expectation, e.g. $P(X \in \mathcal{A}) = E[I_{\mathcal{A}}(X)]$ where $I_{\mathcal{A}}(X)$ is the indicator function for set $\mathcal{A}$, i.e. $X \in \mathcal{A} \to I_{\mathcal{A}}(X) = 1, X \notin \mathcal{A} \to I_{\mathcal{A}}(X) = 0$.

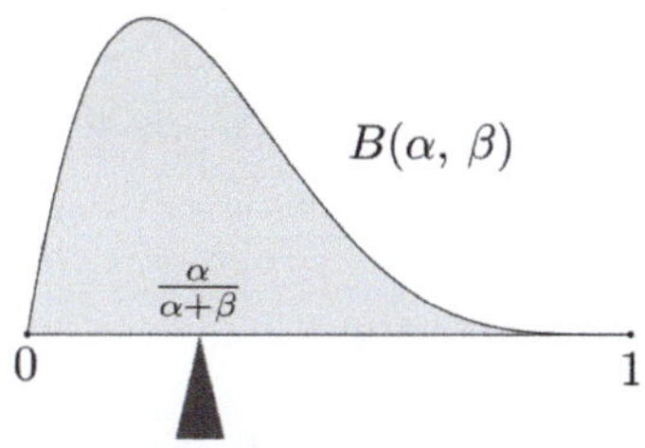

The mass of probability distribution is balanced at the expected value, here a Beta(α,β) distribution with expected value α/(α+β).

In classical mechanics, the center of mass is an analogous concept to expectation. For example, suppose X is a discrete random variable with values x_i and corresponding probabilities p_i. Now consider a weightless rod on which are placed weights, at locations x_i along the rod and having masses p_i (whose sum is one). The point at which the rod balances is E[X].

Expected values can also be used to compute the variance, by means of the computational formula for the variance

$$\operatorname{Var}(X) = E[X^2] - (E[X])^2.$$

A very important application of the expectation value is in the field of quantum mechanics. The expectation value of a quantum mechanical operator $\hat{A}$ operating on a quantum state vector $|\psi\rangle$ is written as $\langle \hat{A} \rangle = \langle \psi | A | \psi \rangle$. The uncertainty in $\hat{A}$ can be calculated using the formula .$(\Delta A)^2 = \langle \hat{A}^2 \rangle - \langle \hat{A} \rangle^2$

Expectation of Matrices

If X is an $m \times n$ matrix, then the expected value of the matrix is defined as the matrix of expected values:

$$E[X] = E\left[\begin{pmatrix} x_{1,1} & x_{1,2} & \cdots & x_{1,n} \\ x_{2,1} & x_{2,2} & \cdots & x_{2,n} \\ \vdots & \vdots & \ddots & \vdots \\ x_{m,1} & x_{m,2} & \cdots & x_{m,n} \end{pmatrix}\right] = \begin{pmatrix} E[x_{1,1}] & E[x_{1,2}] & \cdots & E[x_{1,n}] \\ E[x_{2,1}] & E[x_{2,2}] & \cdots & E[x_{2,n}] \\ \vdots & \vdots & \ddots & \vdots \\ E[x_{m,1}] & E[x_{m,2}] & \cdots & E[x_{m,n}] \end{pmatrix}.$$

This is utilized in covariance matrices.

Formulas for Special Cases

Discrete Distribution Taking Only Non-negative Integer Values

When a random variable takes only values in {0, 1, 2, 3, ...} we can use the following formula for computing its expectation (even when the expectation is infinite):

$$\mathrm{E}[X] = \sum_{i=1}^{\infty} P(X \geq i).$$

Proof.

$$\sum_{i=1}^{\infty} \mathrm{P}(X \geq i) = \sum_{i=1}^{\infty} \sum_{j=i}^{\infty} P(X = j).$$

Interchanging the order of summation, we have

$$\begin{aligned}\sum_{i=1}^{\infty} \sum_{j=i}^{\infty} P(X = j) &= \sum_{j=1}^{\infty} \sum_{i=1}^{j} P(X = j) \\ &= \sum_{j=1}^{\infty} j P(X = j) \\ &= \mathrm{E}[X].\end{aligned}$$

This result can be a useful computational shortcut. For example, suppose we toss a coin where the probability of heads is p. How many tosses can we expect until the first heads (not including the heads itself)? Let X be this number. We are counting only the tails and not the heads which ends the experiment; in particular, we can have $X = 0$. The expectation of X may be computed by $\sum_{i=1}^{\infty} (1-p)^i = \frac{1}{p} - 1$. This is because, when the first i tosses yield tails, the number of tosses is at least i. The last equality used the formula for a geometric progression, $\sum_{i=1}^{\infty} r^i = \sum_{i=0}^{\infty} r^i - 1 = \frac{1}{1-r} - 1$, where $r = 1 - p < 1$.

Continuous Distribution Taking Non-negative Values

Analogously with the discrete case above, when a continuous random variable X takes only non-negative values, we can use the following formula for computing its expectation (even when the expectation is infinite):

$$\mathrm{E}[X] = \int_0^{\infty} P(X \geq x)\, \mathrm{d}x$$

Proof: It is first assumed that X has a density $f_X(x)$. We present two techniques:

- Using integration by parts (a special case of Section 1.4 above):

$$E[X] = \int_0^\infty (-x)(-f_X(x))\,dx = \left[-x(1-F(x))\right]_0^\infty + \int_0^\infty (1-F(x))\,dx$$

and the bracket vanishes because (see Cumulative distribution function#Derived functions)

$$1 - F(x) = o\left(\frac{1}{x}\right) \text{as} \quad \to \infty$$

- Using an interchange in order of integration:

$$\int_0^\infty P(X \geq x)\,dx = \int_0^\infty \int_x^\infty f_X(t)\,dt\,dx$$
$$= \int_0^\infty \int_0^t f_X(t)\,dx\,dt = \int_0^\infty t f_X(t)\,dt = E[X]$$

In case no density exists, it is seen that

$$E[X] = \int_0^\infty \int_0^x dt\,dF(x) = \int_0^\infty \int_t^\infty dF(x)dt = \int_0^\infty (1-F(t))dt.$$

If both negative and non-negative values are taken for *X*, then its expectation is

$$E[X] = \int_0^\infty P(X \geq x)\,dx - \int_{-\infty}^0 P(X \leq x)\,dx$$

History

The idea of the expected value originated in the middle of the 17th century from the study of the so-called problem of points, which seeks to divide the stakes *in a fair way* between two players who have to end their game before it's properly finished. This problem had been debated for centuries, and many conflicting proposals and solutions had been suggested over the years, when it was posed in 1654 to Blaise Pascal by French writer and amateur mathematician Chevalier de Méré. de Méré claimed that this problem couldn't be solved and that it showed just how flawed mathematics was when it came to its application to the real world. Pascal, being a mathematician, was provoked and determined to solve the problem once and for all. He began to discuss the problem in a now famous series of letters to Pierre de Fermat. Soon enough they both independently came up with a solution. They solved the problem in different computational ways but their results were identical because their computations were based on the same fundamental principle. The principle is that the value of a future gain should be directly proportional to the chance of getting it. This principle seemed to have come naturally to both of them. They were very pleased by the fact that they had found essentially the same solution and this in turn made them absolutely convinced they had solved the problem conclusively. However, they did not publish their findings. They only informed a small circle of mutual scientific friends in Paris about it.

Three years later, in 1657, a Dutch mathematician Christiaan Huygens, who had just visited Paris, published a treatise "*De ratiociniis in ludo aleæ*" on probability theory. In this book he considered the problem of points and presented a solution based on the same principle as the solutions of Pascal and Fermat. Huygens also extended the concept of expectation by adding rules for how to calculate expectations in more complicated situations than the original problem (e.g., for three or more players). In this sense this book can be seen as the first successful attempt of laying down the foundations of the theory of probability.

In the foreword to his book, Huygens wrote: "It should be said, also, that for some time some of the best mathematicians of France have occupied themselves with this kind of calculus so that no one should attribute to me the honour of the first invention. This does not belong to me. But these savants, although they put each other to the test by proposing to each other many questions difficult to solve, have hidden their methods. I have had therefore to examine and go deeply for myself into this matter by beginning with the elements, and it is impossible for me for this reason to affirm that I have even started from the same principle. But finally I have found that my answers in many cases do not differ from theirs." (cited by Edwards (2002)). Thus, Huygens learned about de Méré's Problem in 1655 during his visit to France; later on in 1656 from his correspondence with Carcavi he learned that his method was essentially the same as Pascal's; so that before his book went to press in 1657 he knew about Pascal's priority in this subject.

Neither Pascal nor Huygens used the term "expectation" in its modern sense. In particular, Huygens writes: "That my Chance or Expectation to win any thing is worth just such a Sum, as wou'd procure me in the same Chance and Expectation at a fair Lay. ... If I expect a or b, and have an equal Chance of gaining them, my Expectation is worth a+b/2." More than a hundred years later, in 1814, Pierre-Simon Laplace published his tract "*Théorie analytique des probabilités*", where the concept of expected value was defined explicitly:

... this advantage in the theory of chance is the product of the sum hoped for by the probability of obtaining it; it is the partial sum which ought to result when we do not wish to run the risks of the event in supposing that the division is made proportional to the probabilities. This division is the only equitable one when all strange circumstances are eliminated; because an equal degree of probability gives an equal right for the sum hoped for. We will call this advantage *mathematical hope*.

The use of the letter E to denote expected value goes back to W.A. Whitworth in 1901, who used a script E. The symbol has become popular since for English writers it meant "Expectation", for Germans "Erwartungswert", for Spanish "Esperanza matemática" and for French "Espérance mathématique".

Conditional Probability

In probability theory, conditional probability is a measure of the probability of an event given that (by assumption, presumption, assertion or evidence) another event has occurred. If the event of interest is A and the event B is known or assumed to have occurred, "the conditional probability of A given B", or "the probability of A under the condition B", is usually written as $P(A|B)$, or sometimes $P_B(A)$. For example, the probability that any given person has a cough on any given day may be only 5%. But if we know or assume that the person has a cold, then they are much more likely to be coughing. The conditional probability of coughing given that you have a cold might be a much higher 75%.

The concept of conditional probability is one of the most fundamental and one of the most important concepts in probability theory. But conditional probabilities can be quite slippery and require careful interpretation. For example, there need not be a causal or temporal relationship between *A* and *B*.

$P(A|B)$ may or may not be equal to $P(A)$ (the unconditional probability of *A*). If $P(A|B) = P(A)$ and $P(B|A) = P(B)$, then events *A* and *B* are said to be independent: in such a case, having knowledge about either event does not change our knowledge about the other event. Also, in general, $P(A|B)$ (the conditional probability of A given B) is not equal to $P(B|A)$. For example, if you have cancer you might have a 90% chance of testing positive for cancer. In this case what is being measured is that the if event *B* "having cancer" has occurred, the probability *A* - *test is positive* given that *B* *having cancer* occurred is 90%, $P(A|B) = 90\%$. Alternatively, you can test positive for cancer but you may have only a 10% chance of actually having cancer because cancer is very rare. In this case what is being measured is the probability of the event *B* - *having cancer* given that the event *A* - *test is positive* has occurred, $P(B|A) = 10\%$. Falsely equating the two probabilities causes various errors of reasoning such as the base rate fallacy. Conditional probabilities can be correctly reversed using Bayes' theorem.

Definition

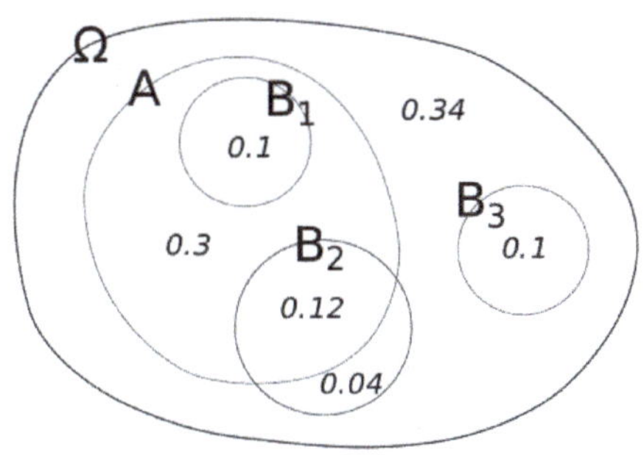

Illustration of conditional probabilities with an Euler diagram. The unconditional probability $P(A) = 0.52$. However, the conditional probability $P(A|B_1) = 1$, $P(A|B_2) = 0.75$, and $P(A|B_3) = 0$.

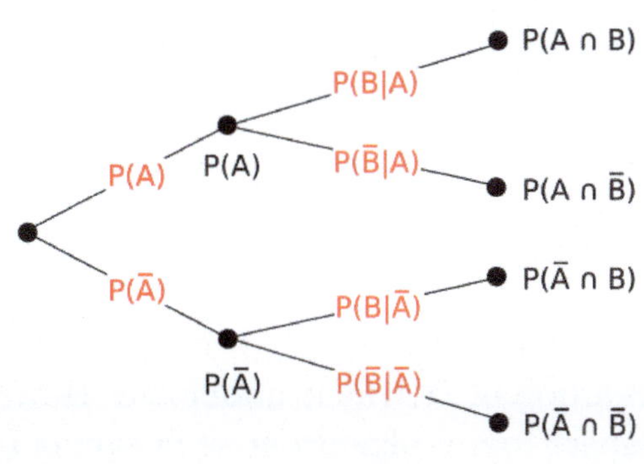

On a tree diagram, branch probabilities are conditional on the event associated with the parent node.

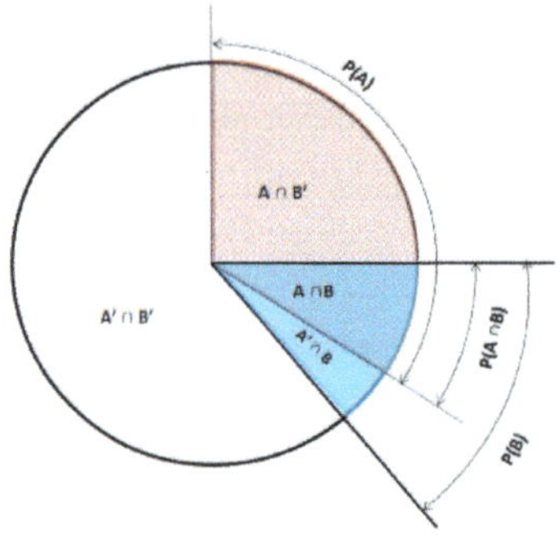

Venn Pie Chart describing conditional probabilities

Conditioning on an Event

Kolmogorov Definition

Given two events *A* and *B* from the sigma-field of a probability space with *P*(*B*) > 0, the conditional probability of *A* given *B* is defined as the quotient of the probability of the joint of events *A* and *B*, and the probability of *B*:

$$P(A \mid B) = \frac{P(A \cap B)}{P(B)}$$

This may be visualized as restricting the sample space to *B*. The logic behind this equation is that if the outcomes are restricted to *B*, this set serves as the new sample space.

Note that this is a definition but not a theoretical result. We just denote the quantity $P(A \cap B) / P(B)$ as $P(A \mid B)$ and call it the conditional probability of *A* given *B*.

As an Axiom of Probability

Some authors, such as De Finetti, prefer to introduce conditional probability as an axiom of probability:

$$P(A \cap B) = P(A \mid B)P(B)$$

Although mathematically equivalent, this may be preferred philosophically; under major probability interpretations such as the subjective theory, conditional probability is considered a primitive entity. Further, this "multiplication axiom" introduces a symmetry with the summation axiom for mutually exclusive events:

$$P(A \cup B) = P(A) + P(B) - \cancel{P(A \cap B)}^{\,0}$$

Measure-theoretic Definition

If *P*(*B*) = 0, then the simple definition of *P*(*A*|*B*) is undefined. However, it is possible to define a conditional probability with respect to a σ-algebra of such events (such as those arising from a continuous random variable).

For example, if X and Y are non-degenerate and jointly continuous random variables with density $f_{X,Y}(x, y)$ then, if B has positive measure,

$$P(X \in A \mid Y \in B) = \frac{\int_{y \in B} \int_{x \in A} f_{X,Y}(x,y)\,dx\,dy}{\int_{y \in B} \int_{x \in \mathbb{R}} f_{X,Y}(x,y)\,dx\,dy}.$$

The case where B has zero measure is problematic. For the case that $B = \{y_0\}$, representing a single point, the conditional probability could be defined as

$$P(X \in A \mid Y = y_0) = \frac{\int_{x \in A} f_{X,Y}(x,y_0)\,dx}{\int_{x \in \mathbb{R}} f_{X,Y}(x,y_0)\,dx},$$

however this approach leads to the Borel–Kolmogorov paradox. The more general case of zero measure is even more problematic, as can be seen by noting that the limit, as all δy_i approach zero, of

$$P(X \in A \mid Y \in \cup_i [y_i, y_i + \delta y_i]) \approxeq \frac{\sum_i \int_{x \in A} f_{X,Y}(x,y_i)\,dx\,\delta y_i}{\sum_i \int_{x \in \mathbb{R}} f_{X,Y}(x,y_i)\,dx\,\delta y_i},$$

depends on their relationship as they approach zero.

Conditioning on a Random Variable

Conditioning on an event may be generalized to conditioning on a random variable. Let X be a random variable; we assume for the sake of presentation that X is discrete, that is, X takes on only finitely many values x. Let A be an event. The conditional probability of A given X is defined as the random variable, written $P(A|X)$, *that takes on the value*

$$P(A \mid X = x)$$

whenever

$$X = x.$$

More formally:

$$P(A \mid X)(\omega) = P(A \mid X = X(\omega)).$$

The conditional probability $P(A|X)$ is a function of X, e.g., if the function g is defined as

$$g(x) = P(A \mid X = x),$$

then

$$P(A \mid X) = g \circ X$$

Note that $P(A|X)$ and X are now both random variables. From the law of total probability, the expected value of $P(A|X)$ is equal to the unconditional probability of A.

Example

Suppose that somebody secretly rolls two fair six-sided dice, and we must predict the outcome (the sum of the two upward faces).

- Let A be the value rolled on die 1.
- Let B be the value rolled on die 2.

What is the Probability that A = 2?

Table 1 shows the sample space of 36 outcomes

Clearly, $A = 2$ in exactly 6 of the 36 outcomes, thus $P(A=2) = 6/36 = 1/6$.

Table 1

+		B					
		1	2	3	4	5	6
A	1	2	3	4	5	6	7
	2	3	4	5	6	7	8
	3	4	5	6	7	8	9
	4	5	6	7	8	9	10
	5	6	7	8	9	10	11
	6	7	8	9	10	11	12

Suppose it is Revealed that A+B ≤ 5

What is the probability $A+B \leq 5$?

Table 2 shows that $A+B \leq 5$ for exactly 10 of the same 36 outcomes, thus $P(A+B \leq 5) = 10/36$

+		B					
		1	2	3	4	5	6
A	1	2	3	4	5	6	7
	2	3	4	5	6	7	8
	3	4	5	6	7	8	9
	4	5	6	7	8	9	10
	5	6	7	8	9	10	11
	6	7	8	9	10	11	12

What is the Probability that A = 2 Given that A+B ≤ 5 ?

Table 3 shows that for 3 of these 10 outcomes, $A = 2$

Thus, the conditional probability $P(A=2 \mid A+B \leq 5) = {}^{3}\!/_{10} = 0.3$.

Table 3

+		B					
		1	2	3	4	5	6
A	1	2	3	4	5	6	7
	2	3	4	5	6	7	8
	3	4	5	6	7	8	9
	4	5	6	7	8	9	10
	5	6	7	8	9	10	11
	6	7	8	9	10	11	12

Use in Inference

In statistical inference, the conditional probability is an update of the probability of an event based on new information. Incorporating the new information can be done as follows

- Let *A* the event of interest be in the sample space, say (*X*,*P*).
- The occurrence of the event *A* knowing that event *B* has or will have occurred, means the occurrence of *A* as it is restricted to *B*, i.e. $A \cap B$.
- Without the knowledge of the occurrence of *B*, the information about the occurrence of *A* would simply be *P*(*A*)
- The probability of *A* knowing that event *B* has or will have occurred, will be the probability of $A \cap B$ compared with *P*(*B*), the probability *B* has occurred.
- This results in $P(A|B) = P(A \cap B)/P(B)$ whenever $P(B)>0$ and 0 otherwise.

Note: This approach results in a probability measure that is consistent with the original probability measure and satisfies all the Kolmogorov Axioms. This conditional probability measure also could have resulted by assuming that the relative magnitude of the probability of A with respect to X will be preserved with respect to B (cf. a Formal Derivation below).

Note: The phraseology "evidence" or "information" is generally used in the Bayesian interpretation of probability. The conditioning event is interpreted as evidence for the conditioned event. That is, *P*(*A*) is the probability of *A* before accounting for evidence *E*, and *P*(*A*|*E*) is the probability of *A* after having accounted for evidence *E* or after having updated *P*(*A*). This is consistent with the frequentist interpretation, which presumably is the first definition given above.

Statistical Independence

Events *A* and *B* are defined to be statistically independent if

$$P(A \cap B) = P(A)P(B).$$

If *P*(*B*) is not zero, then this is equivalent to the statement that

$$P(A \mid B) = P(A).$$

Similarly, if *P*(*A*) is not zero, then

$$P(B \mid A) = P(B).$$

is also equivalent. Although the derived forms may seem more intuitive, they are not the preferred definition as the conditional probabilities may be undefined, and the preferred definition is symmetrical in *A* and *B*.

Common Fallacies

These fallacies should not be confused with Robert K. Shope's 1978 "conditional fallacy", which deals with counterfactual examples that beg the question.

Assuming Conditional Probability is of Similar Size to its Inverse

Relative size	Case B	Case $\bar{B}$	Total
Condition A	w	x	$w+x$
Condition $\bar{A}$	y	z	$y+z$
Total	$w+y$	$x+z$	$w+x+y+z$

$$\mathrm{P}(\mathrm{A}|\mathrm{B}) \times \mathrm{P}(\mathrm{B}) = \frac{w}{w+y} \times \frac{w+y}{w+x+y+z} = \frac{w}{w+x+y+z}$$

$$\mathrm{P}(\mathrm{B}|\mathrm{A}) \times \mathrm{P}(\mathrm{A}) = \frac{w}{w+x} \times \frac{w+x}{w+x+y+z} = \frac{w}{w+x+y+z}$$

A geometric visualisation of Bayes' theorem. In the table, the values w, x, y and z give the relative weights of each corresponding condition and case. The figures denote the cells of the table involved in each metric, the probability being the fraction of each figure that is shaded. This shows that P(A|B) P(B) = P(B|A) P(A) i.e. P(A|B) = P(B|A) P(A)/P(B) . Similar reasoning can be used to show that P(Ā|B) = P(B|Ā) P(Ā)/P(B) etc.

In general, it cannot be assumed that $P(A|B) \approx P(B|A)$. This can be an insidious error, even for those who are highly conversant with statistics. The relationship between $P(A|B)$ and $P(B|A)$ is given by Bayes' theorem:

$$P(B \mid A) = \frac{P(A \mid B)P(B)}{P(A)}$$

$$\Leftrightarrow \frac{P(B \mid A)}{P(A \mid B)} = \frac{P(B)}{P(A)}$$

That is, $P(A|B) \approx P(B|A)$ only if $P(B)/P(A) \approx 1$, or equivalently, $P(A) \approx P(B)$.

Alternatively, noting that $A \cap B = B \cap A$, and applying conditional probability:

$$P(A \mid B)P(B) = P(A \cap B) = P(B \cap A) = P(B \mid A)P(A)$$

Rearranging gives the result.

Assuming Marginal and Conditional Probabilities are of Similar Size

In general, it cannot be assumed that $P(A) \approx P(A|B)$. These probabilities are linked through the law of total probability:

$$P(A) = \sum_n P(A \cap B_n) = \sum_n P(A \mid B_n)P(B_n).$$

where the events (B_n) form a countable partition of A.

This fallacy may arise through selection bias. For example, in the context of a medical claim, let S_C be the event that a sequela (chronic disease) S occurs as a consequence of circumstance (acute condition) C. Let H be the event that an individual seeks medical help. Suppose that in most cases, C does not cause S so $P(S_C)$ is low. Suppose also that medical attention is only sought if S has occurred due to C. From experience of patients, a doctor may therefore erroneously conclude that $P(S_C)$ is high. The actual probability observed by the doctor is $P(S_C|H)$.

Over- or Under-weighting Priors

Not taking prior probability into account partially or completely is called *base rate neglect*. The reverse, insufficient adjustment from the prior probability is *conservatism*.

Formal Derivation

Formally, $P(A|B)$ is defined as the probability of A according to a new probability function on the sample space, such that outcomes not in B have probability 0 and that it is consistent with all original probability measures.

Let Ω be a sample space with elementary events {ω}. Suppose we are told the event B □ Ω has occurred. A new probability distribution (denoted by the conditional notation) is to be assigned on {ω} to reflect this. For events in B, it is reasonable to assume that the relative magnitudes of the probabilities will be preserved. For some constant scale factor α, the new distribution will therefore satisfy:

1. $\omega \in B : P(\omega \mid B) = \alpha P(\omega)$
2. $\omega \notin B : P(\omega \mid B) = 0$
3. $\sum_{\omega \in \Omega} P(\omega \mid B) = 1.$

Substituting 1 and 2 into 3 to select α:

$$\begin{aligned}
1 &= \sum_{\omega\in\Omega} P(\omega|B) \\
&= \sum_{\omega\in B} \alpha P(\omega) + \cancelto{0}{\sum_{\omega\notin B}} \\
&= \alpha \sum_{\omega\in B} P(\omega) \\
&= \alpha \cdot P(B) \\
\Rightarrow \alpha &= \frac{1}{P(B)}
\end{aligned}$$

So the new probability distribution is

$$1.\ \omega \in B : P(\omega \mid B) = \frac{P(\omega)}{P(B)}$$

$$2.\ \omega \notin B : P(\omega \mid B) = 0$$

Now for a general event A,

$$\begin{aligned}
P(A|B) &= \sum_{\omega\in A\cap B} P(\omega|B) + \cancelto{0}{\sum_{\omega\in A\cap B^c} P(\omega|B)} \\
&= \sum_{\omega\in A\cap B} \frac{P(\omega)}{P(B)} \\
&= \frac{P(A\cap B)}{P(B)}
\end{aligned}$$

References

- "Kendall's Advanced Theory of Statistics, Volume 1: Distribution Theory", Alan Stuart and Keith Ord, 6th Ed, (2009), ISBN 9780534243128
- Franklin, James (2001). The Science of Conjecture: Evidence and Probability Before Pascal. Johns Hopkins University Press. ISBN 0801865697.
- Stark, Henry; Woods, John W. (2002). Probability and Random Processes with Applications to Signal Processing (3rd ed.). Pearson. p. 7. ISBN 9788177583564.
- Forbes, Catherine; Evans, Merran; Hastings, Nicholas; Peacock, Brian (2011). Statistical Distributions (4th ed.). Wiley. p. 3. ISBN 9780470390634.
- Larsen, R. J.; Marx, M. L. (2001). An Introduction to Mathematical Statistics and Its Applications (Third ed.). Upper Saddle River, NJ: Prentice Hall. p. 22. ISBN 9780139223037.
- Foerster, Paul A. (2006). Algebra and Trigonometry: Functions and Applications, Teacher's Edition (Classics ed.). Prentice Hall. p. 633. ISBN 0-13-165711-9.
- Yates, Daniel S.; Moore, David S; Starnes, Daren S. (2003). The Practice of Statistics (2nd ed.). New York: Freeman. ISBN 978-0-7167-4773-4.
- Sheldon M Ross (2007). "§2.4 Expectation of a random variable". Introduction to probability models (9th ed.). Academic Press. p. 38 ff. ISBN 0-12-598062-0.
- Richard W Hamming (1991). "§2.5 Random variables, mean and the expected value". The art of probability for scientists and engineers. Addison–Wesley. p. 64 ff. ISBN 0-201-40686-1.

4

Sets : An Integrated Study

Sets are well- defined collections of objects. For example, numbers like 2 or 6 are distinct objects. This section also focuses on topics such as elements, Venn diagram, subset, intersection, complement and Cartesian product. This text will provide an integrated understanding of sets.

Set (Mathematics)

In mathematics, a set is a well-defined collection of distinct objects, considered as an object in its own right. For example, the numbers 2, 4, and 6 are distinct objects when considered separately, but when they are considered collectively they form a single set of size three, written {2,4,6}. Sets are one of the most fundamental concepts in mathematics. Developed at the end of the 19th century, set theory is now a ubiquitous part of mathematics, and can be used as a foundation from which nearly all of mathematics can be derived. In mathematics education, elementary topics such as Venn diagrams are taught at a young age, while more advanced concepts are taught as part of a university degree. The German word *Menge*, rendered as "set" in English, was coined by Bernard Bolzano in his work *The Paradoxes of the Infinite*.

A set of polygons in a Venn diagram

Definition

A set is a well-defined collection of distinct objects. The objects that make up a set (also known as the elements or members of a set) can be anything: numbers, people, letters of the alphabet, other sets, and so on. Georg Cantor, the founder of set theory, gave the following definition of a set at the beginning of his *Beiträge zur Begründung der transfiniten Mengenlehre*:

§ I

The Conception of Power or Cardinal Number

By an "aggregate" (*Menge*) we are to understand any collection into a whole (*Zusammenfassung zu einem Ganzen*) M of definite and separate objects *m* of our intuition or our thought. These objects are called the "elements" of M.

Passage with a translation of the original set definition of Georg Cantor

A set is a gathering together into a whole of definite, distinct objects of our perception [Anschauung] or of our thought—which are called elements of the set.

Sets are conventionally denoted with capital letters. Sets A and B are equal if and only if they have precisely the same elements.

Cantor's definition turned out to be inadequate for formal mathematics; instead, the notion of a "set" is taken as a primitive notion in axiomatic set theory, and the properties of sets are defined by a collection of axioms. The most basic properties are that a set can have elements, and that two sets are equal (one and the same) if and only if every element of each set is an element of the other; this property is called the *extensionality of sets*.

Describing Sets

There are two ways of describing, or specifying the members of, a set. One way is by intensional definition, using a rule or semantic description:

A is the set whose members are the first four positive integers.

B is the set of colors of the French flag.

The second way is by extension – that is, listing each member of the set. An extensional definition is denoted by enclosing the list of members in curly brackets:

$C = \{4, 2, 1, 3\}$

$D = \{\text{blue, white, red}\}$.

One often has the choice of specifying a set either intensionally or extensionally. In the examples above, for instance, $A = C$ and $B = D$.

There are two important points to note about sets. First, in an extensional definition, a set member can be listed two or more times, for example, {11, 6, 6}. However, per extensionality, two definitions of sets which differ only in that one of the definitions lists set members multiple times, define, in fact, the same set. Hence, the set {11, 6, 6} is exactly identical to the set {11, 6}. The second important point is that the order in which the elements of a set are listed is irrelevant (unlike for a sequence or tuple). We can illustrate these two important points with an example:

$\{6, 11\} = \{11, 6\} = \{11, 6, 6, 11\}$.

For sets with many elements, the enumeration of members can be abbreviated. For instance, the set of the first thousand positive integers may be specified extensionally as

$\{1, 2, 3, ..., 1000\}$,

where the ellipsis ("...") indicates that the list continues in the obvious way. Ellipses may also be used where sets have infinitely many members. Thus the set of positive even numbers can be written as $\{2, 4, 6, 8, ... \}$.

The notation with braces may also be used in an intensional specification of a set. In this usage, the braces have the meaning "the set of all ...". So, E = {playing card suits} is the set whose four members are ♠, ♦, ♥, and ♣. A more general form of this is set-builder notation, through which, for instance, the set F of the twenty smallest integers that are four less than perfect squares can be denoted

$F = \{n^2 - 4 : n \text{ is an integer; and } 0 \leq n \leq 19\}$.

In this notation, the colon (":") means "such that", and the description can be interpreted as "F is the set of all numbers of the form $n^2 - 4$, such that n is a whole number in the range from 0 to 19 inclusive." Sometimes the vertical bar ("|") is used instead of the colon.

Membership

If B is a set and x is one of the objects of B, this is denoted $x \in B$, and is read as "x belongs to B", or "x is an element of B". If y is not a member of B then this is written as $y \notin B$, and is read as "y does not belong to B".

For example, with respect to the sets $A = \{1,2,3,4\}$, B = {blue, white, red}, and $F = \{n^2 - 4 : n \text{ is an integer; and } 0 \leq n \leq 19\}$ defined above,

$4 \in A$ and $12 \in F$; but

$9 \notin F$ and green $\notin B$.

Subsets

If every member of set A is also a member of set B, then A is said to be a *subset* of B, written $A \subseteq B$ (also pronounced *A is contained in B*). Equivalently, we can write $B \supseteq A$, read as *B is a superset of A*, *B includes A*, or *B contains A*. The relationship between sets established by $\subseteq$ is called *inclusion* or *containment*.

If A is a subset of, but not equal to, B, then A is called a *proper subset* of B, written $A \subsetneq B$ (*A is a proper subset of B*) or $B \supsetneq A$ (*B is a proper superset of A*).

Note that the expressions $A \subset B$ and $B \supset A$ are used differently by different authors; some authors use them to mean the same as $A \subseteq B$ (respectively $B \supseteq A$), whereas others use them to mean the same as $A \subsetneq B$ (respectively $B \supsetneq A$).

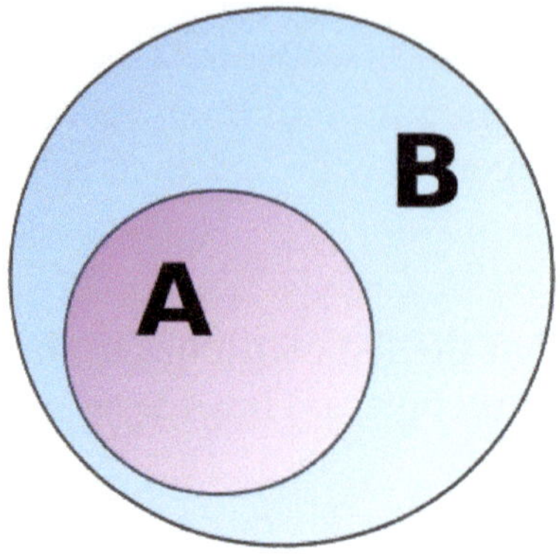

A is a subset of B

Example:

- The set of all men is a proper subset of the set of all people.
- $\{1, 3\} \subseteq \{1, 2, 3, 4\}$.
- $\{1, 2, 3, 4\} \subseteq \{1, 2, 3, 4\}$.

The empty set is a subset of every set and every set is a subset of itself:

- $\emptyset \subseteq A$.
- $A \subseteq A$.

An obvious but useful identity, which can often be used to show that two seemingly different sets are equal:

- $A = B$ if and only if $A \subseteq B$ and $B \subseteq A$.

A partition of a set S is a set of nonempty subsets of S such that every element x in S is in exactly one of these subsets.

Power Sets

The power set of a set S is the set of all subsets of S. Note that the power set contains S itself and the empty set because these are both subsets of S. For example, the power set of the set $\{1, 2, 3\}$ is $\{\{1, 2, 3\}, \{1, 2\}, \{1, 3\}, \{2, 3\}, \{1\}, \{2\}, \{3\}, \emptyset\}$. The power set of a set S is usually written as $P(S)$.

The power set of a finite set with n elements has 2^n elements. This relationship is one of the reasons for the terminology *power set*. For example, the set $\{1, 2, 3\}$ contains three elements, and the power set shown above contains $2^3 = 8$ elements.

The power set of an infinite (either countable or uncountable) set is always uncountable. Moreover, the power set of a set is always strictly "bigger" than the original set in the sense that there is no way to pair every element of S with exactly one element of $P(S)$. (There is never an onto map or surjection from S onto $P(S)$.)

Every partition of a set S is a subset of the powerset of S.

Cardinality

The cardinality $|S|$ of a set S is "the number of members of S." For example, if B = {*blue, white, red*}, $|B| = 3$.

There is a unique set with no members and zero cardinality, which is called the *empty set* (or the *null set*) and is denoted by the symbol ∅. For example, the set of all three-sided squares has zero members and thus is the empty set. Though it may seem trivial, the empty set, like the number zero, is important in mathematics; indeed, the existence of this set is one of the fundamental concepts of axiomatic set theory.

Some sets have infinite cardinality. The set N of natural numbers, for instance, is infinite. Some infinite cardinalities are greater than others. For instance, the set of real numbers has greater cardinality than the set of natural numbers. However, it can be shown that the cardinality of (which is to say, the number of points on) a straight line is the same as the cardinality of any segment of that line, of the entire plane, and indeed of any finite-dimensional Euclidean space.

Special Sets

There are some sets that hold great mathematical importance and are referred to with such regularity that they have acquired special names and notational conventions to identify them. One of these is the empty set, denoted {} or ∅. Another is the unit set {x}, which contains exactly one element, namely x. Many of these sets are represented using blackboard bold or bold typeface. Special sets of numbers include

- **P** or P, denoting the set of all primes: P = {2, 3, 5, 7, 11, 13, 17, ...}.
- **N** or N, denoting the set of all natural numbers: N = {0, 1, 2, 3, . . .} (sometimes defined excluding 0).
- **Z** or Z, denoting the set of all integers (whether positive, negative or zero): Z = {..., −2, −1, 0, 1, 2, ...}.
- **Q** or Q, denoting the set of all rational numbers (that is, the set of all proper and improper fractions): Q = $\{a/b : a, b \in Z, b \neq 0\}$. For example, $1/4 \in Q$ and $11/6 \in Q$. All integers are in this set since every integer a can be expressed as the fraction $a/1$ ($Z \subsetneq Q$).
- **R** or R, denoting the set of all real numbers. This set includes all rational numbers, together with all irrational numbers (that is, numbers that cannot be rewritten as fractions, such as $\sqrt{2}$, as well as transcendental numbers such as π, e and numbers that cannot be defined).
- **C** or C, denoting the set of all complex numbers: C = $\{a + bi : a, b \in R\}$. For example, $1 + 2i \in C$.
- **H** or H, denoting the set of all quaternions: H = $\{a + bi + cj + dk : a, b, c, d \in R\}$. For exam-ple, $1 + i + 2j - k \in H$.

Positive and negative sets are denoted by a superscript - or +. For example, Q^+ represents the set of positive rational numbers.

Each of the above sets of numbers has an infinite number of elements, and each can be considered to be a proper subset of the sets listed below it. The primes are used less frequently than the others outside of number theory and related fields.

Basic Operations

There are several fundamental operations for constructing new sets from given sets.

Unions

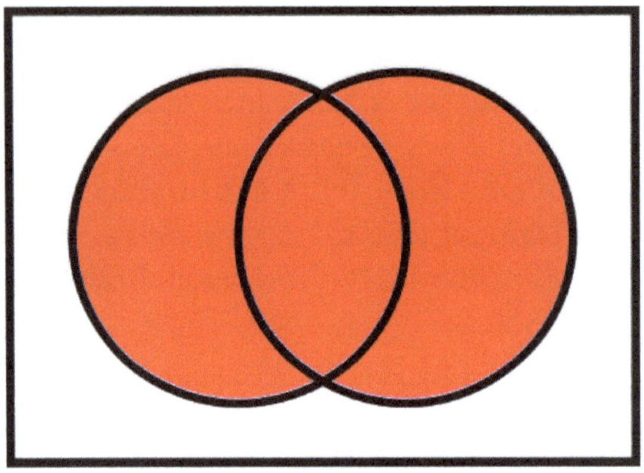

The **union** of A and B, denoted $A \cup B$

Two sets can be "added" together. The *union* of A and B, denoted by $A \cup B$, is the set of all things that are members of either A or B.

Examples:

- $\{1, 2\} \cup \{1, 2\} = \{1, 2\}$.
- $\{1, 2\} \cup \{2, 3\} = \{1, 2, 3\}$.
- $\{1, 2, 3\} \cup \{3, 4, 5\} = \{1, 2, 3, 4, 5\}$

Some basic properties of unions:

- $A \cup B = B \cup A$.
- $A \cup (B \cup C) = (A \cup B) \cup C$.
- $A \subseteq (A \cup B)$.
- $A \cup A = A$.
- $A \cup \emptyset = A$.
- $A \subseteq B$ if and only if $A \cup B = B$.

Intersections

A new set can also be constructed by determining which members two sets have "in common". The *intersection* of A and B, denoted by $A \cap B$, is the set of all things that are members of both A and B. If $A \cap B = \emptyset$, then A and B are said to be *disjoint*.

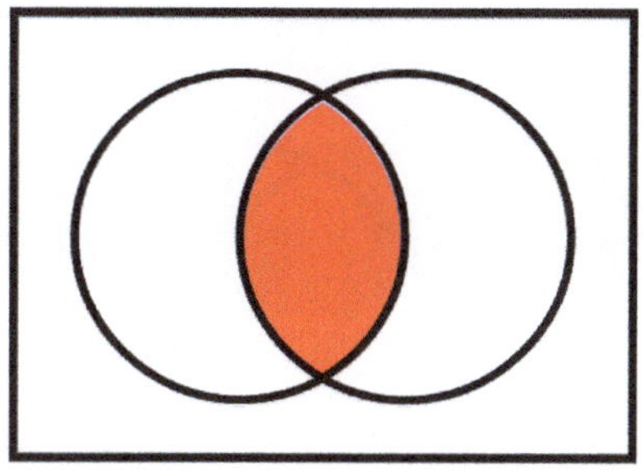

The **intersection** of A and B, denoted $A \cap B$.

Examples:

- $\{1, 2\} \cap \{1, 2\} = \{1, 2\}$.
- $\{1, 2\} \cap \{2, 3\} = \{2\}$.

Some basic properties of intersections:

- $A \cap B = B \cap A$.
- $A \cap (B \cap C) = (A \cap B) \cap C$.
- $A \cap B \subseteq A$.
- $A \cap A = A$.
- $A \cap \emptyset = \emptyset$.
- $A \subseteq B$ if and only if $A \cap B = A$.

Complements

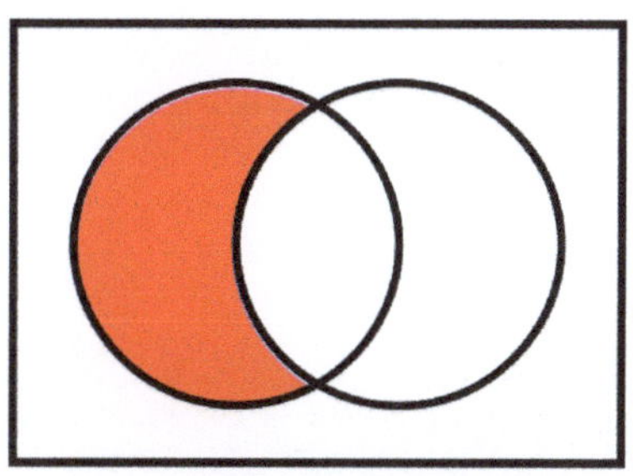

The **relative complement** of B in A

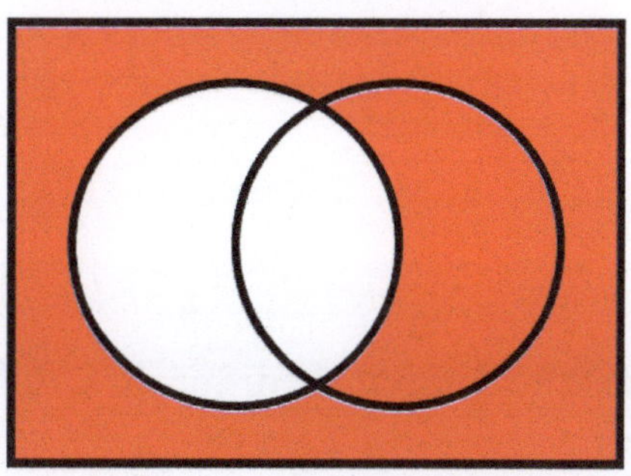

The **complement** of A in U

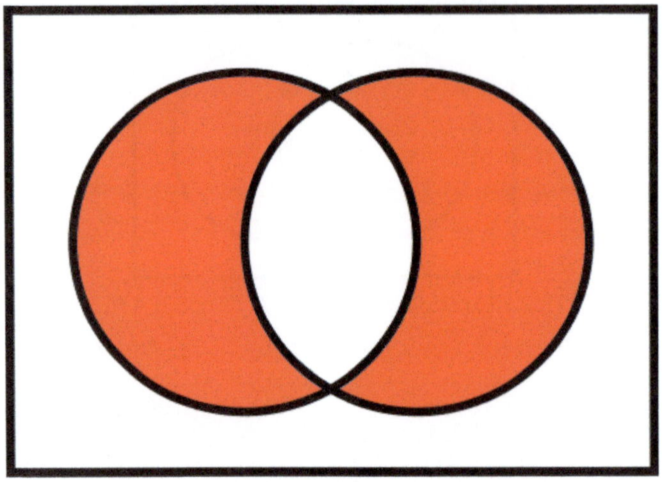

The **symmetric difference** of A and B

Two sets can also be "subtracted". The *relative complement* of B in A (also called the *set-theoretic difference* of A and B), denoted by $A \setminus B$ (or $A - B$), is the set of all elements that are members of A but not members of B. Note that it is valid to "subtract" members of a set that are not in the set, such as removing the element *green* from the set {1, 2, 3}; doing so has no effect.

In certain settings all sets under discussion are considered to be subsets of a given universal set U. In such cases, $U \setminus A$ is called the *absolute complement* or simply *complement* of A, and is denoted by A'.

Examples:

- $\{1, 2\} \setminus \{1, 2\} = \emptyset$.
- $\{1, 2, 3, 4\} \setminus \{1, 3\} = \{2, 4\}$.
- If U is the set of integers, E is the set of even integers, and O is the set of odd integers, then $U \setminus E = E' = O$.

Some basic properties of complements:

- $A \setminus B \neq B \setminus A$ for $A \neq B$.
- $A \cup A' = U$.
- $A \cap A' = \emptyset$.
- $(A')' = A$.
- $A \setminus A = \emptyset$.
- $U' = \emptyset$ and $\emptyset' = U$.
- $A \setminus B = A \cap B'$.

An extension of the complement is the symmetric difference, defined for sets A, B as

- $A \Delta B = (A \setminus B) \cup (B \setminus A)$.

For example, the symmetric difference of {7,8,9,10} and {9,10,11,12} is the set {7,8,11,12}.

Cartesian Product

A new set can be constructed by associating every element of one set with every element of another set. The *Cartesian product* of two sets A and B, denoted by $A \times B$ is the set of all ordered pairs (a, b) such that a is a member of A and b is a member of B.

Examples:

- {1, 2} × {red, white} = {(1, red), (1, white), (2, red), (2, white)}.
- {1, 2} × {red, white, green} = {(1, red), (1, white), (1, green), (2, red), (2, white), (2, green) }.
- $\{1, 2\} \times \{1, 2\} = \{(1, 1), (1, 2), (2, 1), (2, 2)\}$.

Some basic properties of cartesian products:

- $A \times \emptyset = \emptyset$.
- $A \times (B \cup C) = (A \times B) \cup (A \times C)$.
- $(A \cup B) \times C = (A \times C) \cup (B \times C)$.

Let A and B be finite sets. Then

- $|A \times B| = |B \times A| = |A| \times |B|$.

Applications

Set theory is seen as the foundation from which virtually all of mathematics can be derived. For example, structures in abstract algebra, such as groups, fields and rings, are sets closed under one or more operations.

One of the main applications of naive set theory is constructing relations. A relation from a domain A to a codomain B is a subset of the Cartesian product $A \times B$. Given this concept, we are quick to see that the set F of all ordered pairs (x, x^2), where x is real, is quite familiar. It has a domain set R and a codomain set that is also R, because the set of all squares is subset of the set of all reals. If placed in functional notation, this relation becomes $f(x) = x^2$. The reason these two are equivalent is for any given value, y that the function is defined for, its corresponding ordered pair, (y, y^2) is a member of the set F.

Axiomatic Set Theory

Although initially naive set theory, which defines a set merely as *any well-defined* collection, was well accepted, it soon ran into several obstacles. It was found that this definition spawned several paradoxes, most notably:

- Russell's paradox—It shows that the "set of all sets that *do not contain themselves*," i.e. the "set" $\{ x : x \text{ is a set and } x \notin x \}$ does not exist.
- Cantor's paradox—It shows that "the set of all sets" cannot exist.

The reason is that the phrase *well-defined* is not very well defined. It was important to free set theory of these paradoxes because nearly all of mathematics was being redefined in terms of set theory. In an attempt to avoid these paradoxes, set theory was axiomatized based on first-order logic, and thus axiomatic set theory was born.

For most purposes however, naive set theory is still useful.

Principle of Inclusio and Exclusion

This principle gives us the cardinality of the union of sets.

$$|A_1 \cup A_2 \cup A_3 \cup \ldots \cup A_n| = (|A_1| + |A_2| + |A_3| + \ldots |A_n|) -$$
$$(|A_1 \cap A_2| + |A_1 \cap A_3| + \ldots |A_{n\ 1} \cap A_n|) + \ldots +$$
$$(-\ )\quad (|A_1 \cap A_2 \cap A_3 \cap \ldots \cap A\ |)$$

De Morgan's Law

De Morgan stated two laws about sets.

If A and B are any two sets then,

(A ∪ B)′ = A′ ∩ B′

The complement of A union B equals the complement of A intersected with the complement of B.

(A ∩ B)′ = A′ ∪ B′

The complement of A intersected with B is equal to the complement of A union to the complement of B.

Element (Mathematics)

In mathematics, an element, or member, of a set is any one of the distinct objects that make up that set.

Sets

Writing $A = \{1, 2, 3, 4\}$ means that the elements of the set A are the numbers 1, 2, 3 and 4. Sets of elements of A, for example $\{1, 2\}$, are subsets of A.

Sets can themselves be elements. For example, consider the set $B = \{1, 2, \{3, 4\}\}$. The elements of B are *not* 1, 2, 3, and 4. Rather, there are only three elements of B, namely the numbers 1 and 2, and the set $\{3, 4\}$.

The elements of a set can be anything. For example, $C = \{$ red, green, blue $\}$, is the set whose elements are the colors red, green and blue.

Notation and Terminology

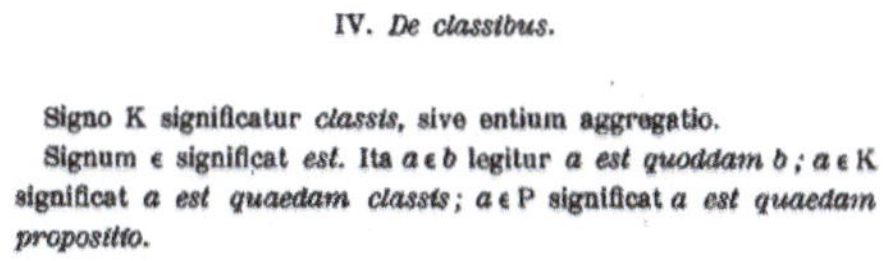
IV. *De classibus.*

Signo K significatur *classis*, sive entium aggregatio.
Signum ϵ significat *est*. Ita *a* ϵ *b* legitur *a est quoddam b* ; *a* ϵ K significat *a est quaedam classis* ; *a* ϵ P significat *a est quaedam propositio*.

First usage of the symbol ϵ in the work Arithmetices principia nova methodo exposita by Giuseppe Peano.

The relation "is an element of", also called set membership, is denoted by the symbol "∈". Writing

$$x \in A$$

means that "x is an element of A". Equivalent expressions are "x is a member of A", "x belongs to A", "x is in A" and "x lies in A". The expressions "A includes x" and "A contains x" are also used to mean set membership, however some authors use them to mean instead "x is a subset of A". Logician George Boolos strongly urged that "contains" be used for membership only and "includes" for the subset relation only.

Another possible notation for the same relation is

$$A \ni x,$$

meaning "A contains x", though it is used less often.

The negation of set membership is denoted by the symbol "∉". Writing

$$x \notin A$$

means that "x is not an element of A".

The symbol ϵ was first used by Giuseppe Peano 1889 in his work Arithmetices principia nova methodo exposita. Here he wrote on page X:

"Signum ϵ significat est. Ita a ϵ b legitur a est quoddam b; ..."

which means

"The symbol ϵ means *is*. So a ϵ b is read as a *is a* b; ..."

The Unicode characters for these symbols are U+2208 ('element of'), U+220B ('contains as member') and U+2209 ('not an element of'). The equivalent LaTeX commands are "\in", "\ni" and "\notin". Mathematica has commands "\[Element]" and "\[NotElement]".

Cardinality of Sets

The number of elements in a particular set is a property known as cardinality; informally, this is the size of a set. In the above examples the cardinality of the set A is 4, while the cardinality of either of the sets B and C is 3. An infinite set is a set with an infinite number of elements, while a finite set is a set with a finite number of elements. The above examples are examples of finite sets. An example of an infinite set is the set of positive integers = { 1, 2, 3, 4, ... }.

Examples

Using the sets defined above, namely $A = \{1, 2, 3, 4\}$, $B = \{1, 2, \{3, 4\}\}$ and $C = \{$ red, green, blue $\}$:

- $2 \in A$
- $\{3,4\} \in B$
- $3,4 \notin B$
- $\{3,4\}$ is a member of B
- Yellow $\notin C$
- The cardinality of $D = \{ 2, 4, 8, 10, 12 \}$ is finite and equal to 5.
- The cardinality of $P = \{ 2, 3, 5, 7, 11, 13, \dots\}$ (the prime numbers) is infinite (this was proven by Euclid).

Venn Diagram

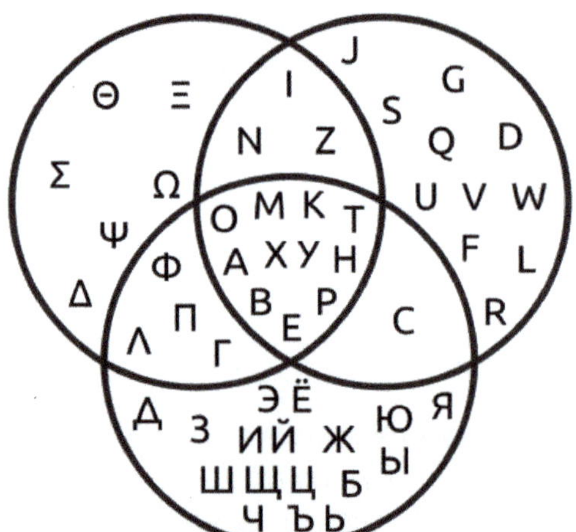

Venn diagram showing which uppercase letter glyphs are shared by the Greek, Latin and Russian alphabets

A Venn diagram (also called a set diagram or logic diagram) is a diagram that shows all possible logical relations between a finite collection of different sets. Typically overlapping shapes, usually circles, are used, and an area-proportional or scaled Venn diagram is one in which the area of the shape is proportional to the number of elements it contains. These diagrams represent elements as points in the plane, and sets as regions inside curves. An element is in a set S just in case the corresponding point is in the region for S. They are thus a special case of Euler diagrams, which do not necessarily show all relations. Venn diagrams were conceived around 1880 by John Venn. They are used to teach elementary set theory, as well as illustrate simple set relationships in probability, logic, statistics, linguistics and computer science.

Example

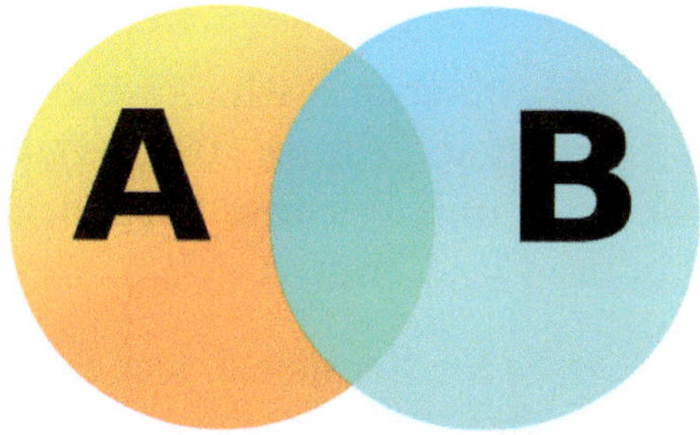

Sets A (creatures with two legs) and B (creatures that can fly)

This example involves two sets, A and B, represented here as coloured circles. The orange circle, set A, represents all living creatures that are two-legged. The blue circle, set B, represents the living creatures that can fly. Each separate type of creature can be imagined as a point somewhere in the diagram. Living creatures that both can fly *and* have two legs—for example, parrots—are then in both sets, so they correspond to points in the region where the blue and orange circles overlap. That region contains all such and only such living creatures.

Humans and penguins are bipedal, and so are then in the orange circle, but since they cannot fly they appear in the left part of the orange circle, where it does not overlap with the blue circle. Mosquitoes have six legs, and fly, so the point for mosquitoes is in the part of the blue circle that does not overlap with the orange one. Creatures that are not two-legged and cannot fly (for example, whales and spiders) would all be represented by points outside both circles.

The combined region of sets A and B is called the *union* of A and B, denoted by $A \cup B$. The union in this case contains all living creatures that are either two-legged or that can fly (or both).

The region in both A and B, where the two sets overlap, is called the *intersection* of A and B, denoted by $A \cap B$. For example, the intersection of the two sets is not empty, because there *are* points that represent creatures that are in *both* the orange and blue circles.

History

Venn diagrams were introduced in 1880 by John Venn in a paper entitled *On the Diagrammatic and Mechanical Representation of Propositions and Reasonings* in the "Philosophical Magazine and Journal of Science", about the different ways to represent propositions by diagrams. The use of these types of diagrams in formal logic, according to Ruskey and M. Weston, is "not an easy history to trace, but it is certain that the diagrams that are popularly associated with Venn, in fact, originated much earlier. They are rightly associated with Venn, however, because he comprehensively surveyed and formalized their usage, and was the first to generalize them".

Venn himself did not use the term "Venn diagram" and referred to his invention as "Eulerian Circles". For example, in the opening sentence of his 1880 article Venn writes, "Schemes of diagrammatic representation have been so familiarly introduced into logical treatises during the last century or so, that many readers, even those who have made no professional study of logic, may be supposed to be acquainted with the general nature and object of such devices. Of these schemes one only, viz. that commonly called 'Eulerian circles,' has met with any general acceptance..." The first to use the term "Venn diagram" was Clarence Irving Lewis in 1918, in his book "A Survey of Symbolic Logic".

Venn diagrams are very similar to Euler diagrams, which were invented by Leonhard Euler in the 18th century. M. E. Baron has noted that Leibniz (1646–1716) in the 17th century produced similar diagrams before Euler, but much of it was unpublished. She also observes even earlier Euler-like diagrams by Ramon Lull in the 13th Century.

In the 20th century, Venn diagrams were further developed. D.W. Henderson showed in 1963 that the existence of an n-Venn diagram with n-fold rotational symmetry implied that n was a prime number. He also showed that such symmetric Venn diagrams exist when n is 5 or 7. In 2002 Peter Hamburger found symmetric Venn diagrams for $n = 11$ and in 2003, Griggs, Killian, and Savage showed that symmetric Venn diagrams exist for all other primes. Thus rotationally symmetric Venn diagrams exist if and only if n is a prime number.

Venn diagrams and Euler diagrams were incorporated as part of instruction in set theory as part of the new math movement in the 1960s. Since then, they have also been adopted in the curriculum of other fields such as reading.

Overview

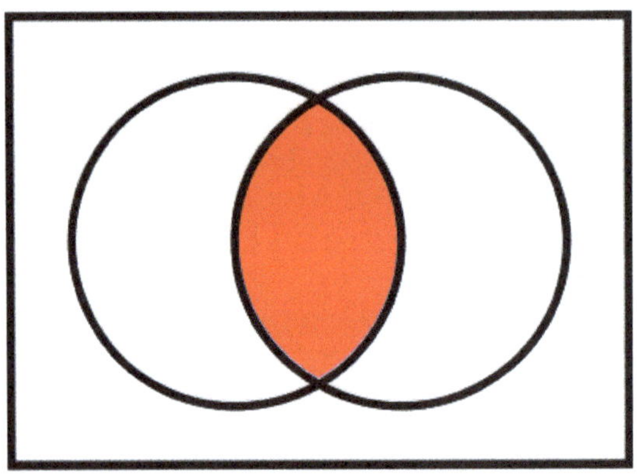

Intersection of two sets $A \cap B$

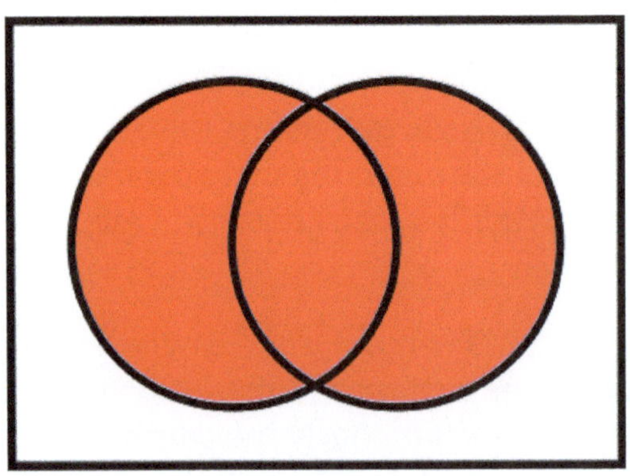

Union of two sets $A \cup B$

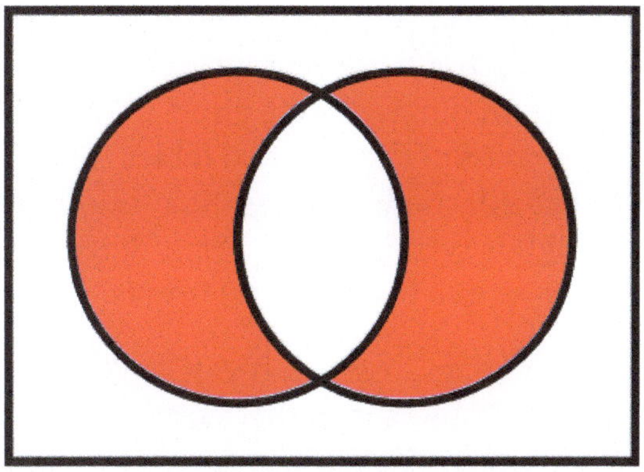

Symmetric difference of two sets
$A \Delta B$

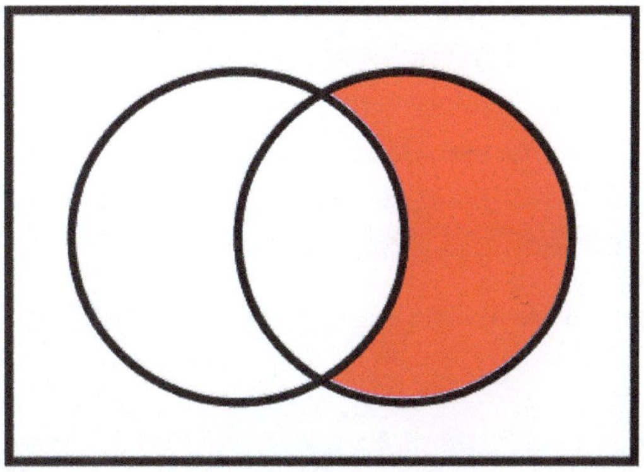

Relative complement of A (left) in B (right)
$A^c \cap B = B \setminus A$

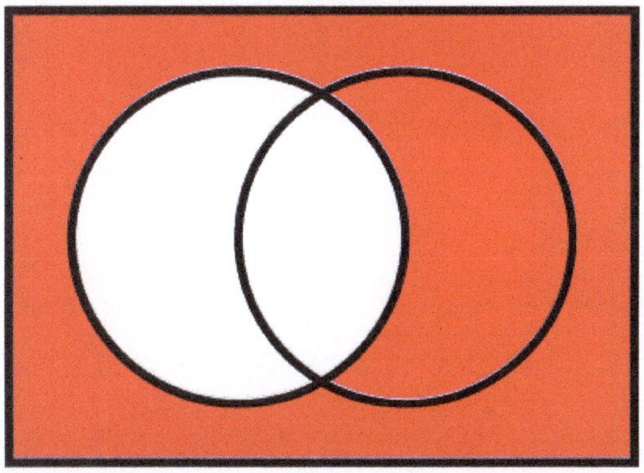

Absolute complement of A in U
$A^c = U \setminus A$

A Venn diagram is constructed with a collection of simple closed curves drawn in a plane. According to Lewis, the "principle of these diagrams is that classes [or *sets*] be represented by regions in such relation to one another that all the possible logical relations of these classes can be indicated in the same diagram. That is, the diagram initially leaves room for any possible relation of the classes, and the actual or given relation, can then be specified by indicating that some particular region is null or is not-null".

Venn diagrams normally comprise overlapping circles. The interior of the circle symbolically represents the elements of the set, while the exterior represents elements that are not members of the set. For instance, in a two-set Venn diagram, one circle may represent the group of all wooden objects, while another circle may represent the set of all tables. The overlapping region or *intersection* would then represent the set of all wooden tables. Shapes other than circles can be employed as shown below by Venn's own higher set diagrams. Venn diagrams do not generally contain information on the relative or absolute sizes (cardinality) of sets; i.e. they are schematic diagrams.

Venn diagrams are similar to Euler diagrams. However, a Venn diagram for n component sets must contain all 2^n hypothetically possible zones that correspond to some combination of inclusion or exclusion in each of the component sets. Euler diagrams contain only the actually possible zones in a given context. In Venn diagrams, a shaded zone may represent an empty zone, whereas in an Euler diagram the corresponding zone is missing from the diagram. For example, if one set represents *dairy products* and another *cheeses*, the Venn diagram contains a zone for cheeses that are not dairy products. Assuming that in the context *cheese* means some type of dairy product, the Euler diagram has the cheese zone entirely contained within the dairy-product zone—there is no zone for (non-existent) non-dairy cheese. This means that as the number of contours increases, Euler diagrams are typically less visually complex than the equivalent Venn diagram, particularly if the number of non-empty intersections is small.

The difference between Euler and Venn diagrams can be seen in the following example. Take the three sets:

- $A = \{1,2,5\}$
- $B = \{1,6\}$
- $C = \{4,7\}$

The Venn and the Euler diagram of those sets are:

- 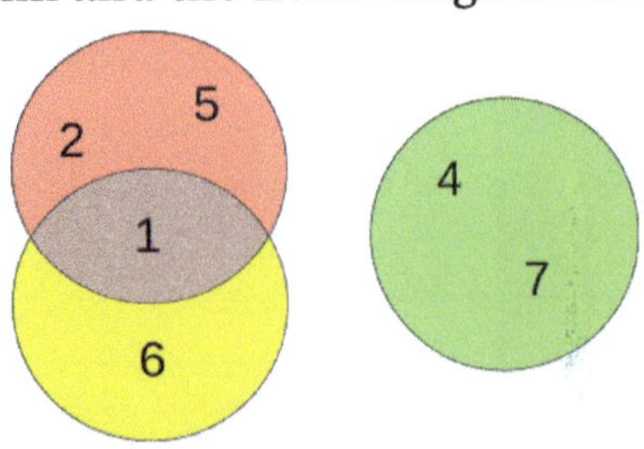

  Euler diagram
- 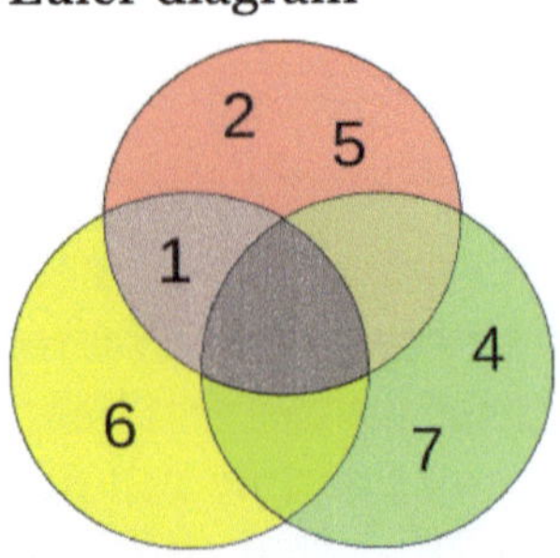

  Venn diagram

Extensions to Higher Numbers of Sets

Venn diagrams typically represent two or three sets, but there are forms that allow for higher numbers. Shown below, four intersecting spheres form the highest order Venn diagram that has the symmetry of a simplex and can be visually represented. The 16 intersections correspond to the vertices of a tesseract (or the cells of a 16-cell respectively).

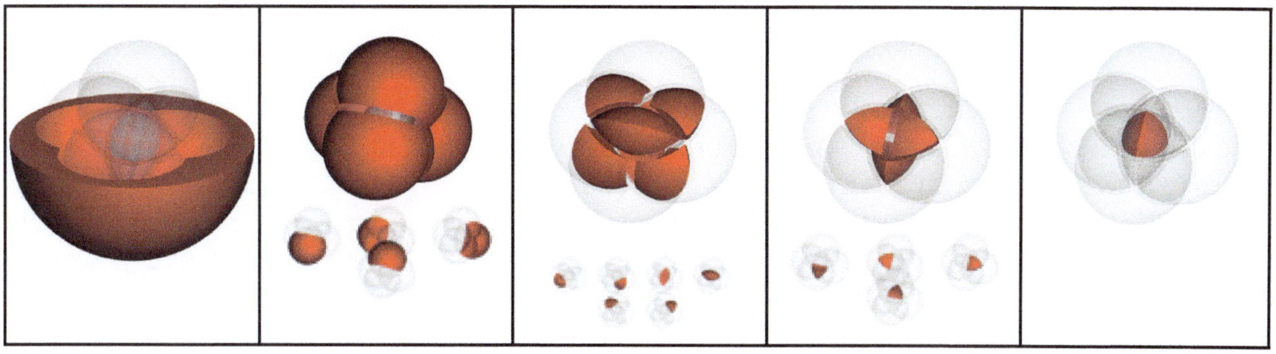

For higher numbers of sets, some loss of symmetry in the diagrams is unavoidable. Venn was keen to find "symmetrical figures...elegant in themselves," that represented higher numbers of sets, and he devised a four-set diagram using ellipses. He also gave a construction for Venn diagrams for *any* number of sets, where each successive curve that delimits a set interleaves with previous curves, starting with the three-circle diagram.

- 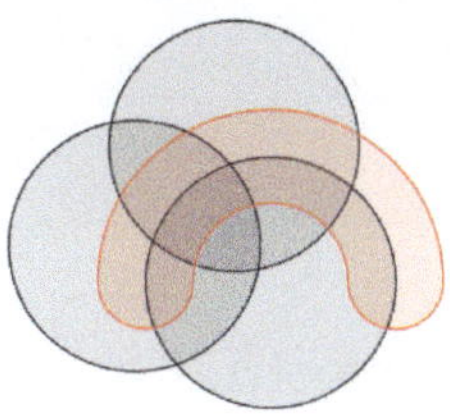

Venn's construction for 4 sets

- 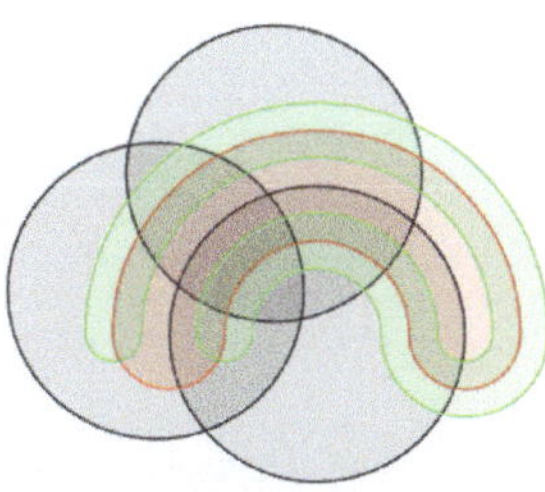

Venn's construction for 5 sets

- 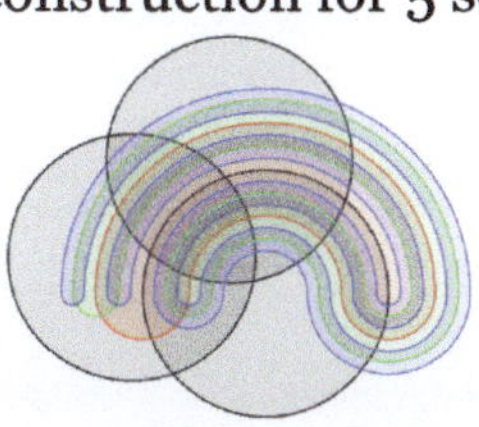

Venn's construction for 6 sets

- 

Venn's four-set diagram using ellipses

- 

Non-example: This Euler diagram is not a Venn diagram for four sets as it has only 13 regions (excluding the outside); there is no region where only the yellow and blue, or only the red and green circles meet.

- 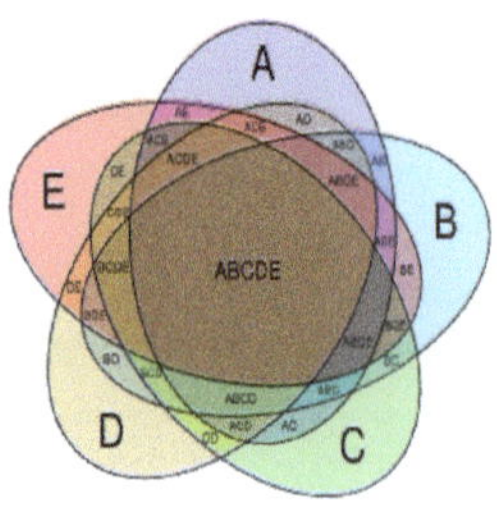

Five-set Venn diagram using congruent ellipses in a 5-fold rotationally symmetrical arrangement devised by Branko Grünbaum. Labels have been simplified for greater readability; for example, A denotes $A \cap B^c \cap C^c \cap D^c \cap E^c$, while BCE denotes $A^c \cap B \cap C \cap D^c \cap E$.

- 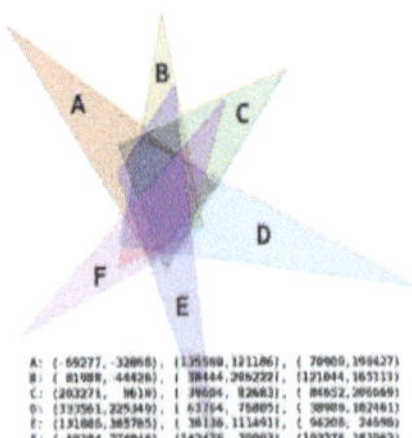

Six-set Venn diagram made of only triangles (interactive version)

Edwards' Venn Diagrams

- 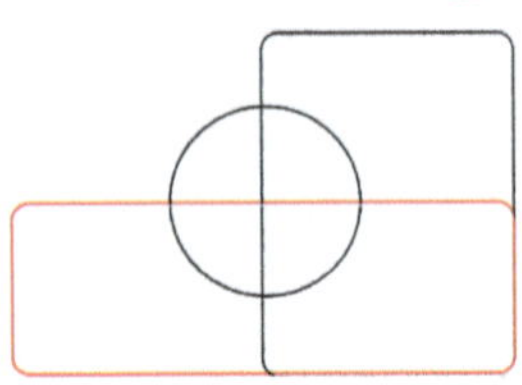

Three sets

- 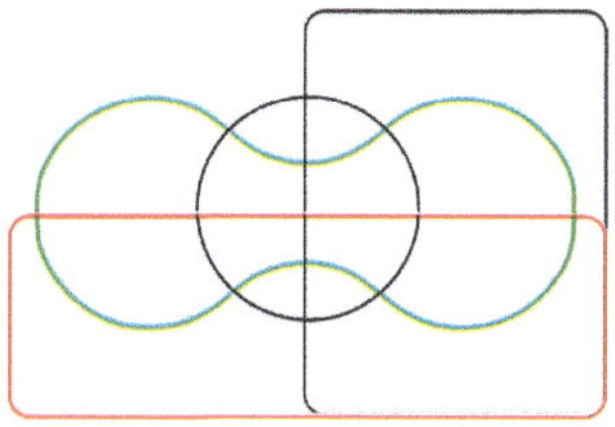

Four sets

- 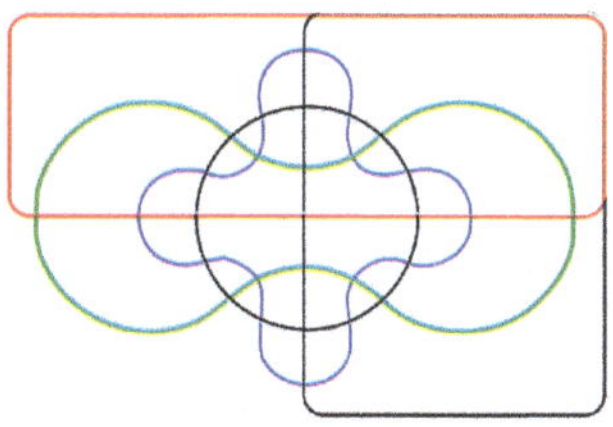

Five sets

- 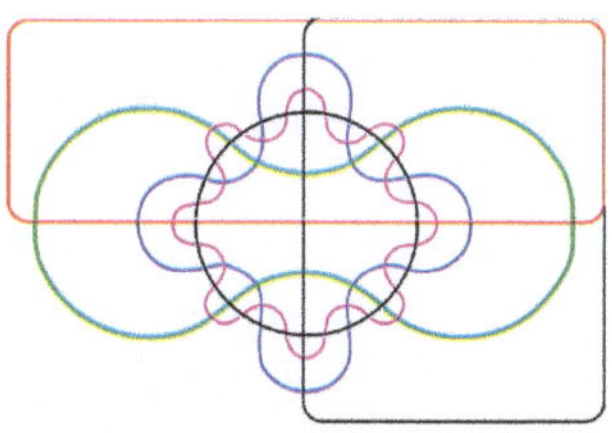

Six sets

A. W. F. Edwards constructed a series of Venn diagrams for higher numbers of sets by segmenting the surface of a sphere. For example, three sets can be easily represented by taking three hemispheres of the sphere at right angles ($x = 0$, $y = 0$ and $z = 0$). A fourth set can be added to the representation by taking a curve similar to the seam on a tennis ball, which winds up and down around the equator, and so on. The resulting sets can then be projected back to a plane to give *cogwheel* diagrams with increasing numbers of teeth, as shown here. These diagrams were devised while designing a stained-glass window in memory of Venn.

Other Diagrams

Edwards' Venn diagrams are topologically equivalent to diagrams devised by Branko Grünbaum, which were based around intersecting polygons with increasing numbers of sides. They are also 2-dimensional representations of hypercubes.

Henry John Stephen Smith devised similar n-set diagrams using sine curves with the series of equations.

$$y_i = \frac{\sin(2^i x)}{2i} \text{ where } 0 \le i \le n-2 \text{ and } i \in \mathbb{N}.$$

Charles Lutwidge Dodgson devised a five-set diagram.

Related Concepts

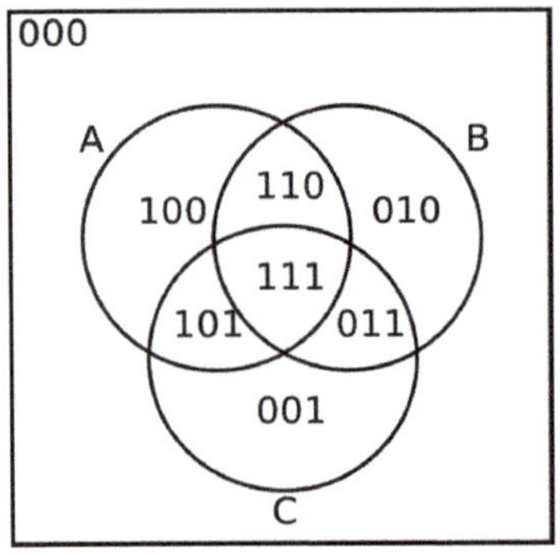

Venn diagram as a truth table

Venn diagrams correspond to truth tables for the propositions $x \in A$, $x \in B$, etc., in the sense that each region of Venn diagram corresponds to one row of the truth table. Another way of representing sets is with R-Diagrams.

Subset

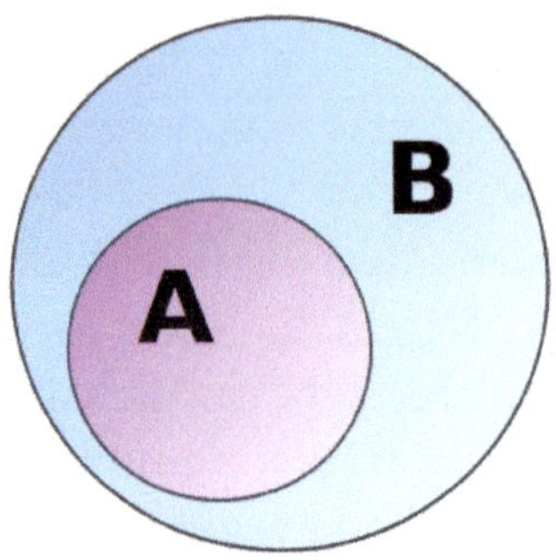

Euler diagram showing
A is a proper subset of *B* and conversely *B* is a proper superset of *A*

In mathematics, especially in set theory, a set *A* is a subset of a set *B*, or equivalently *B* is a superset of *A*, if *A* is "contained" inside *B*, that is, all elements of *A* are also elements of *B*. *A* and *B* may coincide. The relationship of one set being a subset of another is called inclusion or sometimes containment.

The subset relation defines a partial order on sets.

The algebra of subsets forms a Boolean algebra in which the subset relation is called inclusion.

Definitions

If *A* and *B* are sets and every element of *A* is also an element of *B*, then:

- *A* is a subset of (or is included in) *B*, denoted by $A \subseteq B$,

or equivalently

- B is a superset of (or includes) A, denoted by $B \supseteq A$.

If A is a subset of B, but A is not equal to B (i.e. there exists at least one element of B which is not an element of A), then

- A is also a proper (or strict) subset of B; this is written as

or equivalently $A \subsetneq B$.

- B is a proper superset of A; this is written as $B \supsetneq A$.

For any set S, the inclusion relation $\subseteq$ is a partial order on the set $\mathcal{P}(S)$ of all subsets of S (the power set of S) defined by $A \leq B \Leftrightarrow A \subseteq B$. We may also partially order $\mathcal{P}(S)$ by reverse set inclusion by defining $A \leq B \Leftrightarrow B \subseteq A$.

When quantified, $A \subseteq B$ is represented as: $\forall x\{x \in A \rightarrow x \in B\}$.

Property

- A finite set A is a subset of (or is included in) B, denoted by $A \subseteq B$,

If and only if the cardinality of their intersection is equal to the cardinality of A.

Formally:

$$A \subseteq B \Leftrightarrow |A \cap B| = |A|$$

$\subset$ and $\supset$ Symbols

Some authors use the symbols $\subset$ and $\supset$ to indicate *subset* and *superset* respectively; that is, with the same meaning and instead of the symbols, $\subseteq$ and $\supseteq$. So for example, for these authors, it is true of every set A that $A \subset A$.

Other authors prefer to use the symbols $\subset$ and $\supset$ to indicate *proper* subset and superset, respectively, instead of $\subsetneq$ and $\supsetneq$. This usage makes $\subseteq$ and $\subset$ analogous to the inequality symbols $\leq$ and $<$. For example, if $x \leq y$ then x may or may not equal y, but if $x < y$, then x definitely does not equal y, and *is* less than y. Similarly, using the convention that $\subset$ is proper subset, if $A \subseteq B$, then A may or may not equal B, but if $A \subset B$, then A definitely does not equal B.

Examples

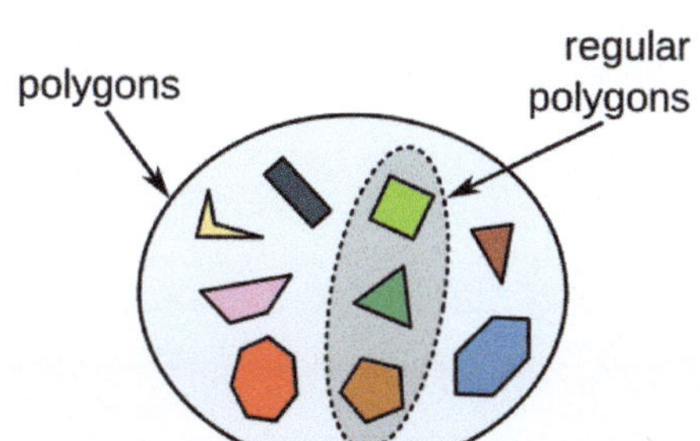

The regular polygons form a subset of the polygons

- The set A = {1, 2} is a proper subset of B = {1, 2, 3}, thus both expressions A ⊆ B and A ⊊ B are true.
- The set D = {1, 2, 3} is a subset of E = {1, 2, 3}, thus D ⊆ E is true, and D ⊊ E is not true (false).
- Any set is a subset of itself, but not a proper subset. (X ⊆ X is true, and X ⊊ X is false for any set X.)
- The empty set { }, denoted by ∅, is also a subset of any given set *X*. It is also always a proper subset of any set except itself.
- The set {*x*: *x* is a prime number greater than 10} is a proper subset of {*x*: *x* is an odd number greater than 10}
- The set of natural numbers is a proper subset of the set of rational numbers; likewise, the set of points in a line segment is a proper subset of the set of points in a line. These are two examples in which both the subset and the whole set are infinite, and the subset has the same cardinality (the concept that corresponds to size, that is, the number of elements, of a finite set) as the whole; such cases can run counter to one's initial intuition.
- The set of rational numbers is a proper subset of the set of real numbers. In this example, both sets are infinite but the latter set has a larger cardinality (or *power*) than the former set.

Another example in an Euler diagram:

- 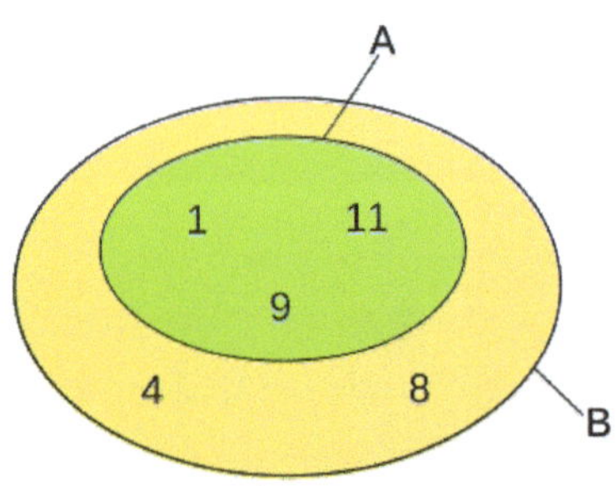

  A is a proper subset of B

- 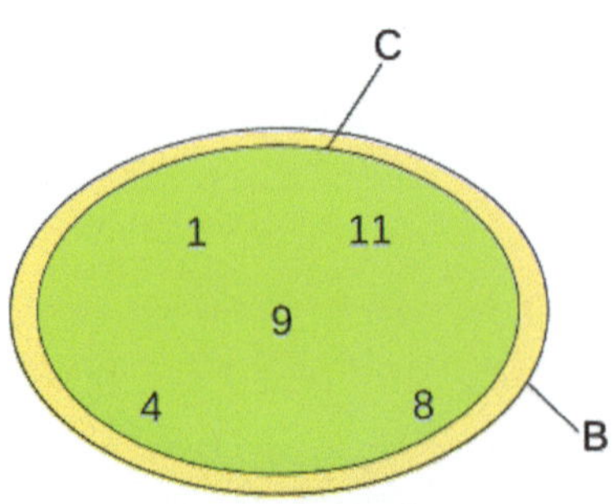

  C is a subset but not a proper subset of B

Other Properties of Inclusion

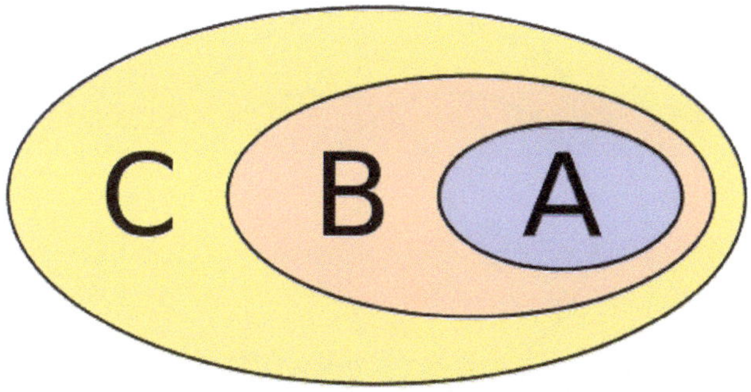

$A \subseteq B$ and $B \subseteq C$ imply $A \subseteq C$

Inclusion is the canonical partial order in the sense that every partially ordered set $(X, \preceq)$ is isomorphic to some collection of sets ordered by inclusion. The ordinal numbers are a simple example—if each ordinal n is identified with the set $[n]$ of all ordinals less than or equal to n, then $a \leq b$ if and only if $[a] \subseteq [b]$.

For the power set $\mathcal{P}(S)$ of a set S, the inclusion partial order is (up to an order isomorphism) the Cartesian product of $k = |S|$ (the cardinality of S) copies of the partial order on $\{0,1\}$ for which $0 < 1$. This can be illustrated by enumerating $S = \{s_1, s_2, ..., s_k\}$ and associating with each subset $T \subseteq S$ (which is to say with each element of 2^S) the k-tuple from $\{0,1\}^k$ of which the ith coordinate is 1 if and only if s_i is a member of T.

Intersection (Set Theory)

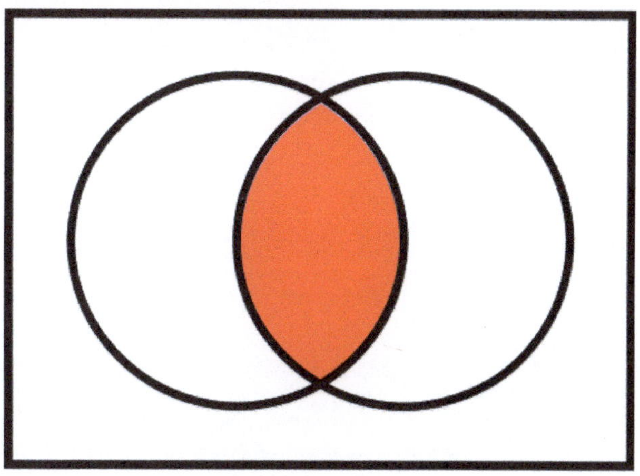

Intersection of two sets:
$A \cap B$

In mathematics, the intersection $A \cap B$ of two sets A and B is the set that contains all elements of A that also belong to B (or equivalently, all elements of B that also belong to A), but no other elements.

For explanation of the symbols used in this article, refer to the table of mathematical symbols.

Basic Definition

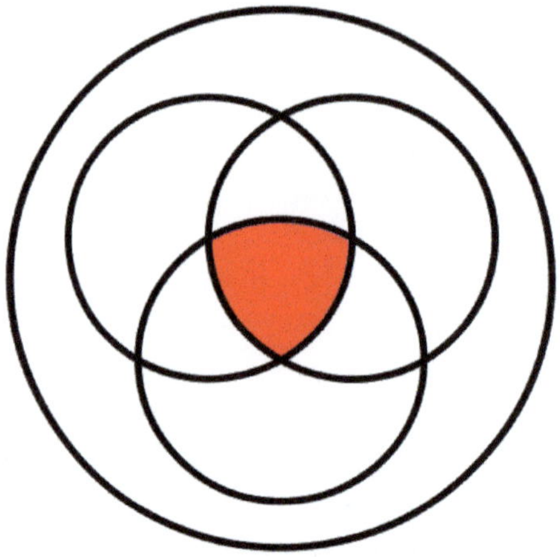

Intersection of three sets:
$A \cap B \cap C$

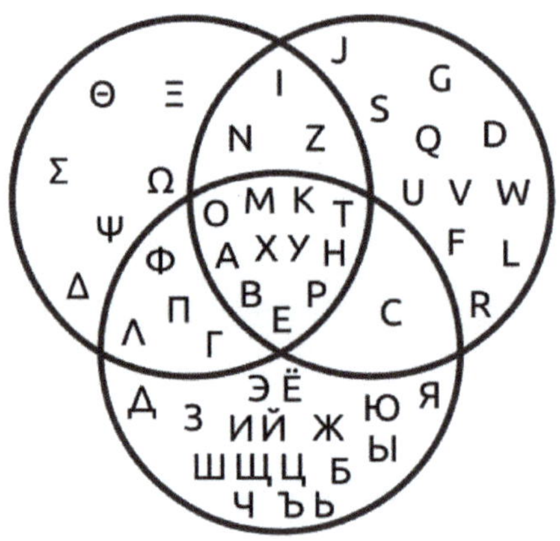

Intersections of the Greek, English and Russian alphabet, considering only the shapes of the letters and ignoring their pronunciation

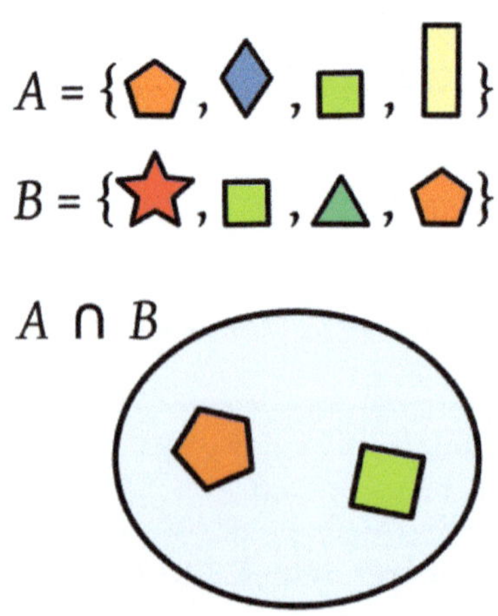

Example of an intersection with sets

The intersection of A and B is written "$A \cap B$". Formally:

$$A \cap B = \{x : x \in A \wedge x \in B\}$$

that is

$x \in A \cap B$ if and only if

$x \in A$ and

$x \in B$.

For example:

- The intersection of the sets {1, 2, 3} and {2, 3, 4} is {2, 3}.
- The number 9 is *not* in the intersection of the set of prime numbers {2, 3, 5, 7, 11, ...} and the set of odd numbers {1, 3, 5, 7, 9, 11, ...}.

More generally, one can take the intersection of several sets at once. The *intersection* of A, B, C, and D, for example, is $A \cap B \cap C \cap D = A \cap (B \cap (C \cap D))$. Intersection is an associative operation; thus,$A \cap (B \cap C) = (A \cap B) \cap C$.

Inside a universe U one may define the complement A^c of A to be the set of all elements of U not in A. Now the intersection of A and B may be written as the complement of the union of their complements, derived easily from De Morgan's laws:$A \cap B = (A^c \cup B^c)^c$

Intersecting and Disjoint Sets

We say that *A intersects (meets) B at an element x* if x belongs to A and B. We say that *A intersects (meets) B* if A intersects B at some element. A intersects B if their intersection is inhabited.

We say that *A and B are disjoint* if A does not intersect B. In plain language, they have no elements in common. A and B are disjoint if their intersection is empty, denoted $A \cap B = \varnothing$.

For example, the sets {1, 2} and {3, 4} are disjoint, the set of even numbers intersects the set of multiples of 3 at 0, 6, 12, 18 and other numbers.

Arbitrary Intersections

The most general notion is the intersection of an arbitrary *nonempty* collection of sets. If M is a nonempty set whose elements are themselves sets, then x is an element of the *intersection* of M if and only if for every element A of M, x is an element of A. In symbols:

$$\left(x \in \bigcap_{A \in M} \mathbf{A}\right) \Leftrightarrow (\forall A \in \mathbf{M}, x \in A).$$

The notation for this last concept can vary considerably. Set theorists will sometimes write "$\cap M$", while others will instead write "$\cap_{A \in M} A$". The latter notation can be generalized to "$\cap_{i \in I} A_i$", which refers to the intersection of the collection $\{A_i : i \in I\}$. Here I is a nonempty set, and A_i is a set for every i in I.

In the case that the index set I is the set of natural numbers, notation analogous to that of an infinite series may be seen:

$$\bigcap_{i=1}^{\infty} A_i.$$

When formatting is difficult, this can also be written "$A_1 \cap A_2 \cap A_3 \cap \ldots$», even though strictly speaking, $A_1 \cap (A_2 \cap (A_3 \cap \ldots$ makes no sense. (This last example, an intersection of countably many sets, is actually very common).

Finally, let us note that whenever the symbol "∩" is placed *before* other symbols instead of *between* them, it should be of a larger size (⋂).

Nullary Intersection

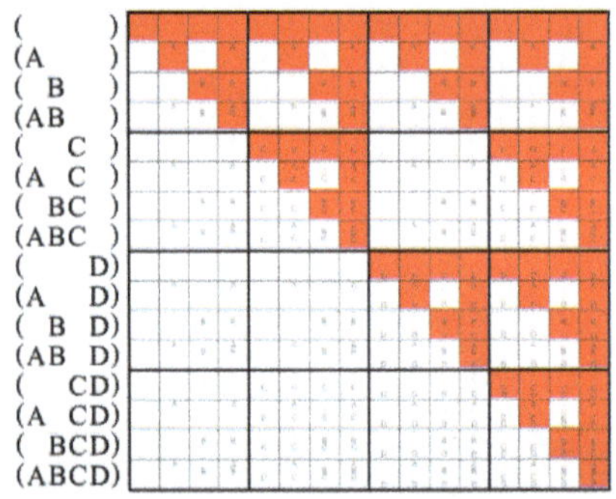

Conjunctions of the arguments in parentheses The conjunction of no argument is the tautology (compare: empty product); accordingly the intersection of no set is the universe.

Note that in the previous section we excluded the case where M was the empty set (∅). The reason is as follows: The intersection of the collection M is defined as the set

$$\bigcap_{A \in M} \mathbf{A} = \{x : \forall A \in \mathbf{M}, x \in A\}.$$

If M is empty there are no sets A in M, so the question becomes "which x's satisfy the stated condition?" The answer seems to be *every possible x*. When M is empty the condition given above is an example of a vacuous truth. So the intersection of the empty family should be the universal set (the identity element for the operation of intersection)

Unfortunately, according to standard (ZFC) set theory, the universal set does not exist. A fix for this problem can be found if we note that the intersection over a set of sets is always a subset of the union over that set of sets. This can symbolically be written as

$$\bigcap_{A \in M} \mathbf{A} \subseteq \bigcup_{A \in M} \mathbf{A}.$$

Therefore, we can modify the definition slightly to

$$\bigcap_{A \in M} \mathbf{A} = \{x \in \bigcup_{A \in M} \mathbf{A} : \forall A \in \mathbf{M}, x \in A\}.$$

Now if M is empty there is no problem. The intersection is the empty set, because the union over the empty set is the empty set. In fact, this is the operation that we would have defined in the first place if we were defining the set in ZFC, as except for the operations defined by the axioms (the power set of a set, for instance), every set must be defined as the subset of some other set or by replacement.

Complement (Set Theory)

In set theory, the complement of a set A refers to elements not in A. The relative complement of A with respect to a set B, written $B \setminus A$, is the set of elements in B but not in A. When all sets under consideration are considered to be subsets of a given set U, the absolute complement of A is the set of elements in U but not in A.

Relative Complement

Definition

If A and B are sets, then the relative complement of A in B, also termed the set-theoretic difference of B and A, is the set of elements in B but not in A.

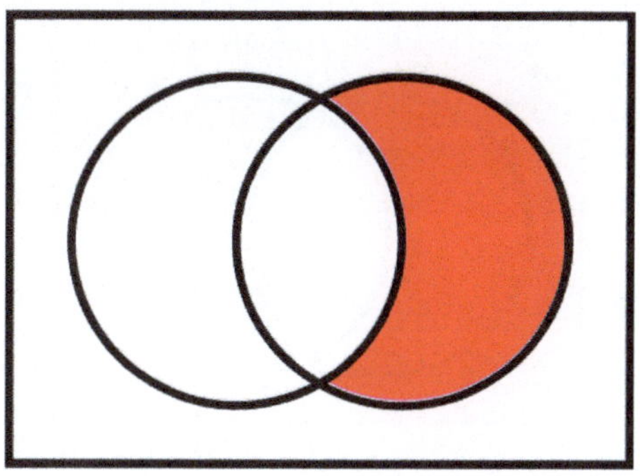

The relative complement of A (left circle) in B (right circle): $B \cap A^c = B \setminus A$

The relative complement of A in B is denoted $B \setminus A$ according to the ISO 31-11 standard. It is sometimes written $B - A$, but this notation is ambiguous, as in some contexts it can be interpreted as the set of all elements $b - a$, where b is taken from B and a from A.

Formally:

$$B \setminus A = \{x \in B \mid x \notin A\}.$$

Examples

- $\{1,2,3\} \setminus \{2,3,4\} = \{1\}$.
- $\{2,3,4\} \setminus \{1,2,3\} = \{4\}$.

If $\mathbb{R}$ is the set of real numbers and $\mathbb{Q}$ is the set of rational numbers, then $\mathbb{R} \setminus \mathbb{Q}$ is the set of irrational numbers.

Properties

Let *A*, *B*, and *C* be three sets. The following identities capture notable properties of relative complements:

- $C \setminus (A \cap B) = (C \setminus A) \cup (C \setminus B)$.
- $C \setminus (A \cup B) = (C \setminus A) \cap (C \setminus B)$.
- $C \setminus (B \setminus A) = (C \cap A) \cup (C \setminus B)$,

with the important special case $C \setminus (C \setminus A) = (C \cap A)$ demonstrating that intersection can be expressed using only the relative complement operation.

- $(B \setminus A) \cap C = (B \cap C) \setminus A = B \cap (C \setminus A)$.
- $(B \setminus A) \cup C = (B \cup C) \setminus (A \setminus C)$.
- $A \setminus A = \varnothing$.
- $\varnothing \setminus A = \varnothing$
- $A \setminus \varnothing = A$.

Absolute Complement

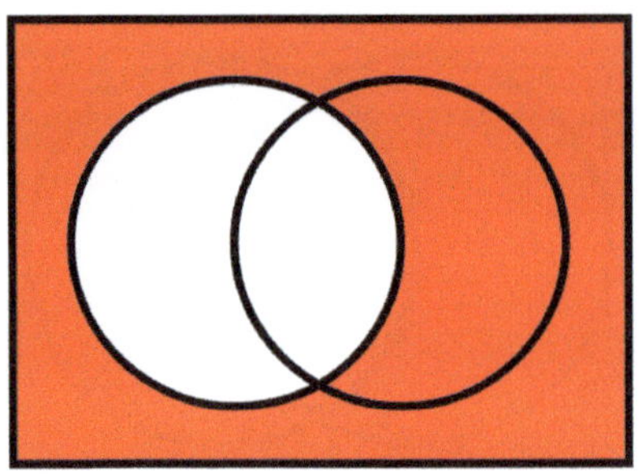

The absolute complement of *A* in *U*: $A^{\complement} = U \setminus A$

Definition

If *A* is a set, then the absolute complement of *A* (or simply the complement of ***A***) is the set of elements not in *A*. In other words, if *U* is the universe that contains all the sets under study, and there is no need to mention it because it is obvious and unique, then the absolute complement of *A* is the relative complement of *A* in *U*:

$$A^{\complement} = U \setminus A.$$

Formally:

$$A^{\complement} = \{x \in U \mid x \notin A\}.$$

The absolute complement of *A* is usually denoted by $A^{\complement}$. Other notations include A^c, $\overline{A}$, $'$, $\complement_U A$, and $\complement A$.

Examples

- Assume that the universe is the set of integers. If A is the set of odd numbers, then the complement of A is the set of even numbers. If B is the set of multiples of 3, then the complement of B is the set of numbers congruent to 1 or 2 modulo 3.
- Assume that the universe is the standard 52-card deck. If the set A is the suit of spades, then the complement of A is the union of the suits of clubs, diamonds, and hearts. If the set B is the union of the suits of clubs and diamonds, then the complement of B is the union of the suits of hearts and spades.

Properties

Let A and B be two sets in a universe U. The following identities capture important properties of absolute complements:

De Morgan's laws:

- $(A \cup B)^C = A^C \cap B^C$.
- $(A \cap B)^C = A^C \cup B^C$.

Complement laws:

- $A \cup A^C = U$.
- $A \cap A^C = \varnothing$.
- $\varnothing^C = U$.
- $U^C = \varnothing$.
- If $A \subset B$, then $B^C \subset A^C$.

(this follows from the equivalence of a conditional with its contrapositive).

Involution or double complement law:

- $(A^C)^C = A$.

Relationships between relative and absolute complements:

- $A \setminus B = A \cap B^C$.
- $(A \setminus B)^C = A^C \cup B$.

Relationship with set difference:

- $A^C \setminus B^C = B \setminus A$.

The first two complement laws above show that if A is a non-empty, proper subset of U, then $\{A, A^c\}$ is a partition of U.

LaTeX Notation

In the LaTeX typesetting language, the command \setminus is usually used for rendering a set difference symbol, which is similar to a backslash symbol. When rendered, the \setminus command looks identical to \backslash except that it has a little more space in front and behind the slash, akin to the LaTeX sequence \mathbin{\backslash}. A variant \smallsetminus is available in the amssymb package.

Complements in Various Programming Languages

Some programming languages allow for manipulation of sets as data structures, using these operators or functions to construct the difference of sets a and b:

.NET Framework

```
b.Except(a);
```

C++

```
set_difference(a.begin(), a.end(), b.begin(), b.end(), result.begin());
```

Clojure

```
(clojure.set/difference a b)
```

Common Lisp

```
set-difference, nset-difference
```

F#

```
Set.difference a b
```

or

```
a - b
```

Falcon

```
diff = a - b
```

Haskell

```
difference a b
a \\ b
```

Java

```
diff = a.clone();
diff.removeAll(b);
```

Julia

setdiff

Mathematica

Complement

MATLAB

setdiff

OCaml

Set.S.diff

Octave

setdiff

PARI/GP

setminus

Pascal

SetDifference := a - b;

Perl 5

for perl version >= 5.10

@a = grep {not $_ ~~ @b} @a;

Perl 6

$A ∖ $B

$A (-) $B # texas version

PHP

array_diff($a, $b);

Prolog

a(X),\+ b(X).

Python

diff = a.difference(b)

diff = a - b

R

setdiff

Racket

```
(set-subtract a b)
```

Ruby

```
diff = a - b
```

Scala

```
a.diff(b)
```

or

```
a -- b
```

Smalltalk (Pharo)

```
a difference: b
```

SQL

```
SELECT * FROM A
EXCEPT
SELECT * FROM B
```

Unix shell

```
comm -23 a b
grep -vf b a # less efficient, but works with small unsorted sets
```

Ordered Pair

In mathematics, an ordered pair (a, b) is a pair of objects. The order in which the objects appear in the pair is significant: the ordered pair (a, b) is different from the ordered pair (b, a) unless $a = b$. (In contrast, the unordered pair $\{a, b\}$ equals the unordered pair $\{b, a\}$.)

Ordered pairs are also called 2-tuples, or sequences (sometimes, lists in a computer science context) of length 2; ordered pairs of scalars are also called 2-dimensional vectors. The entries of an ordered pair can be other ordered pairs, enabling the recursive definition of ordered n-tuples (ordered lists of n objects). For example, the ordered triple (a,b,c) can be defined as $(a, (b,c))$, i.e., as one pair nested in another.

In the ordered pair (a, b), the object a is called the *first entry*, and the object b the *second entry* of the pair. Alternatively, the objects are called the first and second *coordinates*, or the left and right *projections* of the ordered pair.

Cartesian products and binary relations (and hence functions) are defined in terms of ordered pairs.

Generalities

Let (a_1,b_1) and (a_2,b_2) be ordered pairs. Then the *characteristic* (or *defining*) *property* of the ordered pair is:

$$(a_1,b_1)=(a_2,b_2) \text{ if and only if } a_1=a_2 \text{ and } b_1=b_2.$$

The set of all ordered pairs whose first entry is in some set A and whose second entry is in some set B is called the Cartesian product of A and B, and written $A \times B$. A binary relation between sets A and B is a subset of $A \times B$.

The (a, b) notation may be used for other purposes, most notably as denoting open intervals on the real number line. In such situations, the context will usually make it clear which meaning is intended. For additional clarification, the ordered pair may be denoted by the variant notation $< a, b>$, but this notation also has other uses.

The left and right projection of a pair p is usually denoted by $\pi_1(p)$ and $\pi_2(p)$, or by $\pi_l(p)$ and $\pi_r(p)$, respectively. In contexts where arbitrary n-tuples are considered, πn $i(t)$ is a common notation for the i-th component of an n-tuple t.

Informal and Formal Definitions

In some introductory mathematics textbooks an informal (or intuitive) definition of ordered pair is given, such as

For any two objects a and b, the ordered pair (a, b) is a notation specifying the two objects a and b, in that order.

This is usually followed by a comparison to a set of two elements; pointing out that in a set a and b must be different, but in an ordered pair they may be equal and that while the order of listing the elements of a set doesn't matter, in an ordered pair changing the order of distinct entries changes the ordered pair.

This "definition" is unsatisfactory because it is only descriptive and is based on an intuitive understanding of *order*. However, as is sometimes pointed out, no harm will come from relying on this description and almost everyone thinks of ordered pairs in this manner.

A more satisfactory approach is to observe that the characteristic property of ordered pairs given above is all that is required to understand the role of ordered pairs in mathematics. Hence the ordered pair can be taken as a primitive notion, whose associated axiom is the characteristic property. This was the approach taken by the N. Bourbaki group in its *Theory of Sets*, published in 1954. However, this approach also has its drawbacks as both the existence of ordered pairs and their characteristic property must be axiomatically assumed.

Another way to rigorously deal with ordered pairs is to define them formally in the context of set theory. This can be done in several ways and has the advantage that existence and the characteristic property can be proven from the axioms that define the set theory. One of the most cited versions of this definition is due to Kuratowski and his definition was used in the second edition of Bourbaki's *Theory of Sets*, published in 1970. Even those mathematical textbooks that give an informal definition of ordered pairs will often mention the formal definition of Kuratowski in an exercise.

Defining the Ordered Pair Using Set Theory

If one agrees that set theory is an appealing foundation of mathematics, then all mathematical objects must be defined as sets of some sort. Hence if the ordered pair is not taken as primitive, it must be defined as a set. Several set-theoretic definitions of the ordered pair are given below.

Wiener's Definition

Norbert Wiener proposed the first set theoretical definition of the ordered pair in 1914:

$$(a,b) := \{\{\{a\},\varnothing\},\{\{b\}\}\}.$$

He observed that this definition made it possible to define the types of *Principia Mathematica* as sets. *Principia Mathematica* had taken types, and hence relations of all arities, as primitive.

Wiener used $\{\{b\}\}$ instead of $\{b\}$ to make the definition compatible with type theory where all elements in a class must be of the same "type". With nesting b within an additional set its type is made equal to $\{\{a\},\varnothing\}$'s.

Hausdorff's Definition

About the same time as Wiener (1914), Felix Hausdorff proposed his definition:

$$(a,b) := \{\{a,1\},\{b,2\}\}$$

"where 1 and 2 are two distinct objects different from a and b."

Kuratowski Definition

In 1921 Kazimierz Kuratowski offered the now-accepted definition of the ordered pair (a, b):

$$(a, b)_K := \{\{a\}, \{a, b\}\}.$$

Note that this definition is used even when the first and the second coordinates are identical:

$$(x, x)_K = \{\{x\}, \{x, x\}\} = \{\{x\}, \{x\}\} = \{\{x\}\}$$

Given some ordered pair p, the property "x is the first coordinate of p" can be formulated as:

$$\forall Y \in p : x \in Y.$$

The property "x is the second coordinate of p" can be formulated as:

$$(\exists Y \in p : x \in Y) \wedge (\forall Y_1, Y_2 \in p : Y_1 \neq Y_2 \rightarrow (x \notin Y_1 \vee x \notin Y_2)).$$

In the case that the left and right coordinates are identical, the right conjunct $(\forall Y_1, Y_2 \in p : Y_1 \neq Y_2 \to (x \notin Y_1 \vee x \notin Y_2))$ is trivially true, since $Y_1 \neq Y_2$ is never the case.

This is how we can extract the first coordinate of a pair (using the notation for arbitrary intersection and arbitrary union):

$$\pi_1(p) = \bigcup \bigcap p.$$

This is how the second coordinate can be extracted:

$$\pi_2(p) = \bigcup \{x \in \bigcup p \mid \bigcup p \neq \bigcap p \to x \notin \bigcap p\}.$$

Variants

The above Kuratowski definition of the ordered pair is "adequate" in that it satisfies the characteristic property that an ordered pair must satisfy, namely that $(a,b) = (x,y) \leftrightarrow (a = x) \wedge (b = y)$. In particular, it adequately expresses 'order', in that $(a,b) = (b,a)$ is false unless $b = a$. There are other definitions, of similar or lesser complexity, that are equally adequate:

- $(a,b)_{\text{reverse}} := \{\{b\},\{a,b\}\}$;
- $(a,b)_{\text{short}} := \{a,\{a,b\}\}$;
- $(a,b)_1 := \{\{0,a\},\{1,b\}\}$.

The reverse definition is merely a trivial variant of the Kuratowski definition, and as such is of no independent interest. The definition short is so-called because it requires two rather than three pairs of braces. Proving that short satisfies the characteristic property requires the Zermelo–Fraenkel set theory axiom of regularity. Moreover, if one uses von Neumann's set-theoretic construction of the natural numbers, then 2 is defined as the set {0, 1} = {0, {0}}, which is indistinguishable from the pair $(0, 0)_{\text{short}}$. Yet another disadvantage of the short pair is the fact, that even if *a* and *b* are of the same type, the elements of the short pair are not. (However, if $a = b$ then the short version keeps having cardinality 2, which is something one might expect of any "pair", including any "ordered pair". Also note that the short version is used in Tarski–Grothendieck set theory, upon which the Mizar system is founded.)

Proving that Definitions Satisfy the Characteristic Property

Prove: $(a, b) = (c, d)$ if and only if $a = c$ and $b = d$.

Kuratowski:

If. If $a = c$ and $b = d$, then $\{\{a\}, \{a, b\}\} = \{\{c\}, \{c, d\}\}$. Thus $(a, b)_K = (c, d)_K$.

Only if. Two cases: $a = b$, and $a \neq b$.

If $a = b$:

$$(a, b)_K = \{\{a\}, \{a, b\}\} = \{\{a\}, \{a, a\}\} = \{\{a\}\}.$$

$(c, d)_K = \{\{c\}, \{c, d\}\} = \{\{a\}\}$.

Thus $\{c\} = \{c, d\} = \{a\}$, which implies $a = c$ and $a = d$. By hypothesis, $a = b$. Hence $b = d$.

If $a \neq b$, then $(a, b)_K = (c, d)_K$ implies $\{\{a\}, \{a, b\}\} = \{\{c\}, \{c, d\}\}$.

Suppose $\{c, d\} = \{a\}$. Then $c = d = a$, and so $\{\{c\}, \{c, d\}\} = \{\{a\}, \{a, a\}\} = \{\{a\}, \{a\}\} = \{\{a\}\}$. But then $\{\{a\}, \{a, b\}\}$ would also equal $\{\{a\}\}$, so that $b = a$ which contradicts $a \neq b$.

Suppose $\{c\} = \{a, b\}$. Then $a = b = c$, which also contradicts $a \neq b$.

Therefore $\{c\} = \{a\}$, so that $c = a$ and $\{c, d\} = \{a, b\}$.

If $d = a$ were true, then $\{c, d\} = \{a, a\} = \{a\} \neq \{a, b\}$, a contradiction. Thus $d = b$ is the case, so that $a = c$ and $b = d$.

Reverse:

$(a, b)_{reverse} = \{\{b\}, \{a, b\}\} = \{\{b\}, \{b, a\}\} = (b, a)_K$.

If. If $(a, b)_{reverse} = (c, d)_{reverse}$, $(b, a)_K = (d, c)_K$. Therefore $b = d$ and $a = c$.

Only if. If $a = c$ and $b = d$, then $\{\{b\}, \{a, b\}\} = \{\{d\}, \{c, d\}\}$. Thus $(a, b)_{reverse} = (c, d)_{reverse}$.

Short:

If: If $a = c$ and $b = d$, then $\{a, \{a, b\}\} = \{c, \{c, d\}\}$. Thus $(a, b)_{short} = (c, d)_{short}$.

Only if: Suppose $\{a, \{a, b\}\} = \{c, \{c, d\}\}$. Then a is in the left hand side, and thus in the right hand side. Because equal sets have equal elements, one of $a = c$ or $a = \{c, d\}$ must be the case.

If $a = \{c, d\}$, then by similar reasoning as above, $\{a, b\}$ is in the right hand side, so $\{a, b\} = c$ or $\{a, b\} = \{c, d\}$.

If $\{a, b\} = c$ then c is in $\{c, d\} = a$ and a is in c, and this combination contradicts the axiom of regularity, as $\{a, c\}$ has no minimal element under the relation "element of."

If $\{a, b\} = \{c, d\}$, then a is an element of a, from $a = \{c, d\} = \{a, b\}$, again contradicting regularity.

Hence $a = c$ must hold.

Again, we see that $\{a, b\} = c$ or $\{a, b\} = \{c, d\}$.

The option $\{a, b\} = c$ and $a = c$ implies that c is an element of c, contradicting regularity.

So we have $a = c$ and $\{a, b\} = \{c, d\}$, and so: $\{b\} = \{a, b\} \setminus \{a\} = \{c, d\} \setminus \{c\} = \{d\}$, so $b = d$.

Quine–Rosser Definition

Rosser (1953) employed a definition of the ordered pair due to Quine which requires a prior definition of the natural numbers. Let $\mathbb{N}$ be the set of natural numbers and $x \setminus \mathbb{N}$ be the set of the elements of x not in $\mathbb{N}$. Define

$$\varphi(x) = (x \setminus \mathbb{N}) \cup \{n+1 : n \in (x \cap \mathbb{N})\}.$$

Applying this function simply increments every natural number in x. In particular, $\varphi(x)$ does not contain the number 0, so that for any sets x and y,

$$\varphi(x) \neq \{0\} \cup \varphi(y).$$

Define the ordered pair (A, B) as

$$(A, B) = \{\varphi(a) : a \in A\} \cup \{\varphi(b) \cup \{0\} : b \in B\}.$$

Extracting all the elements of the pair that do not contain 0 and undoing φ yields A. Likewise, B can be recovered from the elements of the pair that do contain 0.

In type theory and in outgrowths thereof such as the axiomatic set theory NF, the Quine–Rosser pair has the same type as its projections and hence is termed a "type-level" ordered pair. Hence this definition has the advantage of enabling a function, defined as a set of ordered pairs, to have a type only 1 higher than the type of its arguments. This definition works only if the set of natural numbers is infinite. This is the case in NF, but not in type theory or in NFU. J. Barkley Rosser showed that the existence of such a type-level ordered pair (or even a "type-raising by 1" ordered pair) implies the axiom of infinity.

Morse Definition

Morse–Kelley set theory makes free use of proper classes. Morse defined the ordered pair so that its projections could be proper classes as well as sets. (The Kuratowski definition does not allow this.) He first defined ordered pairs whose projections are sets in Kuratowski's manner. He then *redefined* the pair

$$(x, y) = (\{0\} \times s(x)) \cup (\{1\} \times s(y))$$

where the component Cartesian products are Kuratowski pairs of sets and where

$$s(x) = \{\varnothing\} \cup \{\{t\} \mid t \in x\}$$

This renders possible pairs whose projections are proper classes. The Quine-Rosser definition above also admits proper classes as projections. Similarly the triple is defined as a 3-tuple as follows:

$$(x, y, z) = (\{0\} \times s(x)) \cup (\{1\} \times s(y)) \cup (\{2\} \times s(z))$$

The use of the singleton set $s(x)$ which has an inserted empty set allows tuples to have the uniqueness property that if a is an n-tuple and b is an m-tuple and $a = b$ then $n = m$. Ordered triples which are defined as ordered pairs do not have this property with respect to ordered pairs.

Category Theory

A category-theoretic product $A \times B$ in a category of sets represents the set of ordered pairs, with the first element coming from A and the second coming from B. In this context the characteristic property above is a consequence of the universal property of the product and the fact that elements of a set X can be identified with morphisms from 1 (a one element set) to X. While different objects may have the universal property, they are all naturally isomorphic.

Cartesian Product

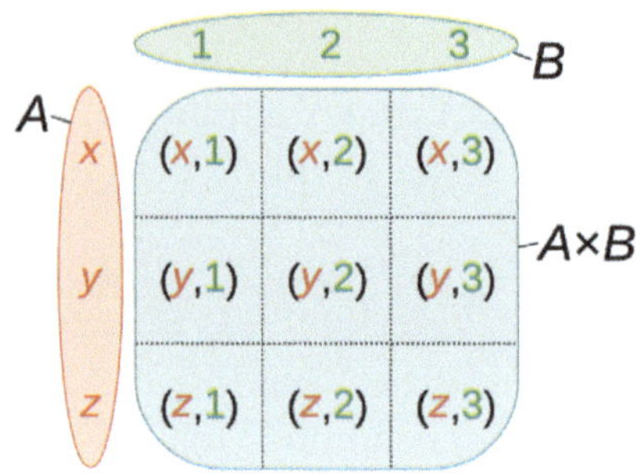

Cartesian product $A \times B$ of the sets $A = \{x, y, z\}$ and $B = \{1, 2, 3\}$

In mathematics, a Cartesian product is a mathematical operation that returns a set (or product set or simply product) from multiple sets. That is, for sets A and B, the Cartesian product $A \times B$ is the set of all ordered pairs (a, b) where $a \in A$ and $b \in B$. Products can be specified using set-builder notation, e.g.

$$A \times B = \{(a,b) \mid a \in A \text{ and } b \in B\}.$$

A table can be created by taking the Cartesian product of a set of rows and a set of columns. If the Cartesian product *rows* × *columns* is taken, the cells of the table contain ordered pairs of the form (row value, column value).

More generally, a Cartesian product of n sets, also known as an $\boldsymbol{n}$-fold Cartesian product, can be represented by an array of n dimensions, where each element is an n-tuple. An ordered pair is a 2-tuple or couple.

The Cartesian product is named after René Descartes, whose formulation of analytic geometry gave rise to the concept, which is further generalized in terms of direct product.

Examples

A Deck of Cards

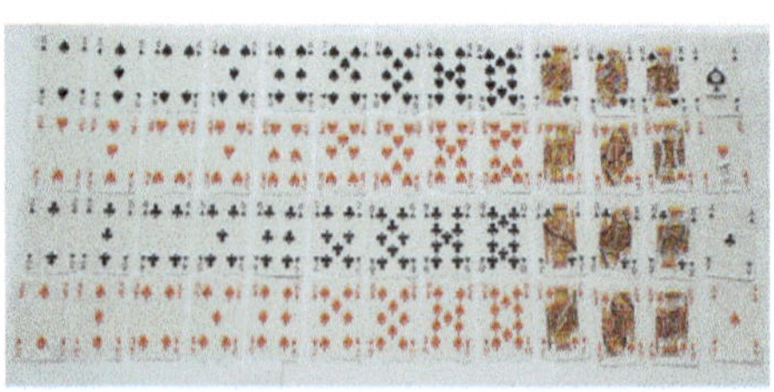

Standard 52-card deck

An illustrative example is the standard 52-card deck. The standard playing card ranks {A, K, Q, J, 10, 9, 8, 7, 6, 5, 4, 3, 2} form a 13-element set. The card suits {♠, ♥, ♦, ♣} form a four-element set. The Cartesian product of these sets returns a 52-element set consisting of 52 ordered pairs, which correspond to all 52 possible playing cards.

Ranks × *Suits* returns a set of the form {(A, ♠), (A, ♥), (A, ♦), (A, ♣), (K, ♠), ..., (3, ♣), (2, ♠), (2, ♥), (2, ♦), (2, ♣)}.

Suits × *Ranks* returns a set of the form {(♠, A), (♠, K), (♠, Q), (♠, J), (♠, 10), ..., (♣, 6), (♣, 5), (♣, 4), (♣, 3), (♣, 2)}.

Both sets are distinct, even disjoint.

A Two-dimensional Coordinate System

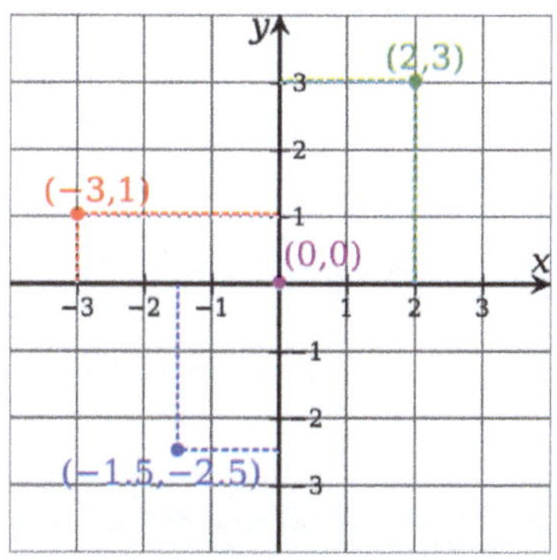

Cartesian coordinates of example points

The main historical example is the Cartesian plane in analytic geometry. In order to represent geometrical shapes in a numerical way and extract numerical information from shapes' numerical representations, René Descartes assigned to each point in the plane a pair of real numbers, called its coordinates. Usually, such a pair's first and second components are called its x and y coordinates, respectively; cf. picture. The set of all such pairs (i.e. the Cartesian product R×R with R denoting the real numbers) is thus assigned to the set of all points in the plane.

Most Common Implementation (Set Theory)

A formal definition of the Cartesian product from set-theoretical principles follows from a definition of ordered pair. The most common definition of ordered pairs, the Kuratowski definition, is $(x, y) = \{\{x\},\{x, y\}\}$. Note that, under this definition, $X \times Y \subseteq \mathcal{P}(\mathcal{P}(X \cup Y))$, where $\mathcal{P}$ represents the power set. Therefore, the existence of the Cartesian product of any two sets in ZFC follows from the axioms of pairing, union, power set, and specification. Since functions are usually defined as a special case of relations, and relations are usually defined as subsets of the Cartesian product, the definition of the two-set Cartesian product is necessarily prior to most other definitions.

Non-commutativity and Non-associativity

Let *A*, *B*, *C*, and *D* be sets.

The Cartesian product *A* × *B* is not commutative,

$$A \times B \neq B \times A,$$

because the ordered pairs are reversed except if at least one of the following conditions is satisfied:

- A is equal to B, or
- A or B is the empty set.

For example:

$A = \{1,2\}; B = \{3,4\}$

$A \times B = \{1,2\} \times \{3,4\} = \{(1,3), (1,4), (2,3), (2,4)\}$

$B \times A = \{3,4\} \times \{1,2\} = \{(3,1), (3,2), (4,1), (4,2)\}$

$A = B = \{1,2\}$

$A \times B = B \times A = \{1,2\} \times \{1,2\} = \{(1,1), (1,2), (2,1), (2,2)\}$

$A = \{1,2\}; B = \emptyset$

$A \times B = \{1,2\} \times \emptyset = \emptyset$

$B \times A = \emptyset \times \{1,2\} = \emptyset$

Strictly speaking, the Cartesian product is not associative (unless one of the involved sets is empty).

$$(A \times B) \times C \neq A \times (B \times C)$$

If for example $A = \{1\}$, then $(A \times A) \times A = \{ ((1,1),1) \} \neq \{ (1,(1,1)) \} = A \times (A \times A)$.

Intersections, Unions, and Subsets

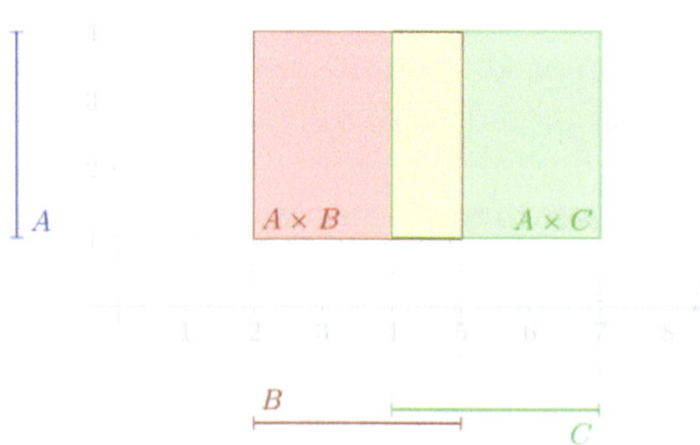

Example sets $A=\{y\in\mathbb{R}:1\leq y\leq 4\}$, $B=\{x\in\mathbb{R}:2\leq x\leq 5\}$, and $C=\{x\in\mathbb{R}:4\leq x\leq 7\}$, demonstrating $A\times(B\cap C)=(A\times B)\cap(A\times C)$, $A\times(B\cup C)=(A\times B)\cup(A\times C)$, and $A\times(B\setminus C)=(A\times B)\setminus(A\times C)$.

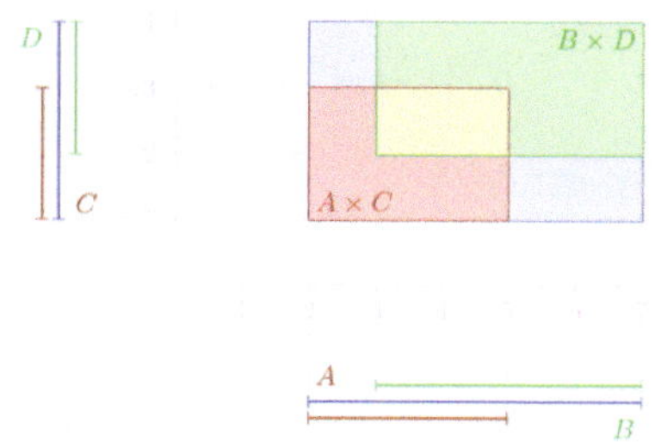

$(A \cup B) \times (C \cup D) \neq (A \times C) \cup (B \times D)$ can be seen from the same example

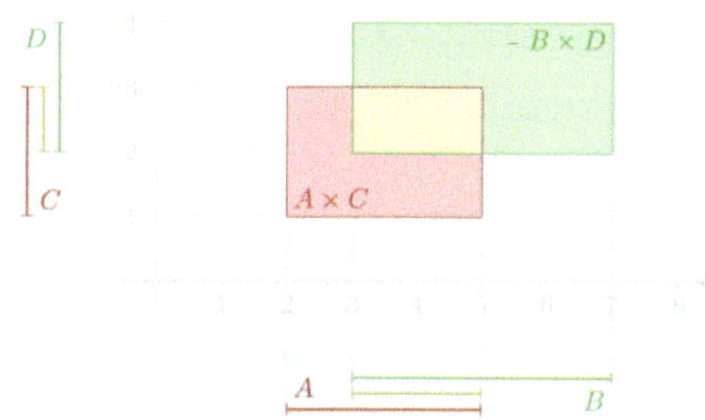

Example sets $A=\{x \in \mathbb{R}: 2 \leq x \leq 5\}$, $B=\{x \in \mathbb{R}: 3 \leq x \leq 7\}$, $C=\{y \in \mathbb{R}: 1 \leq y \leq 3\}$, and $D=\{y \in \mathbb{R}: 2 \leq y \leq 4\}$, demonstrating $(A \cap B) \times (C \cap D) = (A \times C) \cap (B \times D)$.

The Cartesian product behaves nicely with respect to intersections, cf. left picture.

$$(A \cap B) \times (C \cap D) = (A \times C) \cap (B \times D)$$

In most cases the above statement is not true if we replace intersection with union, cf. middle picture.

$$(A \cup B) \times (C \cup D) \neq (A \times C) \cup (B \times D)$$

In fact, we have that:

$$(A \times C) \cup (B \times D) = [(A \setminus B) \times C] \cup [(A \cap B) \times (C \cup D)] \cup [(B \setminus A) \times D]$$

For the set difference we also have the following identity:

$$(A \times C) \setminus (B \times D) = [A \times (C \setminus D)] \cup [(A \setminus B) \times C]$$

Here are some rules demonstrating distributivity with other operators (cf. right picture):

$$A \times (B \cap C) = (A \times B) \cap (A \times C),$$

$$A \times (B \cup C) = (A \times B) \cup (A \times C),$$

$$A \times (B \setminus C) = (A \times B) \setminus (A \times C),$$

$$(A \times B)^{\complement} = (A^{\complement} \times B^{\complement}) \cup (A^{\complement} \times B) \cup (A \times B^{\complement}),$$

where $A^{\complement}$ denotes the absolute complement of A.

Other properties related with subsets are:

if $A \subseteq B$ then $A \times C \subseteq B \times C$,

if both $A, B \neq \varnothing$ then $A \times B \subseteq C \times D \Leftrightarrow A \subseteq C \wedge B \subseteq D$.

Cardinality

The cardinality of a set is the number of elements of the set. For example, defining two sets: A = {a, b} and B = {5, 6}. Both set A and set B consist of two elements each. Their Cartesian product, written as $A \times B$, results in a new set which has the following elements:

$A \times B$ = {(a,5), (a,6), (b,5), (b,6)}.

Each element of A is paired with each element of B. Each pair makes up one element of the output set. The number of values in each element of the resulting set is equal to the number of sets whose cartesian product is being taken; 2 in this case. The cardinality of the output set is equal to the product of the cardinalities of all the input sets. That is,

$|A \times B| = |A| \cdot |B|$.

Similarly

$|A \times B \times C| = |A| \cdot |B| \cdot |C|$

and so on.

The set $A \times B$ is infinite if either A or B is infinite and the other set is not the empty set.

n-ary Product

Cartesian Power

The Cartesian square (or binary Cartesian product) of a set X is the Cartesian product $X^2 = X \times X$. An example is the 2-dimensional plane $R^2 = R \times R$ where R is the set of real numbers: R^2 is the set of all points (x,y) where x and y are real numbers.

The cartesian power of a set X can be defined as:

$$X^n = \underbrace{X \times X \times \cdots \times X}_{n} = \{(x_1, \ldots, x_n) \mid x_i \in X \text{ for all } i = 1, \ldots, n\}.$$

An example of this is $R^3 = R \times R \times R$, with R again the set of real numbers, and more generally R^n.

The n-ary cartesian power of a set X is isomorphic to the space of functions from an n-element set to X. As a special case, the 0-ary cartesian power of X may be taken to be a singleton set, corresponding to the empty function with codomain X.

Finite *n*-ary Product

The Cartesian product can be generalized to the ***n***-ary Cartesian product over n sets $X_1, ..., X_n$:

$$X_1 \times \cdots \times X_n = \{(x_1, \ldots, x_n) : x_i \in X_i\}.$$

It is a set of n-tuples. If tuples are defined as nested ordered pairs, it can be identified to $(X_1 \times ... \times X_{n-1}) \times X_n$.

Infinite Products

It is possible to define the Cartesian product of an arbitrary (possibly infinite) indexed family of sets. If I is any index set, and $\{X_i\}_{i \in I}$ is a family of sets indexed by I, then the Cartesian product of the sets in X is defined to be

$$\prod_{i \in I} X_i = \left\{ f : I \to \bigcup_{i \in I} X_i \;\middle|\; (\forall i)(f(i) \in X_i) \right\},$$

that is, the set of all functions defined on the index set such that the value of the function at a particular index i is an element of X_i. Even if each of the X_i is nonempty, the Cartesian product may be empty if the axiom of choice (which is equivalent to the statement that every such product is nonempty) is not assumed.

For each j in I, the function

$$\pi_j : \prod_{i \in I} X_i \to X_j,$$

defined by $\pi_j(f) = f(j)$ is called the ***j*th** projection map.

An important case is when the index set is $\mathbb{N}$, the natural numbers: this Cartesian product is the set of all infinite sequences with the ith term in its corresponding set X_i. For example, each element of

$$\prod_{n=1}^{\infty} \mathbb{R} = \mathbb{R} \times \mathbb{R} \times \cdots$$

can be visualized as a vector with countably infinite real number components. This set is frequently denoted $\mathbb{R}^{\omega}$, or $\mathbb{R}^{\mathbb{N}}$.

The special case Cartesian exponentiation occurs when all the factors X_i involved in the product are the same set X. In this case,

$$\prod_{i \in I} X_i = \prod_{i \in I} X$$

is the set of all functions from I to X, and is frequently denoted X^I. This case is important in the study of cardinal exponentiation.

The definition of finite Cartesian products can be seen as a special case of the definition for infinite products. In this interpretation, an n-tuple can be viewed as a function on $\{1, 2, ..., n\}$ that takes its value at i to be the ith element of the tuple (in some settings, this is taken as the very definition of an n-tuple).

Other Forms

Abbreviated Form

If several sets are being multiplied together, e.g. $X_1, X_2, X_3, ...$, then some authors choose to abbreviate the Cartesian product as simply $\times X_i$.

Cartesian Product of Functions

If f is a function from A to B and g is a function from X to Y, their Cartesian product $f \times g$ is a function from $A \times X$ to $B \times Y$ with

$$(f \times g)(a, x) = (f(a), g(x)).$$

This can be extended to tuples and infinite collections of functions. Note that this is different from the standard cartesian product of functions considered as sets.

Definitions Outside of Set Theory

Category Theory

Although the Cartesian product is traditionally applied to sets, category theory provides a more general interpretation of the product of mathematical structures. This is distinct from, although related to, the notion of a Cartesian square in category theory, which is a generalization of the fiber product.

Exponentiation is the right adjoint of the Cartesian product; thus any category with a Cartesian product (and a final object) is a Cartesian closed category.

Graph Theory

In graph theory the Cartesian product of two graphs G and H is the graph denoted by $G \times H$ whose vertex set is the (ordinary) Cartesian product $V(G) \times V(H)$ and such that two vertices (u,v) and (u',v') are adjacent in $G \times H$ if and only if $u = u'$ and v is adjacent with v' in H, *or* $v = v'$ and u is adjacent with u' in G. The Cartesian product of graphs is not a product in the sense of category theory. Instead, the categorical product is known as the tensor product of graphs.

Simple Theorems in the Algebra of Sets

The simple theorems in the algebra of sets are some of the elementary properties of the algebra of union (infix ∪), intersection (infix ∩), and set complement (postfix ') of sets.

These properties assume the existence of at least two sets: a given universal set, denoted U, and the empty set, denoted {}. The algebra of sets describes the properties of all possible subsets of U, called the power set of U and denoted *P*(U). *P*(U) is assumed closed under union, intersection, and set complement. The algebra of sets is an interpretation or model of Boolean algebra, with union, intersection, set complement, U, and {} interpreting Boolean sum, product, complement, 1, and 0, respectively.

The properties below are stated without proof, but can be derived from a small number of properties taken as axioms. A "*" follows the algebra of sets interpretation of Huntington's (1904) classic postulate set for Boolean algebra. These properties can be visualized with Venn diagrams. They also follow from the fact that *P*(U) is a Boolean lattice. The properties followed by "L" interpret the lattice axioms.

Elementary discrete mathematics courses sometimes leave students with the impression that the subject matter of set theory is no more than these properties. For more about elementary set theory, see set, set theory, algebra of sets, and naive set theory. For an introduction to set theory at a higher level, ordinal number, Cantor–Bernstein–Schroeder theorem, Cantor's diagonal argument, Cantor's first uncountability proof, Cantor's theorem, well-ordering theorem, axiom of choice, and Zorn's lemma.

The properties below include a defined binary operation, relative complement, denoted by infix "\". The "relative complement of *A* in *B*," denoted $B \setminus A$, is defined as $(A \cup B')'$ and as $A' \cap B$.

PROPOSITION 1. For any U and any subset *A* of U:

- $\{\}' = U$;
- $U' = \{\}$;
- $A \setminus \{\} = A$;
- $\{\} \setminus A = \{\}$;
- $A \cap \{\} = \{\}$;
- $A \cup \{\} = A$; *
- $A \cap U = A$; *
- $A \cup U = U$;
- $A' \cup A = U$; *
- $A' \cap A = \{\}$; *
- $A \setminus A = \{\}$;
- $U \setminus A = A'$;
- $A \setminus U = \{\}$;
- $A'' = A$;
- $A \cap A = A$;

- $A \cup A = A$.

PROPOSITION 2. For any sets A, B, and C:

- $A \cap B = B \cap A$; * L
- $A \cup B = B \cup A$; * L
- $A \cup (A \cap B) = A$; L
- $A \cap (A \cup B) = A$; L
- $(A \cup B) \setminus A = B \setminus A$;
- $A \cap B = \{\}$ if and only if $B \setminus A = B$;
- $(A' \cup B)' \cup (A' \cup B')' = A$;
- $(A \cap B) \cap C = A \cap (B \cap C)$; L
- $(A \cup B) \cup C = A \cup (B \cup C)$; L
- $C \setminus (A \cap B) = (C \setminus A) \cup (C \setminus B)$;
- $C \setminus (A \cup B) = (C \setminus A) \cap (C \setminus B)$;
- $C \setminus (B \setminus A) = (C \setminus B) \cup (C \cap A)$;
- $(B \setminus A) \cap C = (B \cap C) \setminus A = B \cap (C \setminus A)$;
- $(B \setminus A) \cup C = (B \cup C) \setminus (A \setminus C)$.

The distributive laws:

- $A \cap (B \cup C) = (A \cap B) \cup (A \cap C)$; *
- $A \cup (B \cap C) = (A \cup B) \cap (A \cup C)$. *

PROPOSITION 3. Some properties of $\subseteq$:

- $A \subseteq B$ if and only if $A \cap B = A$;
- $A \subseteq B$ if and only if $A \cup B = B$;
- $A \subseteq B$ if and only if $A' \cup B$;
- $A \subseteq B$ if and only if $B' \subseteq A'$;
- $A \subseteq B$ if and only if $A \setminus B = \{\}$;
- $A \cap B \subseteq A \subseteq A \cup B$.

References

- Dauben, Joseph W., Georg Cantor: His Mathematics and Philosophy of the Infinite, Boston: Harvard University Press (1979) ISBN 978-0-691-02447-9.
- Johnson, D. L. (2001). "3.3 Laws". Elements of logic via numbers and sets. Springer Undergraduate Mathematics Series. Berlin: Springer-Verlag. p. 62. ISBN 3-540-76123-3.

- Rosen, Kenneth H. (2012). Discrete Mathematics and Its Applications (7th ed.). New York: McGraw-Hill. p. 119. ISBN 978-0-07-338309-5.
- Megginson, Robert E. (1998), “Chapter 1”, An introduction to Banach space theory, Graduate Texts in Mathematics, 183, New York: Springer-Verlag, pp. xx+596, ISBN 0-387-98431-3
- Lay, Steven R. (2005), Analysis / With an Introduction to Proof (4th ed.), Pearson / Prentice Hall, p. 50, ISBN 978-0-13-148101-5
- Devlin, Keith (2004), Sets, Functions and Logic / An Introduction to Abstract Mathematics (3rd ed.), Chapman & Hall / CRC, p. 79, ISBN 978-1-58488-449-1
- Wolf, Robert S. (1998), Proof, Logic, and Conjecture / The Mathematician’s Toolbox, W. H. Freeman and Co., p. 164, ISBN 978-0-7167-3050-7

5

Number Theory : A Comprehensive Study

Number theory is a branch of pure mathematics. The number theory studies prime numbers and also the properties of integers. The topics explained in this section are analytic number theory, algebraic number theory, Baker's theorem and pentagonal number theorem. The chapter is an overview of the subject matter incorporating the major aspects of number theory.

Number Theory

Number theory or, in older usage, arithmetic is a branch of pure mathematics devoted primarily to the study of the integers. It is sometimes called "The Queen of Mathematics" because of its foundational place in the discipline. Number theorists study prime numbers as well as the properties of objects made out of integers (e.g., rational numbers) or defined as generalizations of the integers (e.g., algebraic integers).

A Lehmer sieve, which is a primitive digital computer once used for finding primes and solving simple Diophantine equations.

Integers can be considered either in themselves or as solutions to equations (Diophantine geometry). Questions in number theory are often best understood through the study of analytical objects (e.g., the Riemann zeta function) that encode properties of the integers, primes or other number-theoretic objects in some fashion (analytic number theory). One may also study real numbers in relation to rational numbers, e.g., as approximated by the latter (Diophantine approximation).

The older term for number theory is *arithmetic*. By the early twentieth century, it had been superseded by "number theory". (The word "arithmetic" is used by the general public to mean "elementary calculations"; it has also acquired other meanings in mathematical logic, as in *Peano arithmetic*, and computer science, as in *floating point arithmetic*.) The use of the term *arithmetic* for *number theory* regained some ground in the second half of the 20th century, arguably in part due to French influence. In particular, *arithmetical* is preferred as an adjective to *number-theoretic*.

History

Dawn of Arithmetic

The first historical find of an arithmetical nature is a fragment of a table: the broken clay tablet Plimpton 322 (Larsa, Mesopotamia, ca. 1800 BCE) contains a list of "Pythagorean triples", i.e., integers (a,b,c) such that $a^2+b^2=c^2$. The triples are too many and too large to have been obtained by brute force. The heading over the first column reads: "The *takiltum* of the diagonal which has been subtracted such that the width..."

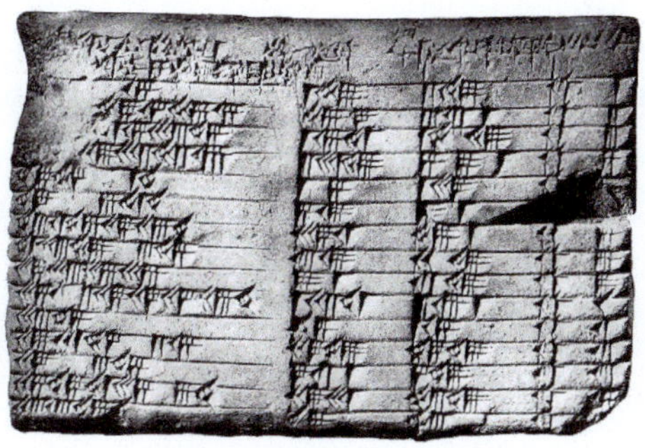

The Plimpton 322 tablet

The table's layout suggests that it was constructed by means of what amounts, in modern language, to the identity

$$\left(\frac{1}{2}\left(x-\frac{1}{x}\right)\right)^2+1=\left(\frac{1}{2}\left(x+\frac{1}{x}\right)\right)^2,$$

which is implicit in routine Old Babylonian exercises. If some other method was used, the triples were first constructed and then reordered by $c>$, presumably for actual use as a "table", i.e., with a view to applications.

It is not known what these applications may have been, or whether there could have been any; Babylonian astronomy, for example, truly flowered only later. It has been suggested instead that the table was a source of numerical examples for school problems.

While Babylonian number theory—or what survives of Babylonian mathematics that can be called thus—consists of this single, striking fragment, Babylonian algebra (in the secondary-school sense of "algebra") was exceptionally well developed. Late Neoplatonic sources state that Pythagoras learned mathematics from the Babylonians. Much earlier sources state that Thales and Pythagoras traveled and studied in Egypt.

Euclid IX 21—34 is very probably Pythagorean; it is very simple material ("odd times even is even", "if an odd number measures [= divides] an even number, then it also measures [= divides] half of it"), but it is all that is needed to prove that $\sqrt{2}$ is irrational. Pythagorean mystics gave great importance to the odd and the even. The discovery that $\sqrt{2}$ is irrational is credited to the early Pythagoreans (pre-Theodorus). By revealing (in modern terms) that numbers could be irrational, this discovery seems to have provoked the first foundational crisis in mathematical history; its proof or its divulgation are sometimes credited to Hippasus, who was expelled or split from the Pythagorean sect. This forced a distinction between *numbers* (integers and the rationals—the subjects of arithmetic), on the one hand, and *lengths* and *proportions* (which we would identify with real numbers, whether rational or not), on the other hand.

The Pythagorean tradition spoke also of so-called polygonal or figurate numbers. While square numbers, cubic numbers, etc., are seen now as more natural than triangular numbers, pentagonal numbers, etc., the study of the sums of triangular and pentagonal numbers would prove fruitful in the early modern period (17th to early 19th century).

We know of no clearly arithmetical material in ancient Egyptian or Vedic sources, though there is some algebra in both. The Chinese remainder theorem appears as an exercise in Sun Zi's *Suan Ching*, also known as *The Mathematical Classic of Sun Zi* (3rd, 4th or 5th century CE.) (There is one important step glossed over in Sun Zi's solution: it is the problem that was later solved by Āryabhaṭa's Kuṭṭaka.)

There is also some numerical mysticism in Chinese mathematics, but, unlike that of the Pythagoreans, it seems to have led nowhere. Like the Pythagoreans' perfect numbers, magic squares have passed from superstition into recreation.

Classical Greece and the Early Hellenistic Period

Aside from a few fragments, the mathematics of Classical Greece is known to us either through the reports of contemporary non-mathematicians or through mathematical works from the early Hellenistic period. In the case of number theory, this means, by and large, *Plato* and *Euclid*, respectively.

Plato had a keen interest in mathematics, and distinguished clearly between arithmetic and calculation. (By *arithmetic* he meant, in part, theorising on number, rather than what *arithmetic* or *number theory* have come to mean.) It is through one of Plato's dialogues—namely, *Theaetetus*— that we know that Theodorus had proven that $\sqrt{3}, \sqrt{5}, \ldots, \sqrt{17}$ are irrational. Theaetetus was, like Plato, a disciple of Theodorus's; he worked on distinguishing different kinds of incommensurables, and was thus arguably a pioneer in the study of number systems. (Book X of Euclid's Elements is described by Pappus as being largely based on Theaetetus's work.)

Euclid devoted part of his *Elements* to prime numbers and divisibility, topics that belong unambiguously to number theory and are basic to it (Books VII to IX of Euclid's Elements). In particular, he gave an algorithm for computing the greatest common divisor of two numbers (the Euclidean algorithm; *Elements*, Prop. VII.2) and the first known proof of the infinitude of primes (*Elements*, Prop. IX.20).

In 1773, Lessing published an epigram he had found in a manuscript during his work as a librarian; it claimed to be a letter sent by Archimedes to Eratosthenes. The epigram proposed what has become known as Archimedes' cattle problem; its solution (absent from the manuscript) requires solving an indeterminate quadratic equation (which reduces to what would later be misnamed Pell's equation). As far as we know, such equations were first successfully treated by the Indian school. It is not known whether Archimedes himself had a method of solution.

Diophantus

Title page of the 1621 edition of Diophantus' *Arithmetica*, translated into Latin by Claude Gaspard Bachet de Méziriac.

Very little is known about Diophantus of Alexandria; he probably lived in the third century CE, that is, about five hundred years after Euclid. Six out of the thirteen books of Diophantus's *Arithmetica* survive in the original Greek; four more books survive in an Arabic translation. The *Arithmetica* is a collection of worked-out problems where the task is invariably to find rational solutions to a system of polynomial equations, usually of the form $f(x,y)=z^2$ or $f(x,y,z)=w^2$. Thus, nowadays, we speak of *Diophantine equations* when we speak of polynomial equations to which rational or integer solutions must be found.

One may say that Diophantus was studying rational points — i.e., points whose coordinates are rational — on curves and algebraic varieties; however, unlike the Greeks of the Classical period, who did what we would now call basic algebra in geometrical terms, Diophantus did what we would now call basic algebraic geometry in purely algebraic terms. In modern language, what Diophantus did was to find rational parametrizations of varieties; that is, given an equation of the form (say) $f(x_1,x_2,x_3)=0$, his aim was to find (in essence) three rational functions g_1,g_2,g_3 such that, for all values of r and s, setting $x_i=g_i(r,s)$ for $i=1,2,3$ gives a solution to $f(x_1,x_2,x_3)=0$.

Diophantus also studied the equations of some non-rational curves, for which no rational parametrisation is possible. He managed to find some rational points on these curves (elliptic curves, as it happens, in what seems to be their first known occurrence) by means of what amounts to a tangent construction: translated into coordinate geometry (which did not exist in Diophantus's time), his method would be visualised as drawing a tangent to a curve at a known rational point, and then finding the other point of intersection of the tangent with the curve; that other point is a new rational point. (Diophantus also resorted to what could be called a special case of a secant construction.)

While Diophantus was concerned largely with rational solutions, he assumed some results on integer numbers, in particular that every integer is the sum of four squares (though he never stated as much explicitly).

Āryabhaṭa, Brahmagupta, Bhāskara

While Greek astronomy probably influenced Indian learning, to the point of introducing trigonometry, it seems to be the case that Indian mathematics is otherwise an indigenous tradition; in particular, there is no evidence that Euclid's Elements reached India before the 18th century.

Āryabhaṭa (476–550 CE) showed that pairs of simultaneous congruences $n \equiv a_1 \bmod m_1$, $n \equiv a_2 \bmod m_2$ could be solved by a method he called *kuṭṭaka*, or *pulveriser*; this is a procedure close to (a generalisation of) the Euclidean algorithm, which was probably discovered independently in India. Āryabhaṭa seems to have had in mind applications to astronomical calculations.

Brahmagupta (628 CE) started the systematic study of indefinite quadratic equations—in particular, the misnamed Pell equation, in which Archimedes may have first been interested, and which did not start to be solved in the West until the time of Fermat and Euler. Later Sanskrit authors would follow, using Brahmagupta's technical terminology. A general procedure (the chakravala, or "cyclic method") for solving Pell's equation was finally found by Jayadeva (cited in the eleventh century; his work is otherwise lost); the earliest surviving exposition appears in Bhāskara II's Bīja-gaṇita (twelfth century).

Indian mathematics remained largely unknown in Europe until the late eighteenth century; Brahmagupta and Bhāskara's work was translated into English in 1817 by Henry Colebrooke.

Arithmetic in the Islamic Golden Age

Al-Haytham seen by the West: frontispice of *Selenographia*, showing Alhasen [*sic*] representing knowledge through reason, and Galileo representing knowledge through the senses.

In the early ninth century, the caliph Al-Ma'mun ordered translations of many Greek mathematical works and at least one Sanskrit work (the *Sindhind*, which may or may not be Brahmagupta's Brāhmasphuṭasiddhānta). Diophantus's main work, the *Arithmetica*, was translated into Arabic by Qusta ibn Luqa (820–912). Part of the treatise *al-Fakhri* (by al-Karajī, 953 – ca. 1029) builds on it to some extent. According to Rashed Roshdi, Al-Karajī's contemporary Ibn al-Haytham knew what would later be called Wilson's theorem.

Western Europe in the Middle Ages

Other than a treatise on squares in arithmetic progression by Fibonacci — who lived and studied in north Africa and Constantinople during his formative years, ca. 1175–1200 — no number theory to speak of was done in western Europe during the Middle Ages. Matters started to change in Europe in the late Renaissance, thanks to a renewed study of the works of Greek antiquity. A catalyst was the textual emendation and translation into Latin of Diophantus's *Arithmetica* (Bachet, 1621, following a first attempt by Xylander, 1575).

Early Modern Number Theory

Fermat

Pierre de Fermat

Pierre de Fermat (1601–1665) never published his writings; in particular, his work on number theory is contained almost entirely in letters to mathematicians and in private marginal notes. He wrote down nearly no proofs in number theory; he had no models in the area. He did make repeated use of mathematical induction, introducing the method of infinite descent.

One of Fermat's first interests was perfect numbers (which appear in Euclid, *Elements* IX) and amicable numbers; this led him to work on integer divisors, which were from the beginning among the subjects of the correspondence (1636 onwards) that put him in touch with the mathematical community of the day. He had already studied Bachet's edition of Diophantus carefully; by 1643, his interests had shifted largely to Diophantine problems and sums of squares (also treated by Diophantus).

Fermat's achievements in arithmetic include:

- Fermat's little theorem (1640), stating that, if a is not divisible by a prime p, then $a^{p-1} \equiv 1 \bmod p$.
- If a and b are coprime, then $a^2 + b^2$ is not divisible by any prime congruent to −1 modulo 4; and every prime congruent to 1 modulo 4 can be written in the form $a^2 + b^2$. These two statements also date from 1640; in 1659, Fermat stated to Huygens that he had proven the latter statement by the method of infinite descent. Fermat and Frenicle also did some work (some of it erroneous) on other quadratic forms.
- Fermat posed the problem of solving $x^2 - Ny^2 = 1$ as a challenge to English mathematicians (1657). The problem was solved in a few months by Wallis and Brouncker. Fermat considered their solution valid, but pointed out they had provided an algorithm without a proof (as had Jayadeva and Bhaskara, though Fermat would never know this.) He states that a proof can be found by descent.
- Fermat developed methods for (doing what in our terms amounts to) finding points on curves of genus 0 and 1. As in Diophantus, there are many special procedures and what amounts to a tangent construction, but no use of a secant construction.
- Fermat states and proves (by descent) in the appendix to *Observations on Diophantus* (Obs. XLV) that $x^4 + y^4 = z^4$ has no non-trivial solutions in the integers. Fermat also mentioned to his correspondents that $x^3 + y^3 = z^3$ has no non-trivial solutions, and that this could be proven by descent. The first known proof is due to Euler (1753; indeed by descent).

Fermat's claim ("Fermat's last theorem") to have shown there are no solutions to $x^n + y^n = z^n$ for all $n \geq 3$ (the only known proof of which is beyond his methods) appears only in his annotations on the margin of his copy of Diophantus; he never claimed this to others and thus would have had no need to retract it if he found any mistake in his supposed proof.

Euler

Leonhard Euler

The interest of Leonhard Euler (1707–1783) in number theory was first spurred in 1729, when a friend of his, the amateur Goldbach, pointed him towards some of Fermat's work on the subject. This has been called the "rebirth" of modern number theory, after Fermat's relative lack of success in getting his contemporaries' attention for the subject. Euler's work on number theory includes the following:

- *Proofs for Fermat's statements.* This includes Fermat's little theorem (generalised by Euler to non-prime moduli); the fact that $p = x^2 + y^2$ if and only if $p \equiv 1 \bmod 4$; initial work towards a proof that every integer is the sum of four squares (the first complete proof is by Joseph-Louis Lagrange (1770), soon improved by Euler himself); the lack of non-zero integer solutions to $x^4 + y^4 = z^2$ (implying the case *n=4* of Fermat's last theorem, the case *n=3* of which Euler also proved by a related method).

- *Pell's equation*, first misnamed by Euler. He wrote on the link between continued fractions and Pell's equation.

- *First steps towards analytic number theory.* In his work of sums of four squares, partitions, pentagonal numbers, and the distribution of prime numbers, Euler pioneered the use of what can be seen as analysis (in particular, infinite series) in number theory. Since he lived before the development of complex analysis, most of his work is restricted to the formal manipulation of power series. He did, however, do some very notable (though not fully rigorous) early work on what would later be called the Riemann zeta function.

- *Quadratic forms.* Following Fermat's lead, Euler did further research on the question of which primes can be expressed in the form $x^2 + Ny^2$, some of it prefiguring quadratic reciprocity.

- *Diophantine equations.* Euler worked on some Diophantine equations of genus 0 and 1. In particular, he studied Diophantus's work; he tried to systematise it, but the time was not yet ripe for such an endeavour – algebraic geometry was still in its infancy. He did notice there was a connection between Diophantine problems and elliptic integrals, whose study he had himself initiated.

Lagrange, Legendre and Gauss

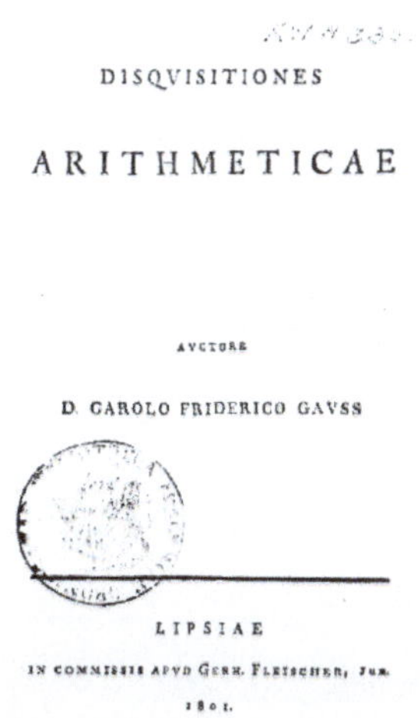
DISQVISITIONES

ARITHMETICAE

AVCTORE

D. CAROLO FRIDERICO GAVSS

LIPSIAE

IN COMMISSIS APVD GERH. FLEISCHER, IVN.

1801.

Carl Friedrich Gauss's Disquisitiones Arithmeticae, first edition

Joseph-Louis Lagrange (1736–1813) was the first to give full proofs of some of Fermat's and Euler's work and observations – for instance, the four-square theorem and the basic theory of the misnamed "Pell's equation" (for which an algorithmic solution was found by Fermat and his contemporaries, and also by Jayadeva and Bhaskara II before them.) He also studied quadratic forms in full generality (as opposed to $mX^2 + nY^2$) — defining their equivalence relation, showing how to put them in reduced form, etc.

Adrien-Marie Legendre (1752–1833) was the first to state the law of quadratic reciprocity. He also conjectured what amounts to the prime number theorem and Dirichlet's theorem on arithmetic progressions. He gave a full treatment of the equation $ax^2 + by^2 + cz^2 = 0$ and worked on quadratic forms along the lines later developed fully by Gauss. In his old age, he was the first to prove "Fermat's last theorem" for $n = 5$ (completing work by Peter Gustav Lejeune Dirichlet, and crediting both him and Sophie Germain).

Carl Friedrich Gauss

In his *Disquisitiones Arithmeticae* (1798), Carl Friedrich Gauss (1777–1855) proved the law of quadratic reciprocity and developed the theory of quadratic forms (in particular, defining their composition). He also introduced some basic notation (congruences) and devoted a section to computational matters, including primality tests. The last section of the *Disquisitiones* established a link between roots of unity and number theory:

The theory of the division of the circle...which is treated in sec. 7 does not belong by itself to arithmetic, but its principles can only be drawn from higher arithmetic.

In this way, Gauss arguably made a first foray towards both Évariste Galois's work and algebraic number theory.

Maturity and Division into Subfields

Ernst Kummer

Peter Gustav Lejeune Dirichlet

Starting early in the nineteenth century, the following developments gradually took place:

- The rise to self-consciousness of number theory (or *higher arithmetic*) as a field of study.
- The development of much of modern mathematics necessary for basic modern number theory: complex analysis, group theory, Galois theory—accompanied by greater rigor in analysis and abstraction in algebra.
- The rough subdivision of number theory into its modern subfields—in particular, analytic and algebraic number theory.

Algebraic number theory may be said to start with the study of reciprocity and cyclotomy, but truly came into its own with the development of abstract algebra and early ideal theory and valuation theory. A conventional starting point for analytic number theory is Dirichlet's theorem on arithmetic progressions (1837), whose proof introduced L-functions and involved some asymptotic analysis and a limiting process on a real variable. The first use of analytic ideas in number theory actually goes back to Euler (1730s), who used formal power series and non-rigorous (or implicit) limiting arguments. The use of *complex* analysis in number theory comes later: the work of Bernhard Riemann (1859) on the zeta function is the canonical starting point; Jacobi's four-square theorem (1839), which predates it, belongs to an initially different strand that has by now taken a leading role in analytic number theory (modular forms).

Main Subdivisions

Elementary Tools

The term *elementary* generally denotes a method that does not use complex analysis. For example, the prime number theorem was first proven using complex analysis in 1896, but an elementary proof was found only in 1949 by Erdős and Selberg. The term is somewhat ambiguous: for example, proofs based on complex Tauberian theorems (e.g. Wiener–Ikehara) are often seen as quite enlightening but not elementary, in spite of using Fourier analysis, rather than complex analysis as such. Here as elsewhere, an *elementary* proof may be longer and more difficult for most readers than a non-elementary one.

Number theory has the reputation of being a field many of whose results can be stated to the layperson. At the same time, the proofs of these results are not particularly accessible, in part because the range of tools they use is, if anything, unusually broad within mathematics.

Analytic Number Theory

Riemann zeta function ζ(s) in the complex plane. The color of a point s gives the value of ζ(s): dark colors denote values close to zero and hue gives the value's argument.

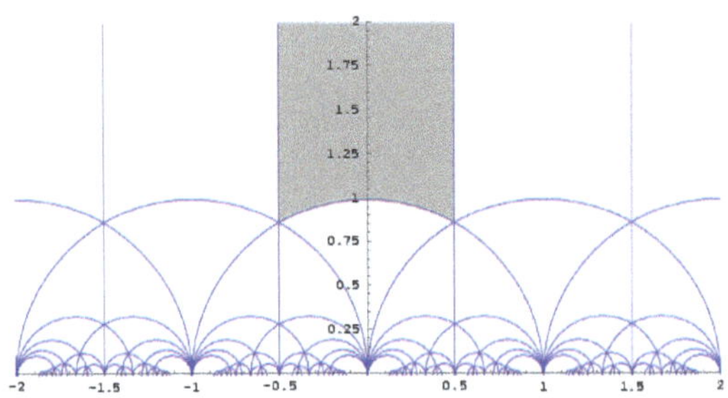

The action of the modular group on the upper half plane. The region in grey is the standard fundamental domain.

Analytic number theory may be defined

- in terms of its tools, as the study of the integers by means of tools from real and complex analysis; or
- in terms of its concerns, as the study within number theory of estimates on size and density, as opposed to identities.

Some subjects generally considered to be part of analytic number theory, e.g., sieve theory, are better covered by the second rather than the first definition: some of sieve theory, for instance, uses little analysis, yet it does belong to analytic number theory.

The following are examples of problems in analytic number theory: the prime number theorem, the Goldbach conjecture (or the twin prime conjecture, or the Hardy–Littlewood conjectures), the Waring problem and the Riemann hypothesis. Some of the most important tools of analytic number theory are the circle method, sieve methods and L-functions (or, rather, the study of their properties). The theory of modular forms (and, more generally, automorphic forms) also occupies an increasingly central place in the toolbox of analytic number theory.

One may ask analytic questions about algebraic numbers, and use analytic means to answer such questions; it is thus that algebraic and analytic number theory intersect. For example, one may define prime ideals (generalizations of prime numbers in the field of algebraic numbers) and ask how many prime ideals there are up to a certain size. This question can be answered by means of an examination of Dedekind zeta functions, which are generalizations of the Riemann zeta function, a key analytic object at the roots of the subject. This is an example of a general procedure in analytic number theory: deriving information about the distribution of a sequence (here, prime ideals or prime numbers) from the analytic behavior of an appropriately constructed complex-valued function.

Algebraic Number Theory

An *algebraic number* is any complex number that is a solution to some polynomial equation $f(x)=0$ with rational coefficients; for example, every solution x of $x^5+(11/2)\,x^3-7x^2+9=0$ (say) is an algebraic number. Fields of algebraic numbers are also called *algebraic number ields*, or shortly *number ields*. Algebraic number theory studies algebraic number fields. Thus, analytic and algebraic number theory can and do overlap: the former is defined by its methods, the latter by its objects of study.

It could be argued that the simplest kind of number fields (viz., quadratic fields) were already studied by Gauss, as the discussion of quadratic forms in *Disquisitiones arithmeticae* can be restated in terms of ideals and norms in quadratic fields. (A *quadratic field* consists of all numbers of the form $a+b\sqrt{d}$, where a and b are rational numbers and d is a fixed rational number whose square root is not rational.) For that matter, the 11th-century chakravala method amounts—in modern terms—to an algorithm for finding the units of a real quadratic number field. However, neither Bhāskara nor Gauss knew of number fields as such.

The grounds of the subject as we know it were set in the late nineteenth century, when *ideal numbers*, the *theory of ideals* and *valuation theory* were developed; these are three complementary ways of dealing with the lack of unique factorisation in algebraic number fields. (For example, in the field generated by the rationals and $\sqrt{-5}$, the number 6 can be factorised both as $6=2\cdot 3$ and $6=(1+\sqrt{-5})(1-\sqrt{-5})$; all of 2, 2, $1+\sqrt{-5}$ and $1-\sqrt{-5}$ are irreducible, and thus, in a naïve sense, analogous to primes among the integers.) The initial impetus for the development of ideal numbers (by Kummer) seems to have come from the study of higher reciprocity laws,i.e., generalisations of quadratic reciprocity.

Number fields are often studied as extensions of smaller number fields: a field *L* is said to be an *extension* of a field *K* if *L* contains *K*. (For example, the complex numbers *C* are an extension of the reals *R*, and the reals *R* are an extension of the rationals *Q*.) Classifying the possible extensions of a given number field is a difficult and partially open problem. Abelian extensions—that is, extensions *L* of *K* such that the Galois group Gal(*L*/*K*) of *L* over *K* is an abelian group—are relatively well understood. Their classification was the object of the programme of class field theory, which was initiated in the late 19th century (partly by Kronecker and Eisenstein) and carried out largely in 1900—1950.

An example of an active area of research in algebraic number theory is Iwasawa theory. The Langlands program, one of the main current large-scale research plans in mathematics, is sometimes described as an attempt to generalise class field theory to non-abelian extensions of number fields.

Diophantine Geometry

The central problem of *Diophantine geometry* is to determine when a Diophantine equation has solutions, and if it does, how many. The approach taken is to think of the solutions of an equation as a geometric object.

For example, an equation in two variables defines a curve in the plane. More generally, an equation, or system of equations, in two or more variables defines a curve, a surface or some other such object in n-dimensional space. In Diophantine geometry, one asks whether there are any *rational points* (points all of whose coordinates are rationals) or *integral points* (points all of whose coordinates are integers) on the curve or surface. If there are any such points, the next step is to ask how many there are and how they are distributed. A basic question in this direction is: are there finitely or infinitely many rational points on a given curve (or surface)? What about integer points?

An example here may be helpful. Consider the Pythagorean equation $x^2+y^2=1$; we would like to study its rational solutions, i.e., its solutions (x,y) such that x and y are both rational. This is the same as asking for all integer solutions to $a^2+b^2=c^2$; any solution to the latter equation gives us a solution $x=a/c$, $y=b/c$ to the former. It is also the same as asking for all points with rational coordinates on the curve described by $x^2+y^2=1$. (This curve happens to be a circle of radius 1 around the origin.)

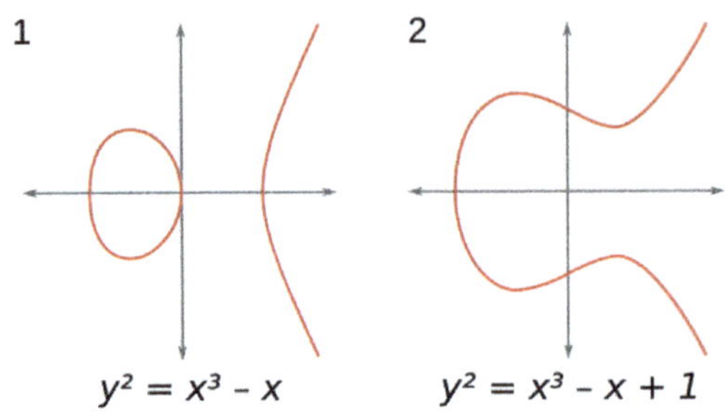

Two examples of an elliptic curve, i.e., a curve of genus 1 having at least one rational point. (Either graph can be seen as a slice of a torus in four-dimensional space.)

The rephrasing of questions on equations in terms of points on curves turns out to be felicitous. The finiteness or not of the number of rational or integer points on an algebraic curve—that is, rational or integer solutions to an equation $f(x,y)=0$, where f is a polynomial in two variables—turns out to depend crucially on the *genus* of the curve. The *genus* can be defined as follows:[note 13] allow the variables in $f(x,y)=0$ to be complex numbers; then $f(x,y)=0$ defines a 2-dimensional surface in (projective) 4-dimensional space (since two complex variables can be decomposed into four real variables, i.e., four dimensions). Count the number of (doughnut) holes in the surface; call this number the *genus* of $f(x,y)=0$. Other geometrical notions turn out to be just as crucial.

There is also the closely linked area of Diophantine approximations: given a number x, how well can it be approximated by rationals? (We are looking for approximations that are good relative to the amount of space that it takes to write the rational: call $a >$(with $\gcd(a,q)=1$) a good approximation to x if $|x-a/q|<\frac{1}{q^c}$, where c is large.) This question is of special interest if x is an algebraic number. If x cannot be well approximated, then some equations do not have integer or rational solutions. Moreover, several concepts (especially that of height) turn out to be crucial both in Diophantine geometry and in the study of Diophantine approximations. This question is also of special interest in transcendental number theory: if a number can be better approximated than any algebraic number, then it is a transcendental number. It is by this argument that π and e have been shown to be transcendental.

Diophantine geometry should not be confused with the geometry of numbers, which is a collection of graphical methods for answering certain questions in algebraic number theory. *Arithmetic geometry*, on the other hand, is a contemporary term for much the same domain as that covered by the term *Diophantine geometry*. The term *arithmetic geometry* is arguably used most often when one wishes to emphasise the connections to modern algebraic geometry (as in, for instance, Faltings' theorem) rather than to techniques in Diophantine approximations.

Recent Approaches and Subfields

The areas below date as such from no earlier than the mid-twentieth century, even if they are based on older material. For example, as is explained below, the matter of algorithms in number theory is very old, in some sense older than the concept of proof; at the same time, the modern study of computability dates only from the 1930s and 1940s, and computational complexity theory from the 1970s.

Probabilistic Number Theory

Take a number at random between one and a million. How likely is it to be prime? This is just another way of asking how many primes there are between one and a million. Further: how many prime divisors will it have, on average? How many divisors will it have altogether, and with what likelihood? What is the probability that it will have many more or many fewer divisors or prime divisors than the average?

Much of probabilistic number theory can be seen as an important special case of the study of variables that are almost, but not quite, mutually independent. For example, the event that a random integer between one and a million be divisible by two and the event that it be divisible by three are almost independent, but not quite.

It is sometimes said that probabilistic combinatorics uses the fact that whatever happens with probability greater than 0 must happen sometimes; one may say with equal justice that many applications of probabilistic number theory hinge on the fact that whatever is unusual must be rare. If certain algebraic objects (say, rational or integer solutions to certain equations) can be shown to be in the tail of certain sensibly defined distributions, it follows that there must be few of them; this is a very concrete non-probabilistic statement following from a probabilistic one.

At times, a non-rigorous, probabilistic approach leads to a number of heuristic algorithms and open problems, notably Cramér's conjecture.

Arithmetic Combinatorics

Let A be a set of N integers. Consider the set $A + A = \{ m + n \mid m, n \in A \}$ consisting of all sums of two elements of A. Is $A + A$ much larger than A? Barely larger? If $A + A$ is barely larger than A, must A have plenty of arithmetic structure, for example, does A resemble an arithmetic progression?

If we begin from a fairly "thick" infinite set A, does it contain many elements in arithmetic progression: a, $a+b, a+2b, a+3b, \ldots, a+10b$, say? Should it be possible to write large integers as sums of elements of A?

These questions are characteristic of *arithmetic combinatorics*. This is a presently coalescing field; it subsumes *additive number theory* (which concerns itself with certain very specific sets A of arithmetic significance, such as the primes or the squares) and, arguably, some of the *geometry of numbers*, together with some rapidly developing new material. Its focus on issues of growth and distribution accounts in part for its developing links with ergodic theory, finite group theory, model theory, and other fields. The term *additive combinatorics* is also used; however, the sets A being studied need not be sets of integers, but rather subsets of non-commutative groups, for which the multiplication symbol, not the addition symbol, is traditionally used; they can also be subsets of rings, in which case the growth of $A+A$ and $A \cdot A$ may be compared.

Computations in Number Theory

While the word *algorithm* goes back only to certain readers of al-Khwārizmī, careful descriptions of methods of solution are older than proofs: such methods (that is, algorithms) are as old as any recognisable mathematics—ancient Egyptian, Babylonian, Vedic, Chinese—whereas proofs appeared only with the Greeks of the classical period. An interesting early case is that of what we now call the Euclidean algorithm. In its basic form (namely, as an algorithm for computing the greatest common divisor) it appears as Proposition 2 of Book VII in *Elements*, together with a proof of correctness. However, in the form that is often used in number theory (namely, as an algorithm for finding integer solutions to an equation $ax+by=c$, or, what is the same, for finding the quantities whose existence is assured by the Chinese remainder theorem) it first appears in the works of Āryabhaṭa (5th–6th century CE) as an algorithm called *kuṭṭaka* ("pulveriser"), without a proof of correctness.

There are two main questions: "can we compute this?" and "can we compute it rapidly?". Anybody can test whether a number is prime or, if it is not, split it into prime factors; doing so rapidly is another matter. We now know fast algorithms for testing primality, but, in spite of much work (both theoretical and practical), no truly fast algorithm for factoring.

The difficulty of a computation can be useful: modern protocols for encrypting messages (e.g., RSA) depend on functions that are known to all, but whose inverses (a) are known only to a chosen few, and (b) would take one too long a time to figure out on one's own. For example, these functions can be such that their inverses can be computed only if certain large integers are factorized. While many difficult computational problems outside number theory are known, most working encryption protocols nowadays are based on the difficulty of a few number-theoretical problems.

On a different note — some things may not be computable at all; in fact, this can be proven in some instances. For instance, in 1970, it was proven, as a solution to Hilbert's 10th problem, that there is no Turing machine which can solve all Diophantine equations. In particular, this means that, given a computably enumerable set of axioms, there are Diophantine equations for which there is no proof, starting from the axioms, of whether the set of equations has or does not have integer solutions. (We would necessarily be speaking of Diophantine equations for which there are no integer solutions, since, given a Diophantine equation with at least one solution, the solution itself provides a proof of the fact that a solution exists. We cannot prove, of course, that a particular Diophantine equation is of this kind, since this would imply that it has no solutions.)

Applications

The number-theorist Leonard Dickson (1874–1954) said "Thank God that number theory is unsullied by any application". Such a view is no longer applicable to number theory. In 1974, Donald Knuth said "...virtually every theorem in elementary number theory arises in a natural, motivated way in connection with the problem of making computers do high-speed numerical calculations". Elementary number theory is taught in discrete mathematics courses for computer scientists; and; on the other hand, number theory also has applications to the continuous in numerical analysis As well as the well-known applications to cryptography, there are also applications to many other areas of mathematics.

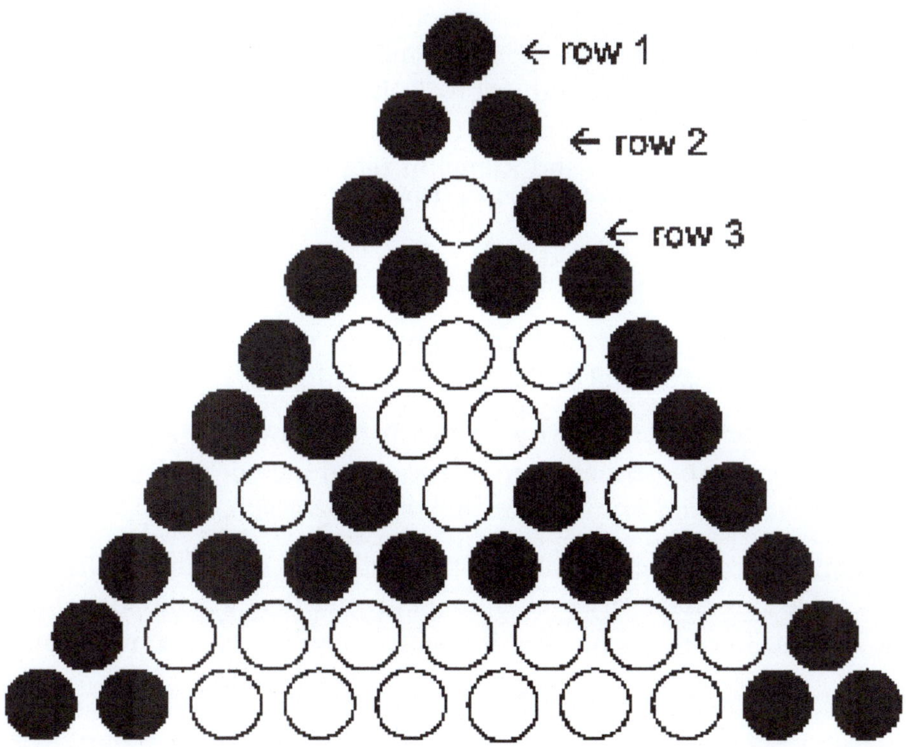

Prizes

The American Mathematical Society awards the *Cole Prize in Number Theory*. Moreover number theory is one of the three mathematical subdisciplines rewarded by the *Fermat Prize*.

Analytic Number Theory

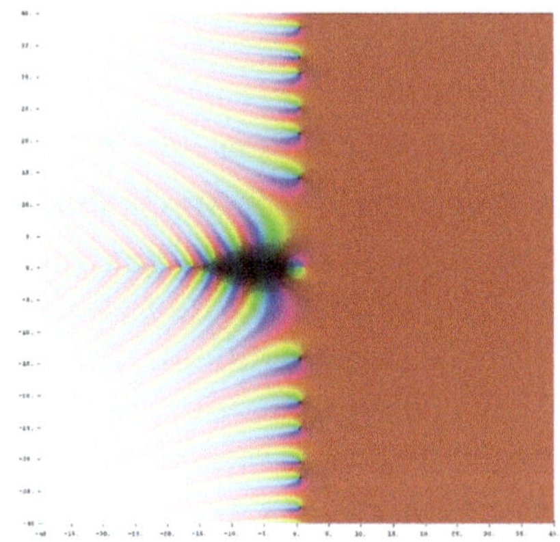

Riemann zeta function ζ(s) in the complex plane. The color of a point s encodes the value of ζ(s): colors close to black denote values close to zero, while hue encodes the value's argument.

In mathematics, analytic number theory is a branch of number theory that uses methods from mathematical analysis to solve problems about the integers. It is often said to have begun with Peter Gustav Lejeune Dirichlet's 1837 introduction of Dirichlet L-functions to give the first proof of Dirichlet's theorem on arithmetic progressions. It is well known for its results on prime numbers (involving the Prime Number Theorem and Riemann zeta function) and additive number theory (such as the Goldbach conjecture and Waring's problem).

Branches of Analytic Number Theory

Analytic number theory can be split up into two major parts, divided more by the type of problems they attempt to solve than fundamental differences in technique.

- Multiplicative number theory deals with the distribution of the prime numbers, such as estimating the number of primes in an interval, and includes the prime number theorem and Dirichlet's theorem on primes in arithmetic progressions.
- Additive number theory is concerned with the additive structure of the integers, such as Goldbach's conjecture that every even number greater than 2 is the sum of two primes. One of the main results in additive number theory is the solution to Waring's problem.

History

Precursors

Much of analytic number theory was inspired by the prime number theorem. Let $\pi(x)$ be the prime-counting function that gives the number of primes less than or equal to x, for any real number x.

For example, $\pi(10) = 4$ because there are four prime numbers (2, 3, 5 and 7) less than or equal to 10. The prime number theorem then states that $x / \ln(x)$ is a good approximation to $\pi(x)$, in the sense that the limit of the *quotient* of the two functions $\pi(x)$ and $x / \ln(x)$ as x approaches infinity is 1:

$$\lim_{x\to\infty} > \frac{\pi(x)}{x/\ln(x)} = 1$$

known as the asymptotic law of distribution of prime numbers.

Adrien-Marie Legendre conjectured in 1797 or 1798 that $\pi(a)$ is approximated by the function $a/(A \ln(a) + B)$, where A and B are unspecified constants. In the second edition of his book on number theory (1808) he then made a more precise conjecture, with $A = 1$ and $B = -1.08366$. Carl Friedrich Gauss considered the same question: "Im Jahr 1792 oder 1793", according to his own recollection nearly sixty years later in a letter to Encke (1849), he wrote in his logarithm table (he was then 15 or 16) the short note "Primzahlen unter $a(=\infty)\frac{a}{\ln a}$". But Gauss never published this conjecture. In 1838 Peter Gustav Lejeune Dirichlet came up with his own approximating function, the logarithmic integral li(x) (under the slightly different form of a series, which he communicated to Gauss). Both Legendre's and Dirichlet's formulas imply the same conjectured asymptotic equivalence of $\pi(x)$ and $x / \ln(x)$ stated above, although it turned out that Dirichlet's approximation is considerably better if one considers the differences instead of quotients.

Dirichlet

Johann Peter Gustav Lejeune Dirichlet is credited with the creation of analytic number theory, a field in which he found several deep results and in proving them introduced some fundamental tools, many of which were later named after him. In 1837 he published Dirichlet's theorem on arithmetic progressions, using mathematical analysis concepts to tackle an algebraic problem and thus creating the branch of analytic number theory. In proving the theorem, he introduced the Dirichlet characters and L-functions. In 1841 he generalized his arithmetic progressions theorem from integers to the ring of Gaussian integers $\mathbb{Z}[i]$.

Chebyshev

In two papers from 1848 and 1850, the Russian mathematician Pafnuty L'vovich Chebyshev attempted to prove the asymptotic law of distribution of prime numbers. His work is notable for the use of the zeta function $\zeta(s)$ (for real values of the argument "s", as are works of Leonhard Euler, as early as 1737) predating Riemann's celebrated memoir of 1859, and he succeeded in proving a slightly weaker form of the asymptotic law, namely, that if the limit of $\pi(x)/(x/\ln(x))$ as x goes to infinity exists at all, then it is necessarily equal to one. He was able to prove unconditionally that this ratio is bounded above and below by two explicitly given constants near to 1 for all x. Although Chebyshev's paper did not prove the Prime Number Theorem, his estimates for $\pi(x)$ were strong enough for him to prove Bertrand's postulate that there exists a prime number between n and $2n$ for any integer $n \geq 2$.

Riemann

"...es ist sehr wahrscheinlich, dass alle Wurzeln reell sind. Hiervon wäre allerdings ein strenger

Beweis zu wünschen; ich habe indess die Aufsuchung desselben nach einigen flüchtigen vergeblichen Versuchen vorläufig bei Seite gelassen, da er für den nächsten Zweck meiner Untersuchung entbehrlich schien.""...it is very probable that all roots are real. Of course one would wish for a rigorous proof here; I have for the time being, after some fleeting vain attempts, provisionally put aside the search for this, as it appears dispensable for the next objective of my investigation."

Riemann's statement of the Riemann hypothesis, from his 1859 paper. (He was discussing a version of the zeta function, modified so that its roots are real rather than on the critical line.)

Bernhard Riemann made some famous contributions to modern analytic number theory. In a single short paper (the only one he published on the subject of number theory), he investigated the Riemann zeta function and established its importance for understanding the distribution of prime numbers. He made a series of conjectures about properties of the zeta function, one of which is the well-known Riemann hypothesis.

Hadamard and De La Vallée-Poussin

xtending the ideas of Riemann, two proofs of the prime number theorem were obtained independently by Jacques Hadamard and Charles Jean de la Vallée-Poussin and appeared in the same year (1896). Both proofs used methods from complex analysis, establishing as a main step of the proof that the Riemann zeta function $\zeta(s)$ is non-zero for all complex values of the variable s that have the form $s = 1 + it$ with $t > 0$.

Modern Times

The biggest technical change after 1950 has been the development of *sieve methods*, particularly in multiplicative problems. These are combinatorial in nature, and quite varied. The extremal branch of combinatorial theory has in return been greatly influenced by the value placed in analytic number theory on quantitative upper and lower bounds. Another recent development is *probabilistic number theory*, which uses methods from probability theory to estimate the distribution of number theoretic functions, such as how many prime divisors a number has.

Developments within analytic number theory are often refinements of earlier techniques, which reduce the error terms and widen their applicability. For example, the *circle method* of Hardy and Littlewood was conceived as applying to power series near the unit circle in the complex plane; it is now thought of in terms of finite exponential sums (that is, on the unit circle, but with the power series truncated). The needs of diophantine approximation are for auxiliary functions that are not generating functions—their coefficients are constructed by use of a pigeonhole principle—and involve several complex variables. The fields of diophantine approximation and transcendence theory have expanded, to the point that the techniques have been applied to the Mordell conjecture.

Problems and Results

Theorems and results within analytic number theory tend not to be exact structural results about the integers, for which algebraic and geometrical tools are more appropriate. Instead, they give approximate bounds and estimates for various number theoretical functions, as the following examples illustrate.

Multiplicative Number Theory

Euclid showed that there are an infinite number of primes but it is very difficult to find an efficient method for determining whether or not a number is prime, especially a large number. A related but easier problem is to determine the asymptotic distribution of the prime numbers; that is, a rough description of how many primes are smaller than a given number. Gauss, amongst others, after computing a large list of primes, conjectured that the number of primes less than or equal to a large number N is close to the value of the integral

$$\int_2^N \frac{1}{\log t} dt.$$

In 1859 Bernhard Riemann used complex analysis and a special meromorphic function now known as the Riemann zeta function to derive an analytic expression for the number of primes less than or equal to a real number x. Remarkably, the main term in Riemann's formula was exactly the above integral, lending substantial weight to Gauss's conjecture. Riemann found that the error terms in this expression, and hence the manner in which the primes are distributed, are closely related to the complex zeros of the zeta function. Using Riemann's ideas and by getting more information on the zeros of the zeta function, Jacques Hadamard and Charles Jean de la Vallée-Poussin managed to complete the proof of Gauss's conjecture. In particular, they proved that if

$$\pi(x) = (\text{number of primes } \leq x),$$

then

$$\lim_{x\to\infty} \frac{\pi(x)}{x/\log x} = 1.$$

This remarkable result is what is now known as the *Prime Number Theorem*. It is a central result in analytic number theory. Loosely speaking, it states that given a large number N, the number of primes less than or equal to N is about $N/\log(N)$.

More generally, the same question can be asked about the number of primes in any arithmetic progression $a+nq$ for any integer n. In one of the first applications of analytic techniques to number theory, Dirichlet proved that any arithmetic progression with a and q coprime contains infinitely many primes. The prime number theorem can be generalised to this problem; letting

$$\pi(x,a,q) = (\text{number of primes } \leq x \text{ such that } p \text{ is in the arithmetic progression } a+nq, n \in \mathbf{Z}),$$

then if a and q are coprime,

$$\lim_{x\to\infty} \frac{\pi(x,a,q)\phi(q)}{x/\log x} = 1.$$

There are also many deep and wide ranging conjectures in number theory whose proofs seem too difficult for current techniques, such as the Twin prime conjecture which asks whether there are infinitely many primes p such that $p + 2$ is prime. On the assumption of the Elliott–Halberstam conjecture it has been proven recently that there are infinitely many primes p such that $p + k$ is prime for some positive even k at most 12. Also, it has been proven unconditionally (i.e. not depending on unproven conjectures) that there are infinitely many primes p such that $p + k$ is prime for some positive even k at most 246.

Additive Number Theory

One of the most important problems in additive number theory is Waring's problem, which asks whether it is possible, for any $k \geq 2$, to write any positive integer as the sum of a bounded number of kth powers,

$$n = x_1^k + \cdots + x_\ell^k.$$

The case for squares, $k = 2$, was answered by Lagrange in 1770, who proved that every positive integer is the sum of at most four squares. The general case was proved by Hilbert in 1909, using algebraic techniques which gave no explicit bounds. An important breakthrough was the application of analytic tools to the problem by Hardy and Littlewood. These techniques are known as the circle method, and give explicit upper bounds for the function $G(k)$, the smallest number of kth powers needed, such as Vinogradov's bound

$$G(k) \leq k(3\log k + 11).$$

Diophantine Problems

Diophantine problems are concerned with integer solutions to polynomial equations: one may study the distribution of solutions, that is, counting solutions according to some measure of "size" or *height*.

An important example is the Gauss circle problem, which asks for integers points (x y) which satisfy

$$x^2 + y^2 \leq r^2.$$

In geometrical terms, given a circle centered about the origin in the plane with radius r, the problem asks how many integer lattice points lie on or inside the circle. It is not hard to prove that the answer is $\pi r^2 + E(r)$, where $E(r)/r^2 \to 0$ as $r \to \infty$. Again, the difficult part and a great achievement of analytic number theory is obtaining specific upper bounds on the error term $E(r)$.

It was shown by Gauss that $E(r) = O(r)$. In general, an $O(r)$ error term would be possible with the unit circle (or, more properly, the closed unit disk) replaced by the dilates of any bounded planar region with piecewise smooth boundary. Furthermore, replacing the unit circle by the unit square, the error term for the general problem can be as large as a linear function of r. Therefore, getting an error bound of the form $O(r^\delta)$ for some $\delta < 1$ in the case of the circle is a significant improvement. The first to attain this was Sierpiński in 1906, who showed $E(r) = O(r^{2/3})$. In 1915, Hardy and Landau each showed that one does *not* have $E(r) = O(r^{1/2})$. Since then the goal has been to show that for each fixed $\epsilon > 0$ there exists a real number $C(\epsilon)$ such that $E(r) \leq C(\epsilon) r^{1/2+\epsilon}$.

In 2000 Huxley showed that $E(r) = O(r^{131/208})$, which is the best published result.

Methods of Analytic Number Theory

Dirichlet Series

One of the most useful tools in multiplicative number theory are Dirichlet series, which are functions of a complex variable defined by an infinite series of the form

$$f(s) = \sum_{n=1}^{\infty} a_n n^{-s}.$$

Depending on the choice of coefficients a_n, this series may converge everywhere, nowhere, or on some half plane. In many cases, even where the series does not converge everywhere, the holomorphic function it defines may be analytically continued to a meromorphic function on the entire complex plane. The utility of functions like this in multiplicative problems can be seen in the formal identity

$$\left(\sum_{n=1}^{\infty} a_n n^{-s}\right)\left(\sum_{n=1}^{\infty} b_n n^{-s}\right) = \sum_{n=1}^{\infty}\left(\sum_{k\ell=n} a_k b_\ell\right)n^{-s};$$

hence the coefficients of the product of two Dirichlet series are the multiplicative convolutions of the original coefficients. Furthermore, techniques such as partial summation and Tauberian theorems can be used to get information about the coefficients from analytic information about the Dirichlet series. Thus a common method for estimating a multiplicative function is to express it as a Dirichlet series (or a product of simpler Dirichlet series using convolution identities), examine this series as a complex function and then convert this analytic information back into information about the original function.

Riemann Zeta Function

Euler showed that the fundamental theorem of arithmetic implies (at least formally) the *Euler product*

$$\sum_{n=1}^{\infty} \frac{1}{n^s} = \prod_{p}^{\infty} \frac{1}{1-p^{-s}} \text{ for } s > 1 \text{ } (p \text{ is prime number})$$

Euler's proof of the infinity of prime numbers makes use of the divergence of the term at the left hand side for $s = 1$ (the so-called harmonic series), a purely analytic result. Euler was also the first to use analytical arguments for the purpose of studying properties of integers, specifically by constructing generating power series. This was the beginning of analytic number theory.

Later, Riemann considered this function for complex values of s and showed that this function can be extended to a meromorphic function on the entire plane with a simple pole at $s = 1$. This function is now known as the Riemann Zeta function and is denoted by $\zeta(s)$. There is a plethora of literature on this function and the function is a special case of the more general Dirichlet L-functions.

Analytic number theorists are often interested in the error of approximations such as the prime number theorem. In this case, the error is smaller than $x/\log x$. Riemann's formula for $\pi(x)$ shows that the error term in this approximation can be expressed in terms of the zeros of the zeta function. In his 1859 paper, Riemann conjectured that all the "non-trivial" zeros of ζ lie on the line $\Re(s) = 1/2$ but never provided a proof of this statement. This famous and long-standing conjecture is known as the *Riemann Hypothesis* and has many deep implications in number theory; in fact, many important theorems have been proved under the assumption that the hypothesis is true. For example, under the assumption of the Riemann Hypothesis, the error term in the prime number theorem is $O(x^{1/2+\varepsilon})$.

In the early 20th century G. H. Hardy and Littlewood proved many results about the zeta function in an attempt to prove the Riemann Hypothesis. In fact, in 1914, Hardy proved that there were infinitely many zeros of the zeta function on the critical line

$$\Re(z) = 1/2.$$

This led to several theorems describing the density of the zeros on the critical line.

Algebraic Number Theory

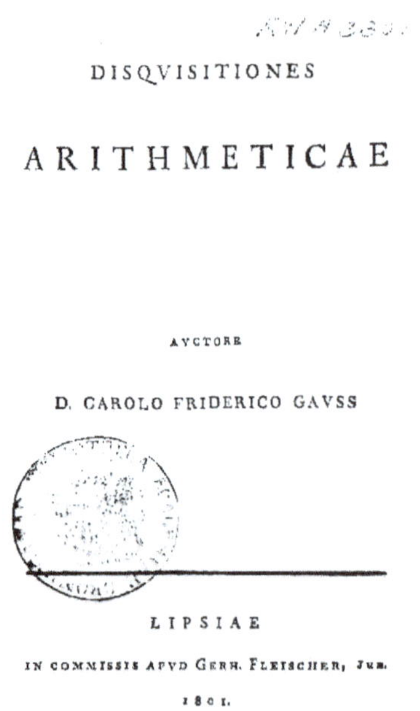

DISQVISITIONES

ARITHMETICAE

AVCTORE

D. CAROLO FRIDERICO GAVSS

LIPSIAE

IN COMMISSIS APVD GERH. FLEISCHER, IUN.

1801.

Title page of the first edition of Disquisitiones Arithmeticae, one of the founding works of modern algebraic number theory.

Algebraic number theory is a major branch of number theory that studies algebraic structures related to algebraic integers. This is generally accomplished by considering a ring of algebraic integers O in an algebraic number field K/Q, and studying their algebraic properties such as factorization, the behaviour of ideals, and field extensions. In this setting, the familiar features of the integers—such as unique factorization—need not hold. The virtue of the primary machinery employed—Galois theory, group cohomology, group representations, and L-functions—is that it allows one to deal with new phenomena and yet partially recover the behaviour of the usual integers.

History of Algebraic Number Theory

Diophantus

The beginnings of algebraic number theory can be traced to Diophantine equations, named after the 3rd-century Alexandrian mathematician, Diophantus, who studied them and developed methods for the solution of some kinds of Diophantine equations. A typical Diophantine problem is to find two integers x and y such that their sum, and the sum of their squares, equal two given numbers A and B, respectively:

$$A = x + y$$

$$B = x^2 + y^2.$$

Diophantine equations have been studied for thousands of years. For example, the solutions to the quadratic Diophantine equation $x^2 + y^2 = z^2$ are given by the Pythagorean triples, originally solved by the Babylonians (c. 1800 BC). Solutions to linear Diophantine equations, such as $26x + 65y = 13$, may be found using the Euclidean algorithm (c. 5th century BC).

Diophantus's major work was the *Arithmetica*, of which only a portion has survived.

Fermat

Fermat's last theorem was first conjectured by Pierre de Fermat in 1637, famously in the margin of a copy of *Arithmetica* where he claimed he had a proof that was too large to fit in the margin. No successful proof was published until 1995 despite the efforts of countless mathematicians during the 358 intervening years. The unsolved problem stimulated the development of algebraic number theory in the 19th century and the proof of the modularity theorem in the 20th century.

Gauss

One of the founding works of algebraic number theory, the *Disquisitiones Arithmeticae* (Latin: *Arithmetical Investigations*) is a textbook of number theory written in Latin by Carl Friedrich Gauss in 1798 when Gauss was 21 and first published in 1801 when he was 24. In this book Gauss brings together results in number theory obtained by mathematicians such as Fermat, Euler, Lagrange and Legendre and adds important new results of his own. Before the *Disquisitiones* was published, number theory consisted of a collection of isolated theorems and conjectures. Gauss brought the work of his predecessors together with his own original work into a systematic framework, filled in gaps, corrected unsound proofs, and extended the subject in numerous ways.

The *Disquisitiones* was the starting point for the work of other nineteenth century European mathematicians including Ernst Kummer, Peter Gustav Lejeune Dirichlet and Richard Dedekind. Many of the annotations given by Gauss are in effect announcements of further research of his own, some of which remained unpublished. They must have appeared particularly cryptic to his contemporaries; we can now read them as containing the germs of the theories of L-functions and complex multiplication, in particular.

Dirichlet

In a couple of papers in 1838 and 1839 Peter Gustav Lejeune Dirichlet proved the first class number formula, for quadratic forms (later refined by his student Kronecker). The formula, which Jacobi called a result "touching the utmost of human acumen", opened the way for similar results regarding more general number fields. Based on his research of the structure of the unit group of quadratic fields, he proved the Dirichlet unit theorem, a fundamental result in algebraic number theory.

He first used the pigeonhole principle, a basic counting argument, in the proof of a theorem in diophantine approximation, later named after him Dirichlet's approximation theorem. He published important contributions to Fermat's last theorem, for which he proved the cases $n = 5$ and $n = 14$, and to the biquadratic reciprocity law. The Dirichlet divisor problem, for which he found the first results, is still an unsolved problem in number theory despite later contributions by other researchers.

Dedekind

Richard Dedekind's study of Lejeune Dirichlet's work was what led him to his later study of algebraic number fields and ideals. In 1863, he published Lejeune Dirichlet's lectures on number theory as *Vorlesungen über Zahlentheorie* ("Lectures on Number Theory") about which it has been written that:

"Although the book is assuredly based on Dirichlet's lectures, and although Dedekind himself referred to the book throughout his life as Dirichlet's, the book itself was entirely written by Dedekind, for the most part after Dirichlet's death." (Edwards 1983)

1879 and 1894 editions of the *Vorlesungen* included supplements introducing the notion of an ideal, fundamental to ring theory. (The word "Ring", introduced later by Hilbert, does not appear in Dedekind's work.) Dedekind defined an ideal as a subset of a set of numbers, composed of algebraic integers that satisfy polynomial equations with integer coefficients. The concept underwent further development in the hands of Hilbert and, especially, of Emmy Noether. Ideals generalize Ernst Eduard Kummer's ideal numbers, devised as part of Kummer's 1843 attempt to prove Fermat's Last Theorem.

Hilbert

David Hilbert unified the field of algebraic number theory with his 1897 treatise *Zahlbericht* (literally "report on numbers"). He also resolved a significant number-theory problem formulated by Waring in 1770. As with the finiteness theorem, he used an existence proof that shows there must be solutions for the problem rather than providing a mechanism to produce the answers. He then had little more to publish on the subject; but the emergence of Hilbert modular forms in the dissertation of a student means his name is further attached to a major area.

He made a series of conjectures on class field theory. The concepts were highly influential, and his own contribution lives on in the names of the Hilbert class field and of the Hilbert symbol of local class field theory. Results were mostly proved by 1930, after work by Teiji Takagi.

Artin

Emil Artin established the Artin reciprocity law in a series of papers (1924; 1927; 1930). This law is a general theorem in number theory that forms a central part of global class field theory. The term "reciprocity law" refers to a long line of more concrete number theoretic statements which it generalized, from the quadratic reciprocity law and the reciprocity laws of Eisenstein and Kummer to Hilbert's product formula for the norm symbol. Artin's result provided a partial solution to Hilbert's ninth problem.

Modern Theory

Around 1955, Japanese mathematicians Goro Shimura and Yutaka Taniyama observed a possible link between two apparently completely distinct, branches of mathematics, elliptic curves and modular forms. The resulting modularity theorem (at the time known as the Taniyama–Shimura conjecture) states that every elliptic curve is modular, meaning that it can be associated with a unique modular form.

It was initially dismissed as unlikely or highly speculative, and was taken more seriously when number theorist André Weil found evidence supporting it, but no proof; as a result the "astounding" conjecture was often known as the Taniyama–Shimura-Weil conjecture. It became a part of the Langlands programme, a list of important conjectures needing proof or disproof.

From 1993 to 1994, Andrew Wiles provided a proof of the modularity theorem for semistable elliptic curves, which, together with Ribet's theorem, provides a proof for Fermat's Last Theorem. Both Fermat's Last Theorem and the Modularity Theorem were almost universally considered inaccessible to proof by contemporaneous mathematicians (meaning, impossible or virtually impossible to prove using current knowledge). Wiles first announced his proof in June 1993 in a version that was soon recognized as having a serious gap in a key point. The proof was corrected by Wiles, in part via collaboration with Richard Taylor, and the final, widely accepted, version was released in September 1994, and formally published in 1995. The proof uses many techniques from algebraic geometry and number theory, and has many ramifications in these branches of mathematics. It also uses standard constructions of modern algebraic geometry, such as the category of schemes and Iwasawa theory, and other 20th-century techniques not available to Fermat.

Basic Notions

Unique Factorization and the Ideal Class Group

One of the first properties of Z that can fail in the ring of integers O of an algebraic number field K is that of the unique factorization of integers into prime numbers. The prime numbers in Z are generalized to irreducible elements in O, and though the unique factorization of elements of O into irreducible elements may hold in some cases (such as for the Gaussian integers Z[i]), it may also fail, as in the case of Z[√-5] where

$$6 = 2\cdot 3 = (1+\sqrt{-5})\cdot(1-\sqrt{-5}).$$

The ideal class group of O is a measure of how much unique factorization of elements fails; in particular, the ideal class group is trivial if, and only if, O is a unique factorization domain.

Factoring Prime Ideals in Extensions

Unique factorization can be partially recovered for O in that it has the property of unique factorization of *ideals* into prime ideals (i.e. it is a Dedekind domain). This makes the study of the prime ideals in O particularly important. This is another area where things change from Z to O: the prime numbers, which generate prime ideals of Z (in fact, every single prime ideal of Z is of the form $(p):=p\mathrm{Z}$ for some prime number p,) may no longer generate prime ideals in O. For example, in the ring of Gaussian integers, the ideal 2Z[i] is no longer a prime ideal; in fact

$$2\mathbf{Z}[i] = \left((1+i)\mathbf{Z}[i]\right)^2.$$

On the other hand, the ideal 3Z[i] is a prime ideal. The complete answer for the Gaussian integers is obtained by using a theorem of Fermat's, with the result being that for an odd prime number p

$$p\mathbf{Z}[i] \text{ is a prime ideal if } p \equiv 3(\mathrm{mod}4)$$

$$p\mathbf{Z}[i] \text{ is not a prime ideal if } p \equiv 1(\mathrm{mod}4).$$

Generalizing this simple result to more general rings of integers is a basic problem in algebraic number theory. Class field theory accomplishes this goal when K is an abelian extension of Q (i.e. a Galois extension with abelian Galois group).

Primes and Places

An important generalization of the notion of prime ideal in O is obtained by passing from the so-called *ideal-theoretic* approach to the so-called *valuation-theoretic* approach. The relation between the two approaches arises as follows. In addition to the usual absolute value function $|\cdot| : \mathrm{Q} \to \mathrm{R}$, there are absolute value functions $|\cdot|_p : \mathrm{Q} \to \mathrm{R}$ defined for each prime number p in Z, called p-adic absolute values. Ostrowski's theorem states that these are all possible absolute value functions on Q (up to equivalence). This suggests that the usual absolute value could be considered as another prime. More generally, a prime of an algebraic number field ***K*** (also called a place) is an equivalence class of absolute values on K. The primes in K are of two sorts: $\mathfrak{p}$-adic absolute values like $|\cdot|_{\mathfrak{p}}$, one for each prime ideal $\mathfrak{p}$ of O, and absolute values like $|\cdot|$ obtained by considering K as a subset of the complex numbers in various possible ways and using the absolute value $|\cdot| : \mathrm{C} \to \mathrm{R}$. A prime of the first kind is called a finite prime (or finite place) and one of the second kind is called an infinite prime (or infinite place). Thus, the set of primes of Q is generally denoted $\{ 2, 3, 5, 7, ..., \infty \}$, and the usual absolute value on Q is often denoted $|\cdot|_\infty$ in this context.

The set of infinite primes of K can be described explicitly in terms of the embeddings $K \to \mathrm{C}$ (i.e. the non-zero ring homomorphisms from K to C). Specifically, the set of embeddings can be split up into two disjoint subsets, those whose image is contained in R, and the rest. To each embedding $\sigma : K \to \mathrm{R}$, there corresponds a unique prime of K coming from the absolute value obtained by composing σ with the usual absolute value on R; a prime arising in this fashion is called a real prime (or real place). To an embedding $\tau : K \to \mathrm{C}$ whose image is *not* contained in R, one can construct a distinct embedding τ, called the *conjugate embedding*, by composing τ with the complex conjugation map $\mathrm{C} \to \mathrm{C}$. Given such a pair of embeddings τ and τ, there corresponds a unique prime of K again obtained by composing τ with the usual absolute value (composing τ instead gives the same absolute value function since $|z| = |\bar{z}|$ for any complex number z, where $\bar{z}$ denotes the complex conjugate of z). Such a prime is called a complex prime (or complex place). The description of the set of infinite primes is then as follows: each infinite prime corresponds either to a unique embedding $\sigma : K \to \mathbb{R}$, or a pair of conjugate embeddings $\tau, \bar{\tau} : K \to \mathbb{C}$. The number of real (respectively, complex) primes is often denoted r_1 (respectively, r_2). Then, the total number of embeddings $K \to \mathbb{C}$ is r_1+2r_2 (which, in fact, equals the degree of the extension $K/\mathbb{Q}$).

Units

The fundamental theorem of arithmetic describes the multiplicative structure of Z. It states that every non-zero integer can be written (essentially) uniquely as a product of prime powers and ±1. The unique factorization of ideals in the ring *O* recovers part of this description, but fails to address the factor ±1. The integers 1 and -1 are the invertible elements (i.e. units) of Z. More generally, the invertible elements in *O* form a group under multiplication called the unit group of *O*, denoted $O^\times$. This group can be much larger than the cyclic group of order 2 formed by the units of Z. Dirichlet's unit theorem describes the abstract structure of the unit group as an abelian group. A more precise statement giving the structure of $O^\times \otimes_Z Q$ as a Galois module for the Galois group of K/Q is also possible. The size of the unit group, and its lattice structure give important numerical information about *O*, as can be seen in the class number formula.

Local Fields

Completing a number field *K* at a place *w* gives a complete field. If the valuation is archimedean, one gets R or C, if it is non-archimedean and lies over a prime *p* of the rationals, one gets a finite extension K_w / Q_p: a complete, discrete valued field with finite residue field. This process simplifies the arithmetic of the field and allows the local study of problems. For example, the Kronecker–Weber theorem can be deduced easily from the analogous local statement. The philosophy behind the study of local fields is largely motivated by geometric methods. In algebraic geometry, it is common to study varieties locally at a point by localizing to a maximal ideal. Global information can then be recovered by gluing together local data. This spirit is adopted in algebraic number theory. Given a prime in the ring of algebraic integers in a number field, it is desirable to study the field locally at that prime. Therefore, one localizes the ring of algebraic integers to that prime and then completes the fraction field much in the spirit of geometry.

Major Results

Finiteness of the Class Group

One of the classical results in algebraic number theory is that the ideal class group of an algebraic number field *K* is finite. The order of the class group is called the class number, and is often denoted by the letter *h*.

Dirichlet's Unit Theorem

Dirichlet's unit theorem provides a description of the structure of the multiplicative group of units $O^\times$ of the ring of integers *O*. Specifically, it states that $O^\times$ is isomorphic to $G \times Z^r$, where *G* is the finite cyclic group consisting of all the roots of unity in *O*, and $r = r_1 + r_2 - 1$ (where r_1 (respectively,

r_2) denotes the number of real embeddings (respectively, pairs of conjugate non-real embeddings) of K). In other words, $O^\times$ is a finitely generated abelian group of rank $r_1 + r_2 - 1$ whose torsion consists of the roots of unity in O.

Reciprocity Laws

In terms of the Legendre symbol, the law of quadratic reciprocity for positive odd primes states

$$\left(\frac{p}{q}\right)\left(\frac{q}{p}\right)=(-1)^{\frac{p-1}{2}\frac{q-1}{2}}.$$

A reciprocity law is a generalization of the law of quadratic reciprocity.

There are several different ways to express reciprocity laws. The early reciprocity laws found in the 19th century were usually expressed in terms of a power residue symbol (p/q) generalizing the quadratic reciprocity symbol, that describes when a prime number is an nth power residue modulo another prime, and gave a relation between (p/q) and (q/p). Hilbert reformulated the reciprocity laws as saying that a product over p of Hilbert symbols $(a,b/p)$, taking values in roots of unity, is equal to 1. Artin's reformulated reciprocity law states that the Artin symbol from ideals (or ideles) to elements of a Galois group is trivial on a certain subgroup. Several more recent generalizations express reciprocity laws using cohomology of groups or representations of adelic groups or algebraic K-groups, and their relationship with the original quadratic reciprocity law can be hard to see.

Class Number Formula

The class number formula relates many important invariants of a number field to a special value of its Dedekind zeta function.

Related Areas

Algebraic number theory interacts with many other mathematical disciplines. It uses tools from homological algebra. Via the analogy of function fields vs. number fields, it relies on techniques and ideas from algebraic geometry. Moreover, the study of higher-dimensional schemes over Z instead of number rings is referred to as arithmetic geometry. Algebraic number theory is also used in the study of arithmetic hyperbolic 3-manifolds.

Algebraic Structure

In mathematics, and more specifically in abstract algebra, an algebraic structure is a set (called carrier set or underlying set) with one or more finitary operations defined on it that satisfies a list of axioms.

Examples of algebraic structures include groups, rings, fields, and lattices. More complex structures can be defined by introducing multiple operations, different underlying sets, or by altering the defining axioms. Examples of more complex algebraic structures include vector spaces, modules, and algebras.

The properties of specific algebraic structures are studied in abstract algebra. The general theory of algebraic structures has been formalized in universal algebra. Category theory is used to study the relationships between two or more classes of algebraic structures, often of different kinds. For example, Galois theory studies the connection between certain fields and groups, algebraic structures of two different kinds.

Introduction

Addition and multiplication on numbers are the prototypical example of an operation that combines two elements of a set to produce a third. These operations obey several algebraic laws. For example, $a+ (b + c) = (a + b) + c$ and $a(bc) = (ab)c$, both examples of the *associative law*. Also $a + b = b + a$, and $ab = ba$, the *commutative law*. Many systems studied by mathematicians have operations that obey some, but not necessarily all, of the laws of ordinary arithmetic. For example, rotations of objects in three-dimensional space can be combined by performing the first rotation and then applying the second rotation to the object in its new orientation. This operation on rotations obeys the associative law, but can fail the commutative law.

Mathematicians give names to sets with one or more operations that obey a particular collection of laws, and study them in the abstract as algebraic structures. When a new problem can be shown to follow the laws of one of these algebraic structures, all the work that has been done on that category in the past can be applied to the new problem.

In full generality, algebraic structures may involve an arbitrary number of sets and operations that can combine more than two elements (higher arity), but this article focuses on binary operations on one or two sets. The examples here are by no means a complete list, but they are meant to be a representative list and include the most common structures. Longer lists of algebraic structures may be found in the external links and within *Category:Algebraic structures*. Structures are listed in approximate order of increasing complexity.

Examples

One Set with Operations

Simple structures: no binary operation:

- Set: a degenerate algebraic structure S having no operations.
- Pointed set: S has one or more distinguished elements, often 0, 1, or both.
- Unary system: S and a single unary operation over S.
- Pointed unary system: a unary system with S a pointed set.

Group-like structures: one binary operation. The binary operation can be indicated by any symbol, or with no symbol (juxtaposition) as is done for ordinary multiplication of real numbers.

- Magma or groupoid: S and a single binary operation over S.
- Semigroup: an associative magma.
- Monoid: a semigroup with identity.

- Group: a monoid with a unary operation (inverse), giving rise to inverse elements.
- Abelian group: a group whose binary operation is commutative.
- Semilattice: a semigroup whose operation is idempotent and commutative. The binary operation can be called either meet or join.
- Quasigroup: a magma obeying the latin square property. A quasigroup may also be represented using three binary operations.
- Loop: a quasigroup with identity.

Ring-like structures or Ringoids: two binary operations, often called addition and multiplication, with multiplication distributing over addition.

- Semiring: a ringoid such that S is a monoid under each operation. Addition is typically assumed to be commutative and associative, and the monoid product is assumed to distribute over the addition on both sides, and the additive identity satisfies $0\,x = 0$ for all x.
- Near-ring: a semiring whose additive monoid is a (not necessarily abelian) group.
- Ring: a semiring whose additive monoid is an abelian group.
- Lie ring: a ringoid whose additive monoid is an abelian group, but whose multiplicative operation satisfies the Jacobi identity rather than associativity.
- Boolean ring: a commutative ring with idempotent multiplication operation.
- Field: a commutative ring which contains a multiplicative inverse for every nonzero element
- Kleene algebras: a semiring with idempotent addition and a unary operation, the Kleene star, satisfying additional properties.
- *-algebra: a ring with an additional unary operation (*) satisfying additional properties.

Lattice structures: two or more binary operations, including operations called meet and join, connected by the absorption law.

- Complete lattice: a lattice in which arbitrary meet and joins exist.
- Bounded lattice: a lattice with a greatest element and least element.
- Complemented lattice: a bounded lattice with a unary operation, complementation, denoted by postfix □. The join of an element with its complement is the greatest element, and the meet of the two elements is the least element.
- Modular lattice: a lattice whose elements satisfy the additional *modular identity*.
- Distributive lattice: a lattice in which each of meet and join distributes over the other. Distributive lattices are modular, but the converse does not hold.
- Boolean algebra: a complemented distributive lattice. Either of meet or join can be defined in terms of the other and complementation. This can be shown to be equivalent with the ring-like structure of the same name above.

- Heyting algebra: a bounded distributive lattice with an added binary operation, relative pseudo-complement, denoted by infix $\rightarrow$, and governed by the axioms $x \rightarrow x = 1$, $x\,(x \rightarrow y) = x\,y$, $y\,(x \rightarrow y) = y$, $x \rightarrow (y\,z) = (x \rightarrow y)\,(x \rightarrow z)$.

Arithmetics: two binary operations, addition and multiplication. S is an infinite set. Arithmetics are pointed unary systems, whose unary operation is injective successor, and with distinguished element 0.

- Robinson arithmetic. Addition and multiplication are recursively defined by means of successor. 0 is the identity element for addition, and annihilates multiplication. Robinson arithmetic is listed here even though it is a variety, because of its closeness to Peano arithmetic.
- Peano arithmetic. Robinson arithmetic with an axiom schema of induction. Most ring and field axioms bearing on the properties of addition and multiplication are theorems of Peano arithmetic or of proper extensions thereof.

Two Sets with Operations

Module-like structures: composite systems involving two sets and employing at least two binary operations.

- Group with operators: a group G with a set Ω and a binary operation $\Omega \times G \rightarrow G$ satisfying certain axioms.
- Module: an Abelian group M and a ring R acting as operators on M. The members of R are sometimes called scalars, and the binary operation of *scalar multiplication* is a function $R \times M \rightarrow M$, which satisfies several axioms. Counting the ring operations these systems have at least three operations.
- Vector space: a module where the ring R is a division ring or field.
- Graded vector space: a vector space with a direct sum decomposition breaking the space into "grades".
- Quadratic space: a vector space V over a field F with a function from V into F satisfying certain properties. Every quadratic space is also an inner product space.

Algebra-like structures: composite system defined over two sets, a ring R and a R module M equipped with an operation called multiplication. This can be viewed as a system with five binary operations: two operations on R, two on M and one involving both R and M.

- Algebra over a ring (also *R-algebra*): a module over a commutative ring R, which also carries a multiplication operation that is compatible with the module structure. This includes distributivity over addition and linearity with respect to multiplication by elements of R. The theory of an algebra over a field is especially well developed.
- Associative algebra: an algebra over a ring such that the multiplication is associative.
- Nonassociative algebra: a module over a commutative ring, equipped with a ring multiplication operation that is not necessarily associative. Often associativity is replaced with a

different identity, such as alternation, the Jacobi identity, or the Jordan identity.

- Coalgebra: a vector space with a "comultiplication" defined dually to that of associative algebras.
- Lie algebra: a special type of nonassociative algebra whose product satisfies the Jacobi identity.
- Lie coalgebra: a vector space with a "comultiplication" defined dually to that of Lie algebras.
- Graded algebra: a graded vector space with an algebra structure compatible with the grading. The idea is that if the grades of two elements a and b are known, then the grade of ab is known, and so the location of the product ab is determined in the decomposition.
- Inner product space: an F vector space V with a bilinear binary operation from $V \times V \to F$.

Four or more binary operations:

- Bialgebra: an associative algebra with a compatible coalgebra structure.
- Lie bialgebra: a Lie algebra with a compatible bialgebra structure.
- Hopf algebra: a bialgebra with a connection axiom (antipode).
- Clifford algebra: a graded associative algebra equipped with an exterior product from which may be derived several possible inner products. Exterior algebras and geometric algebras are special cases of this construction.

Hybrid Structures

Algebraic structures can also coexist with added structure of non-algebraic nature, such as partial order or a topology. The added structure must be compatible, in some sense, with the algebraic structure.

- Topological group: a group with a topology compatible with the group operation.
- Lie group: a topological group with a compatible smooth manifold structure.
- Ordered groups, ordered rings and ordered fields: each type of structure with a compatible partial order.
- Archimedean group: a linearly ordered group for which the Archimedean property holds.
- Topological vector space: a vector space whose M has a compatible topology.
- Normed vector space: a vector space with a compatible norm. If such a space is complete (as a metric space) then it is called a Banach space.
- Hilbert space: an inner product space over the real or complex numbers whose inner product gives rise to a Banach space structure.
- Vertex operator algebra
- Von Neumann algebra: a *-algebra of operators on a Hilbert space equipped with the weak operator topology.

Universal Algebra

Algebraic structures are defined through different configurations of axioms. Universal algebra abstractly studies such objects. One major dichotomy is between structures that are axiomatized entirely by *identities* and structures that are not. If all axioms defining a class of algebras are identities, then the class of objects is a variety.

Identities are equations formulated using only the operations the structure allows, and variables that are tacitly universally quantified over the relevant universe. Identities contain no connectives, existentially quantified variables, or relations of any kind other than the allowed operations. The study of varieties is an important part of universal algebra. An algebraic structure in a variety may be understood as the quotient algebra of term algebra (also called "absolutely free algebra") divided by the equivalence relations generated by a set of identities. So, a collection of functions with given signatures generate a free algebra, the term algebra T. Given a set of equational identities (the axioms), one may consider their symmetric, transitive closure E. The quotient algebra T/E is then the algebraic structure or variety. Thus, for example, groups have a signature containing two operators: the multiplication operator m, taking two arguments, and the inverse operator i, taking one argument, and the identity element e, a constant, which may be considered an operator that takes zero arguments. Given a (countable) set of variables x, y, z, etc. the term algebra is the collection of all possible terms involving m, i, e and the variables; so for example, $m(i(x), m(x,m(y,e)))$ would be an element of the term algebra. One of the axioms defining a group is the identity $m(x, i(x)) = e$; another is $m(x,e) = x$. The axioms can be represented as trees. These equations induce equivalence classes on the free algebra; the quotient algebra then has the algebraic structure of a group.

Several non-variety structures fail to be varieties, because either:

1. It is necessary that $0 \neq 1$, 0 being the additive identity element and 1 being a multiplicative identity element, but this is a nonidentity;
2. Structures such as fields have some axioms that hold only for nonzero members of S. For an algebraic structure to be a variety, its operations must be defined for *all* members of S; there can be no partial operations.

Structures whose axioms unavoidably include nonidentities are among the most important ones in mathematics, e.g., fields and division rings. Although structures with nonidentities retain an undoubted algebraic flavor, they suffer from defects varieties do not have. For example, the product of two fields is not a field.

Category Theory

Category theory is another tool for studying algebraic structures (see, for example, Mac Lane 1998). A category is a collection of *objects* with associated *morphisms*. Every algebraic structure has its own notion of homomorphism, namely any function compatible with the operation(s) defining the structure. In this way, every algebraic structure gives rise to a category. For example, the category of groups has all groups as objects and all group homomorphisms as morphisms. This concrete category may be seen as a category of sets with added category-theoretic structure. Likewise, the category of topological groups (whose morphisms are the continuous group homomorphisms) is a category of topological spaces with extra structure. A forgetful functor between categories of algebraic structures "forgets" a part of a structure.

There are various concepts in category theory that try to capture the algebraic character of a context, for instance

- Algebraic Category
- essentially algebraic category
- presentable category
- locally presentable category
- monadic functors and categories
- universal property.

Different Meanings of "Structure"

In a slight abuse of notation, the word "structure" can also refer to just the operations on a structure, instead of the underlying set itself. For example, the sentence, "We have defined a ring *structure* on the set A," means that we have defined ring *operations* on the set A. For another example, the group $(\mathbb{Z},+)$ can be seen as a set $\mathbb{Z}$ that is equipped with an *algebraic structure,* namely the *operation* $+$.

Abstract Algebra

The permutations of Rubik's Cube form a group, a fundamental concept within abstract algebra.

In algebra, which is a broad division of mathematics, abstract algebra (occasionally called modern algebra) is the study of algebraic structures. Algebraic structures include groups, rings, fields, modules, vector spaces, lattices, and algebras. The term *abstract algebra* was coined in the early 20th century to distinguish this area of study from the other parts of algebra.

Algebraic structures, with their associated homomorphisms, form mathematical categories. Category theory is a formalism that allows a unified way for expressing properties and constructions that are similar for various structures.

Universal algebra is a related subject that studies types of algebraic structures as single objects. For example, the structure of groups is a single object in universal algebra, which is called *variety of groups*.

History

As in other parts of mathematics, concrete problems and examples have played important roles in the development of abstract algebra. Through the end of the nineteenth century, many -- perhaps most -- of these problems were in some way related to the theory of algebraic equations. Major themes include:

- Solving of systems of linear equations, which led to linear algebra
- Attempts to find formulae for solutions of general polynomial equations of higher degree that resulted in discovery of groups as abstract manifestations of symmetry
- Arithmetical investigations of quadratic and higher degree forms and diophantine equations, that directly produced the notions of a ring and ideal.

Numerous textbooks in abstract algebra start with axiomatic definitions of various algebraic structures and then proceed to establish their properties. This creates a false impression that in algebra axioms had come first and then served as a motivation and as a basis of further study. The true order of historical development was almost exactly the opposite. For example, the hypercomplex numbers of the nineteenth century had kinematic and physical motivations but challenged comprehension. Most theories that are now recognized as parts of algebra started as collections of disparate facts from various branches of mathematics, acquired a common theme that served as a core around which various results were grouped, and finally became unified on a basis of a common set of concepts. An archetypical example of this progressive synthesis can be seen in the history of group theory.

Early Group Theory

There were several threads in the early development of group theory, in modern language loosely corresponding to *number theory*, *theory of equations*, and *geometry*.

Leonhard Euler considered algebraic operations on numbers modulo an integer, modular arithmetic, in his generalization of Fermat's little theorem. These investigations were taken much further by Carl Friedrich Gauss, who considered the structure of multiplicative groups of residues mod n and established many properties of cyclic and more general abelian groups that arise in this way. In his investigations of composition of binary quadratic forms, Gauss explicitly stated the associative law for the composition of forms, but like Euler before him, he seems to have been more interested in concrete results than in general theory. In 1870, Leopold Kronecker gave a definition of an abelian group in the context of ideal class groups of a number field, generalizing Gauss's work; but it appears he did not tie his definition with previous work on groups, particularly permutation groups. In 1882, considering the same question, Heinrich M. Weber realized the connection and gave a similar definition that involved the cancellation property but omitted the existence of the inverse element, which was sufficient in his context (finite groups).

Permutations were studied by Joseph-Louis Lagrange in his 1770 paper *Réflexions sur la résolution algébrique des équations (Thoughts on the algebraic solution of equations)* devoted to solutions of algebraic equations, in which he introduced Lagrange resolvents. Lagrange's goal was to understand why equations of third and fourth degree admit formulae for solutions, and he identified as key objects permutations of the roots. An important novel step taken by Lagrange in this paper was the abstract view of the roots, i.e. as symbols and not as numbers. However, he did not consider composition of permutations. Serendipitously, the first edition of Edward Waring's *Meditationes Algebraicae* (*Meditations on Algebra*) appeared in the same year, with an expanded version published in 1782. Waring proved the main theorem on symmetric functions, and specially considered the relation between the roots of a quartic equation and its resolvent cubic. *Mémoire sur la résolution des équations* (*Memoire on the Solving of Equations*) of Alexandre Vandermonde (1771) developed the theory of symmetric functions from a slightly different angle, but like Lagrange, with the goal of understanding solvability of algebraic equations.

> *Kronecker claimed in 1888 that the study of modern algebra began with this first paper of Vandermonde. Cauchy states quite clearly that Vandermonde had priority over Lagrange for this remarkable idea, which eventually led to the study of group theory.*

Paolo Ruffini was the first person to develop the theory of permutation groups, and like his predecessors, also in the context of solving algebraic equations. His goal was to establish the impossibility of an algebraic solution to a general algebraic equation of degree greater than four. En route to this goal he introduced the notion of the order of an element of a group, conjugacy, the cycle decomposition of elements of permutation groups and the notions of primitive and imprimitive and proved some important theorems relating these concepts, such as

> *if G is a subgroup of* S_5 *whose order is divisible by 5 then G contains an element of order 5.*

Note, however, that he got by without formalizing the concept of a group, or even of a permutation group. The next step was taken by Évariste Galois in 1832, although his work remained unpublished until 1846, when he considered for the first time what is now called the *closure property* of a group of permutations, which he expressed as

> ... if in such a group one has the substitutions S and T then one has the substitution ST.

The theory of permutation groups received further far-reaching development in the hands of Augustin Cauchy and Camille Jordan, both through introduction of new concepts and, primarily, a great wealth of results about special classes of permutation groups and even some general theorems. Among other things, Jordan defined a notion of isomorphism, still in the context of permutation groups and, incidentally, it was he who put the term *group* in wide use.

The abstract notion of a group appeared for the first time in Arthur Cayley's papers in 1854. Cayley realized that a group need not be a permutation group (or even *finite*), and may instead consist of matrices, whose algebraic properties, such as multiplication and inverses, he systematically investigated in succeeding years. Much later Cayley would revisit the question whether abstract groups were more general than permutation groups, and establish that, in fact, any group is isomorphic to a group of permutations.

Modern Algebra

The end of the 19th and the beginning of the 20th century saw a tremendous shift in the methodology of mathematics. Abstract algebra emerged around the start of the 20th century, under the name *modern algebra*. Its study was part of the drive for more intellectual rigor in mathematics. Initially, the assumptions in classical algebra, on which the whole of mathematics (and major parts of the natural sciences) depend, took the form of axiomatic systems. No longer satisfied with establishing properties of concrete objects, mathematicians started to turn their attention to general theory. Formal definitions of certain algebraic structures began to emerge in the 19th century. For example, results about various groups of permutations came to be seen as instances of general theorems that concern a general notion of an *abstract group*. Questions of structure and classification of various mathematical objects came to forefront.

These processes were occurring throughout all of mathematics, but became especially pronounced in algebra. Formal definition through primitive operations and axioms were proposed for many basic algebraic structures, such as groups, rings, and fields. Hence such things as group theory and ring theory took their places in pure mathematics. The algebraic investigations of general fields by Ernst Steinitz and of commutative and then general rings by David Hilbert, Emil Artin and Emmy Noether, building up on the work of Ernst Kummer, Leopold Kronecker and Richard Dedekind, who had considered ideals in commutative rings, and of Georg Frobenius and Issai Schur, concerning representation theory of groups, came to define abstract algebra. These developments of the last quarter of the 19th century and the first quarter of 20th century were systematically exposed in Bartel van der Waerden's *Moderne algebra*, the two-volume monograph published in 1930–1931 that forever changed for the mathematical world the meaning of the word *algebra* from *the theory of equations* to the *theory of algebraic structures*.

Basic Concepts

By abstracting away various amounts of detail, mathematicians have defined various algebraic structures that are used in many areas of mathematics. For instance, almost all systems studied are sets, to which the theorems of set theory apply. Those sets that have a certain binary operation defined on them form magmas, to which the concepts concerning magmas, as well those concerning sets, apply. We can add additional constraints on the algebraic structure, such as associativity (to form semigroups); identity, and inverses (to form groups); and other more complex structures. With additional structure, more theorems could be proved, but the generality is reduced. The "hierarchy" of algebraic objects (in terms of generality) creates a hierarchy of the corresponding theories: for instance, the theorems of group theory may be used when studying rings (algebraic objects that have two binary operations with certain axioms) since a ring is a group over one of its operations. In general there is a balance between the amount of generality and the richness of the theory: more general structures have usually less nontrivial theorems and less applications.

Examples of algebraic structures with a single binary operation are:

- Magmas
- Quasigroups
- Monoids

- Semigroups
- Groups

Examples involving several operations include:

- Rings
- Fields
- Modules
- Vector spaces
- Algebras over fields
- Associative algebras
- Lie algebras
- Lattices
- Boolean algebras

Applications

Because of its generality, abstract algebra is used in many fields of mathematics and science. For instance, algebraic topology uses algebraic objects to study topologies. The Poincaré conjecture, proved in 2003, asserts that the fundamental group of a manifold, which encodes information about connectedness, can be used to determine whether a manifold is a sphere or not. Algebraic number theory studies various number rings that generalize the set of integers. Using tools of algebraic number theory, Andrew Wiles proved Fermat's Last Theorem.

In physics, groups are used to represent symmetry operations, and the usage of group theory could simplify differential equations. In gauge theory, the requirement of local symmetry can be used to deduce the equations describing a system. The groups that describe those symmetries are Lie groups, and the study of Lie groups and Lie algebras reveals much about the physical system; for instance, the number of force carriers in a theory is equal to dimension of the Lie algebra, and these bosons interact with the force they mediate if the Lie algebra is nonabelian.

Algebraic Integer

In Algebraic number theory, an algebraic integer is a complex number that is a root of some monic polynomial (a polynomial whose leading coefficient is 1) with coefficients in $\mathbb{Z}$ (the set of integers). The set of all algebraic integers, A, is closed under addition and multiplication and therefore is a commutative subring of the complex numbers. The ring A is the integral closure of regular integers $\mathbb{Z}$ in complex numbers.

The ring of integers of a number field K, denoted by O_K, is the intersection of K and A: it can also be characterised as the maximal order of the field K. Each algebraic integer belongs to the ring of integers of some number field. A number x is an algebraic integer if and only if the ring $\mathbb{Z}[x]$ is finitely generated as an abelian group, which is to say, as a $\mathbb{Z}$-module.

Definitions

The following are equivalent definitions of an algebraic integer. Let K be a number field (i.e., a finite extension of $\mathbb{Q}$, the set of rational numbers), in other words, $K = \mathbb{Q}(\theta)$ for some algebraic number $\theta \in \mathbb{C}$ by the primitive element theorem.

- $\alpha \in K$ is an algebraic integer if there exists a monic polynomial $f(x) \in \mathbb{Z}[x]$ such that $f(\alpha) = 0$.
- $\alpha \in K$ is an algebraic integer if the minimal monic polynomial of α over $\mathbb{Q}$ is in $\mathbb{Z}[x]$.
- $\alpha \in K$ is an algebraic integer if $\mathbb{Z}[\alpha]$ is a finitely generated $\mathbb{Z}$-module.
- $\alpha \in K$ is an algebraic integer if there exists a finitely generated $\mathbb{Z}$-submodule $M \subset \mathbb{C}$ such that $\alpha M \subseteq M$.

Algebraic integers are a special case of integral elements of a ring extension. In particular, an algebraic integer is an integral element of a finite extension $K / \mathbb{Q}$.

Examples

- The only algebraic integers which are found in the set of rational numbers are the integers. In other words, the intersection of $\mathbb{Q}$ and A is exactly $\mathbb{Z}$. The rational number a/b is not an algebraic integer unless b divides a. Note that the leading coefficient of the polynomial $bx - a$ is the integer b. As another special case, the square root $\sqrt{n}$ of a non-negative integer n is an algebraic integer, and so is irrational unless n is a perfect square.
- If d is a square free integer then the extension $K = \mathbb{Q}(\sqrt{d})$ is a quadratic field of rational numbers. The ring of algebraic integers O_K contains $\sqrt{d}$ since this is a root of the monic polynomial $x^2 - d$. Moreover, if $d \equiv 1 \pmod 4$ the element $(1 + \sqrt{d})/2$ is also an algebraic integer. It satisfies the polynomial $x^2 - x + (1 - d)/4$ where the constant term $(1 - d)/4$ is an integer. The full ring of integers is generated by $\sqrt{d}$ or $(1 + \sqrt{d})/2$ respectively.
- The ring of integers of the field $F = \mathbb{Q}[\alpha], \alpha = \sqrt[3]{m}$ has the following integral basis, writing $m = hk^2$ for two square-free coprime integers h and k:

$$\begin{cases} 1, \alpha, \dfrac{\alpha^2 \pm k^2\alpha + k^2}{3k} & m \equiv \pm 1 \mod 9 \\ 1, \alpha, \dfrac{\alpha^2}{k} & \text{else} \end{cases}$$

- If ζ_n is a primitive n-th root of unity, then the ring of integers of the cyclotomic field $\mathbb{Q}(\zeta_n)$ is precisely $\mathbb{Z}[\zeta_n]$.
- If α is an algebraic integer then $\beta = \sqrt[n]{\alpha}$ is another algebraic integer. A polynomial for β is obtained by substituting x^n in the polynomial for α.

Non-example

- If $P(x)$ is a primitive polynomial which has integer coefficients but is not monic, and P is irreducible over $\mathbb{Q}$, then none of the roots of P are algebraic integers. (Here *primitive* is used in the sense that the highest common factor of the set of coefficients of P is 1; this is weaker than requiring the coefficients to be pairwise relatively prime.)

- is a Bézout domain, as a consequence of the principal ideal theorem.

Baker's Theorem

In transcendental number theory, a mathematical discipline, Baker's theorem gives a lower bound for the absolute value of linear combinations of logarithms of algebraic numbers. The result, proved by Alan Baker (1966, 1967a, 1967b), subsumed many earlier results in transcendental number theory and solved a problem posed by Alexander Gelfond nearly fifteen years earlier. Baker used this to prove the transcendence of many numbers, to derive effective bounds for the solutions of some Diophantine equations, and to solve the class number problem of finding all imaginary quadratic fields with class number 1.

History

To simplify notation, we introduce the set L of logarithms of nonzero algebraic numbers, that is

$$L = \left\{\lambda \in \mathbf{C} : e^{\lambda} \in \overline{\mathbf{Q}}\right\}.$$

Using this notation, several results in transcendental number theory become much easier to state. For example the Hermite–Lindemann theorem becomes the statement that any nonzero element of L is transcendental.

In 1934, Alexander Gelfond and Theodor Schneider independently proved the Gelfond–Schneider theorem. This result is usually stated as: if a is algebraic and not equal to 0 or 1, and if b is algebraic and irrational, then a^b is transcendental. Equivalently, though, it says that if λ_1 and λ_2 are elements of L that are linearly independent over the rational numbers, then they are linearly independent over the algebraic numbers. So if λ_1 and λ_2 are elements of L and λ_2 isn't zero, then the quotient λ_1/λ_2 is either a rational number or transcendental. It can't be an algebraic irrational number like $\sqrt{2}$.

Although proving this result of "rational linear independence implies algebraic linear independence" for two elements of L was sufficient for his and Schneider's result, Gelfond felt that it was crucial to extend this result to arbitrarily many elements of L. Indeed, from Gel'fond (1960, p. 177):

...one may assume ... that the most pressing problem in the theory of transcendental numbers is the investigation of the measures of transcendence of finite sets of logarithms of algebraic numbers.

This problem was solved fourteen years later by Alan Baker and has since had numerous applications not only to transcendence theory but in algebraic number theory and the study of Diophantine equations as well. Baker received the Fields medal in 1970 for both this work and his applications of it to Diophantine equations.

Statement

With the above notation, Baker's theorem is a nonhomogeneous generalisation of the Gelfond–Schneider theorem. Specifically it states:

Baker's Theorem. If $\lambda_1,\ldots,\lambda_n$ are elements of L that are linearly independent over the rational numbers, then for any algebraic numbers $\beta_0, \ldots, \beta_n$, not all zero, we have

$$|\beta_0 + \beta_1\lambda_1 + \cdots + \beta_n\lambda_n| > H^{-C}$$

where H is the maximum of the heights of the β's and C is an effectively computable number depending on n, the λ's, and the maximum d of the degrees of the β's. (If β_0 is nonzero then the assumption that the λ's are linearly independent can be dropped.) In particular this number is nonzero, so 1 and the λ's are linearly independent over the algebraic numbers.

Just as the Gelfond–Schneider theorem is equivalent to the statement about the transcendence of numbers of the form a^b, so too Baker's theorem implies the transcendence of numbers of the form

$$a_1^{b_1}\cdots a_n^{b_n},$$

where the b_i are all algebraic, irrational, and 1, $b_1,\ldots,b_n$ are linearly independent over the rationals, and the a_i are all algebraic and not 0 or 1.

Baker (1977) also gave several versions with explicit constants. For example, if $e^{\lambda_j} = \alpha_j$ has height at most $A_j \geq 4$ and all the numbers β_j have height at most $B \geq 4$ then the linear form

$$\Lambda = \beta_0 + \beta_1\lambda_1 + \cdots + \beta_n\lambda_n$$

is either 0 or satisfies

$$\log|\Lambda| > (16nd)^{200n}\Omega(\log\Omega - \log\log A_n)(\log B + \log\Omega)$$

where

$$\Omega = \log A_1 \log A_2 \cdots \log A_n$$

and the field generated by all the α's and β's over the rationals has degree at most d. In the special case when $\beta_0=0$ and all the β_j are rational integers, the rightmost term log Ω can be deleted.

An explicit result by Baker and Wüstholz for a linear form Λ with integer coefficients yields a lower bound of the form

$$\log|\Lambda| > -Ch(\alpha_1)h(\alpha_2)\cdots h(\alpha_n)\log\left(\max\{|\beta_1|,\ldots,|\beta_n|\}\right),$$

with a constant C

$$C = 18(n+1)! \cdot n^{n+1} \cdot (32d)^{n+2} \log(2nd),$$

where d is the degree of the number field generated by the α_i.

Baker's Method

Baker's proof of his theorem is an extension of the argument given by Gel'fond (1960, chapter III, section 4). The main ideas of the proof are illustrated by the proof of the following qualitative version of the theorem of Baker (1966) described by Serre (1971): if the numbers $2\pi i$ and $\log a_1, \ldots, \log a_n$ are linearly independent over the rational numbers, for nonzero algebraic numbers $a_1, \ldots, a_n$, then they are linearly independent over the algebraic numbers. The precise quantitative version of Bakers theory can be proved by replacing the conditions that things are zero by conditions that things are sufficiently small throughout the proof.

The main idea of Bakers proof is to construct an auxiliary function $\Phi(z_1,\ldots,z_{n-1})$ of several variables that vanishes to high order at many points of the form $\Phi(l,l,\ldots,l)$, then repeatedly show that it vanishes to lower order at even more points of this form. Finally the fact that it vanishes (to order 1) at enough points of this form implies using Vandermonde determinants that there is a multiplicative relation between the numbers a_i.

Construction of the Function Φ

Assume there is a relation

$$\beta_1 \log \alpha_1 + \cdots + \beta_{n-1} \log \alpha_{n-1} = \log \alpha_n$$

for algebraic numbers $\alpha_1, \ldots, \alpha_n, \beta_1, \ldots, \beta_{n-1}$ The function Φ is of the form

$$\Phi(z_1,\ldots,z_{n-1}) = \sum_{\lambda_1=0}^{L} \cdots \sum_{\lambda_n=0}^{L} p(\lambda_1 \ldots, \lambda_n) \alpha_1^{(\lambda_1+\lambda_n\beta_1)z_1} \cdots \alpha_{n-1}^{(\lambda_{n-1}+\lambda_n\beta_{n-1})z_{n-1}}$$

The integer coefficients p are chosen so that they are not all zero and Φ and its derivatives of order at most some constant M vanish at $z_1 = \ldots = z_{n-1} = l$, for integers l with $0 \le l \le h$ for some constant h. This is possible because these conditions are homogeneous linear equations in the coefficients p, which have a non-zero solution provided the number of unknown variables p is larger than the number of equations. The linear relation between the logs of the α's is needed to cut down the number of linear equations that have to be satisfied. Moreover, using Siegel's lemma, the sizes of the coefficients p can be chosen to be not too large. The constants L, h, and M have to be carefully adjusted to that the next part of the proof works, and are subject to some constraints, which are roughly:

- L must be somewhat smaller than M to make the argument about extra zeros below work.
- A small power of h must be larger than L to make the final step of the proof work.
- L^n must be larger than about $M^{n-1}h$ in order that it is possible to solve for the coefficients p.

The constraints can be satisfied by taking h to be sufficiently large, M to be some fixed power of h, and L to be a slightly smaller power of h. Baker took M to be about h^2 and L to be about $h^{2-1/2n}$.

The linear relation between the logarithms of the α's is used to reduce L slightly; roughly speaking, without it the condition L^n must be larger than about $M^{n-1}h$ would become L^n must be larger than about $M^n h$, which is incompatible with the condition that L is somewhat smaller than M.

Zeros of Φ(l, ..., l)

The next step is to show that Φ vanishes to slightly smaller order at many more points of the form $z_1 = ... = z_{n-1} = l$ for integers l. This idea was Baker's key innovation: previous work on this problem involved trying to increase the number of derivatives that vanish while keeping the number of points fixed, which does not seem to work in the multivariable case. This is done by combining two ideas; First one shows that the derivatives at these points are quite small, by using the fact that many derivatives of Φ vanish at many nearby points. Then one shows that derivatives of Φ at this point are given by algebraic integers times known constants. If an algebraic integer has all its conjugates bounded by a known constant, then it cannot be too small unless it is zero, because the product of all conjugates of a nonzero algebraic integer is at least 1 in absolute value. Combining these two ideas implies that Φ vanishes to slightly smaller order at many more points $z_1 = ... = z_{n-1} = l$. This part of the argument requires that Φ does not increase too rapidly; the growth of Φ depends on the size of L, so requires a bound on the size of L, which turns out to be roughly that L must be somewhat smaller than M. More precisely, Baker showed that since Φ vanishes to order M at h consecutive integers, it also vanishes to order $M/2$ at $h^{1+1/8n}$ consecutive integers 1, 2, 3, ... Repeating this argument J times shows that Φ vanishes to order $M/2^J$ at $h^{1+J/8n}$ points, provided that h is sufficiently large and L is somewhat smaller than $M/2^J$.

One then takes J large enough that $h^{1+J/8n} > (L+1)^n$ (J larger than about $16n$ will do if $h^2 > L$) so that that $\Phi(l, ..., l) = 0$ for all integers l with $1 \leq l \leq (L+1)^n$.

Completion of the Proof

The condition that $\Phi(l, ..., l)=0$ for all integers l with $1 \leq l \leq (L+1)^n$ can be written as

$$\sum_{\lambda_1=0}^{L} \cdots \sum_{\lambda_n=0}^{L} p(\lambda_1 \ldots, \lambda_n)\alpha_1^{\lambda_1 l} \cdots \alpha_n^{\lambda_n l} = 0$$

This consists of $(L+1)^n$ homogeneous linear equations in the $(L+1)^n$ unknowns p, and by assumption has a non-zero solution p, so the determinant of the matrix of coefficients must vanish. However this matrix is a Vandermonde matrix, so the formula for the determinant of such a matrix forces an equality two of the values

$$\alpha_1^{\lambda_1} \cdots \alpha_n^{\lambda_n}$$

so the numbers $\alpha_1, ..., \alpha_n$ are multiplicatively dependent. Taking logs then shows that $2\pi i$, log $\alpha_1, ...,$ log α_n are linearly dependent over the rationals.

Extensions and Generalizations

Baker (1966) in fact gave a quantitative version of the theorem, giving effective lower bounds for the linear form in logarithms. This is done by a similar argument, except statements about something being zero are replaced by statements giving a small upper bound for it, and so on.

Baker (1967a) showed how to eliminate the assumption about $2\pi i$ in the theorem. This requires a modification of the final step of the proof. One shows that many derivatives of the function $\varphi(z) = \Phi(z, ..., z)$ vanish at $z=0$, by an argument similar to the one above. But these equations for the first $(L+1)^n$ derivatives again give a homogeneous set of linear equations for the coefficients p, so the determinant is zero, and is again a Vandermonde determinant, this time for the numbers $\lambda_1 \log \alpha_1 + ... + \lambda_n \log \alpha_n$. So two of these expressions must be the same which shows that $\log \alpha_1,...,\log \alpha_n$ are linearly dependent over the rationals.

Baker (1967b) gave an inhomogeneous version of the theorem, showing that $\beta_0+\beta_1\log(\alpha_1) + ... + \beta_n \log(\alpha_n)$ is nonzero for nonzero algebraic numbers $\beta_0, ..., \beta_n, \alpha_1, ..., \alpha_n$, and moreover giving an effective lower bound for it. The proof is similar to the homogeneous case: one can assume that

$$\beta_0 + \beta_1 \log \alpha_1 + \cdots + \beta_{n-1} \log \alpha_{n-1} = \log \alpha_n$$

and one inserts an extra variable z_0 into Φ as follows:

$$\Phi(z_0,\ldots,z_{n-1}) = \sum_{\lambda_0=0}^{L} \cdots \sum_{\lambda_n=0}^{L} p(\lambda_0 \ldots, \lambda_n) z_0^{\lambda_0} e^{\lambda_n \beta_0 z_0} \alpha_1^{(\lambda_1+\lambda_n\beta_1)z_1} \cdots \alpha_{n-1}^{(\lambda_{n-1}+\lambda_n\beta_{n-1})z_{n-1}}$$

Corollaries

As mentioned above, the theorem includes numerous earlier transcendence results concerning the exponential function, such as the Hermite–Lindemann theorem and Gelfond–Schneider theorem. It is not quite as encompassing as the still unproven Schanuel's conjecture, and does not imply the six exponentials theorem nor, clearly, the still open four exponentials conjecture.

The main reason Gelfond desired an extension of his result was not just for a slew of new transcendental numbers. In 1935 he used the tools he had developed to prove the Gelfond–Schneider theorem to derive a lower bound for the quantity

$$|\beta_1\lambda_1 + \beta_2\lambda_2|$$

where β_1 and β_2 are algebraic and λ_1 and λ_2 are in L. Baker's proof gave lower bounds for quantities like the above but with arbitrarily many terms, and he could use these bounds to develop effective means of tackling Diophantine equations and to solve Gauss' class number problem.

Extensions

Baker's theorem grants us the linear independence over the algebraic numbers of logarithms of algebraic numbers. This is weaker than proving their *algebraic* independence. So far no progress has been made on this problem at all. It has been conjectured that if $\lambda_1,...,\lambda_n$ are elements of L that are linearly independent over the rational numbers, then they are algebraically independent too.

This is a special case of Schanuel's conjecture, but so far it remains to be proved that there even exist two algebraic numbers whose logarithms are algebraically independent. Indeed, Baker's theorem rules out linear relations between logarithms of algebraic numbers unless there are trivial reasons for them; the next most simple case, that of ruling out homogeneous quadratic relations, is the still open four exponentials conjecture.

Similarly, extending the result to algebraic independence but in the p-adic setting, and using the p-adic logarithm function, remains an open problem. It is known that proving algebraic independence of linearly independent p-adic logarithms of algebraic p-adic numbers would prove Leopoldt's conjecture on the p-adic ranks of units of a number field.

Chinese Remainder Theorem

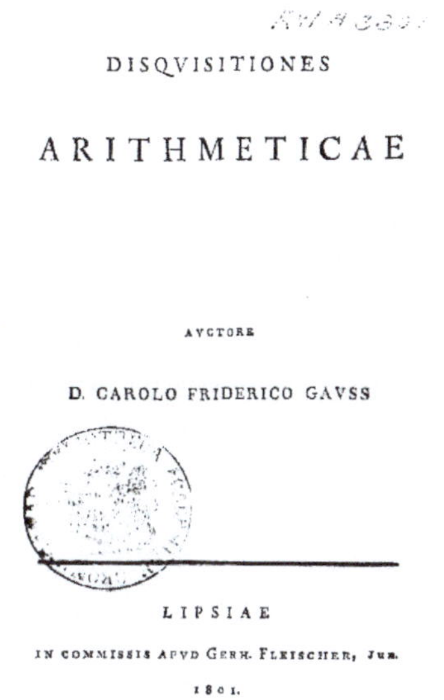
DISQVISITIONES

ARITHMETICAE

AVCTORE

D. CAROLO FRIDERICO GAVSS

LIPSIAE

IN COMMISSIS APVD GERH. FLEISCHER, Jun.

1801.

The Chinese remainder theorem appears in Gauss's 1801 book *Disquisitiones Arithmeticae*.

The Chinese remainder theorem is a theorem of number theory, which states that, if one knows the remainders of the division of an integer n by several integers, then one can determine uniquely the remainder of the division of n by the product of these integers, under the condition that the divisors are pairwise coprime.

This theorem has this name because it is a theorem about *remainders*, which was first discovered in the 3rd century AD by the Chinese mathematician Sunzi.

The Chinese remainder theorem is widely used for computing with large integers, as it allows replacing a computation for which one knows a bound on the size of the result by several similar computations on small integers.

The Chinese remainder theorem (expressed in terms of congruences) is true over every principal ideal domain. It has been generalized to any commutative ring, with a formulation involving ideals.

History

The earliest known statement of the theorem, as a problem with specific numbers, appears in the 3rd-century book *Sunzi's Mathematical Classic* by the Chinese mathematician Sun Tzu:

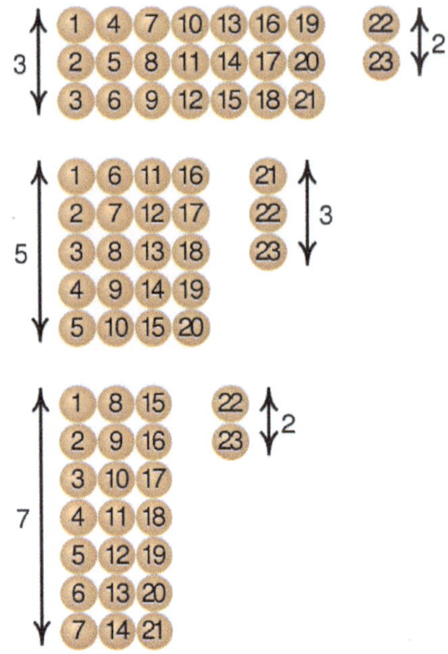

Sun Tzu's original formulation: $x \equiv 2 \pmod 3 \equiv 3 \pmod 5 \equiv 2 \pmod 7$ with the solution $x = 23 + 105k$ where $k \in \mathrm{Z}$

Sun Tzu's work contains neither a proof nor a full algorithm. What amounts to an algorithm for solving this problem was described by Aryabhata (6th century). Special cases of the Chinese remainder theorem were also known to Brahmagupta (7th century), and appear in Fibonacci's Liber Abaci (1202). The result was later generalized with a complete solution called *Dayanshu* (大衍術) in Qin Jiushao's 1247 *Mathematical Treatise in Nine Sections* (數書九章, *Shushu Jiuzhang*).

The notion of congruences was first introduced and used by Gauss in his *Disquisitiones Arithmeticae* of 1801. Gauss illustrates the Chinese remainder theorem on a problem involving calendars, namely, "to find the years that have a certain period number with respect to the solar and lunar cycle and the Roman indiction." Gauss introduces a procedure for solving the problem that had already been used by Euler but was in fact an ancient method that had appeared several times.

Theorem Statement

Let $n_1, \ldots, n_k$ be integers greater than 1, which are often called *moduli* or *divisors*. Let us denote by N the product of the n_i.

The Chinese remainder theorem asserts that if the n_i are pairwise coprime, and if $a_1, \ldots, a_k$ are integers such that $0 \le a_i < n_i$ for every i, then there is one and only one integer x, such that $0 \le x < N$ and the remainder of the Euclidean division of x by n_i is a_i for every i.

This may be restated as follows in term of congruences: If the n_i are pairwise coprime, and if $a_1, \ldots,$

a_k are any integers, then there exists an integer x such that

$$\begin{aligned} x &\equiv a_1 \pmod{n_1} \\ &\vdots \\ x &\equiv a_k \pmod{n_k}, \end{aligned}$$

and any two such x are congruent modulo N.

In abstract algebra, the theorem is often restated as: if the n_i are pairwise coprime, the map

$$x \bmod N \quad \mapsto \quad (x \bmod n_1, \ldots, x \bmod n_k)$$

defines a ring isomorphism

$$\mathbb{Z}/N\mathbb{Z} \cong \mathbb{Z}/n_1\mathbb{Z} \times \cdots \times \mathbb{Z}/n_k\mathbb{Z}$$

between the ring of integers modulo N and the direct product of the rings of integers modulo the n_i. This means that for doing a sequence of arithmetic operations in $\mathbb{Z}/N\mathbb{Z}$, one may do the same computation independently in each $\mathbb{Z}/n_i\mathbb{Z}$ and then get the result by applying the isomorphism (from the right to the left). This may be much faster than the direct computation if N and the number of operations are large. This is widely used, under the name *multi-modular computation*, for linear algebra over the integers or the rational numbers.

The theorem can also be restated in the language of combinatorics as the fact that the infinite arithmetic progressions of integers form a Helly family.

Proof

The existence and the uniqueness of the solution may be proved independently. However the first proof of existence, given below, uses the uniqueness.

Uniqueness

Suppose that x and y are both solutions to all the congruences. As x and y give the same remainder, when divided by n_i, their difference $x - y$ is a multiple of each n_i. As the n_i are pairwise coprime, their product N divides also $x - y$, and thus x and y are congruent modulo N. If x and y are supposed to be non negative and less than N (as in the first statement of the theorem), then their difference may be a multiple of N only if $x = y$.

Existence (First Proof)

The map

$$x \mapsto (x \bmod n_1, \ldots, x \bmod n_k)$$

maps congruence classes modulo N to sequences of congruence classes modulo n_i. The proof of uniqueness shows that this map is injective. As the domain and the codomain of this map have the same number of elements, the map is also surjective, which proves the existence of the solution.

This proof is very simple, but does not provide any direct way for computing a solution. Moreover, it cannot be generalized to other situations where the following proof can.

Existence (Constructive Proof)

Existence may be established by an explicit construction of x. This construction may be split in two steps, firstly by solving the problem in the case of two moduli, and the second one by extending this solution to the general case by induction on the number of moduli.

Case of Two Moduli

We want to solve the system

$$\begin{aligned} x &\equiv a_1 \pmod{n_1} \\ x &\equiv a_2 \pmod{n_2}, \end{aligned}$$

where n_1 and n_2 are coprime.

Bézout's identity asserts the existence of two integers m_1 and m_2 such that

$$m_1n_1 + m_2n_2 = 1.$$

The integers m_1 and m_2 may be computed by Extended Euclidean algorithm.

A solution is given by

$$x = a_1m_2n_2 + a_2m_1n_1.$$

In fact

$$\begin{aligned} x &= a_1m_2n_2 + a_2m_1n_1 \\ &= a_1m_2n_2 + a_2m_1n_1 + (a_1m_1n_1 - a_1m_1n_1) \\ &= a_1(m_1n_1 + m_2n_2) - a_1m_1n_1 + a_2m_1n_1 \\ &= a_1 + n_1m_1(a_2 - a_1). \end{aligned}$$

This shows that $x \equiv a_1 \pmod{n_1}$. The second congruence is proved similarly, by exchanging the indices.

General Case

Let us consider a sequence of congruence equations

$$\begin{aligned} x &\equiv a_1 \pmod{n_1} \\ &\vdots \\ x &\equiv a_k \pmod{n_k}, \end{aligned}$$

where the n_i are pairwise coprime. The two first equations have a solution $a_{1,2}$ provided by the method of the previous section. The set of the solutions of these two first equations is the set of all solutions of the equation

$$x \equiv a_{1,2} \pmod{n_1 n_2}.$$

As the other n_i are coprime with $n_1 n_2$, this reduces solving the initial problem of k equations to a similar problem with $k-1$ equations. Iterating the process, one gets eventually the solutions of the initial problem

Existence (Direct Construction)

For constructing a solution, it is not necessary to make an induction on the number of moduli. However, such a direct construction involves more computation with large numbers, which makes it less efficient and less used. Nevertheless, Lagrange interpolation is a special case of this construction, applied to polynomials instead of integers.

Let $N_i = N / n_i$ be the product of all moduli but one. As the n_i are pairwise coprime, N_i and n_i are coprime. Thus Bézout's identity applies, and there exist integers M_i and m_i such that

$$M_i N_i + m_i n_i = 1.$$

A solution of the system of congruences is

$$x = \sum_{i=1}^{k} a_i M_i N_i.$$

In fact, as N_i is a multiple of n_j for $i \neq j$, we have

$$x \equiv a_i M_i N_i \equiv a_i (M_i N_i + m_i n_i) - a_i m_i n_i \equiv a_i \pmod{n_i},$$

for every i.

Computation

Let us consider a system of congruences

$$\begin{aligned} x &\equiv a_1 \pmod{n_1} \\ &\vdots \\ x &\equiv a_k \pmod{n_k}, \end{aligned}$$

where the n_i are pairwise coprime, and let $N = n_1, \ldots, n_k$. In this section several methods are described for computing the unique solution for x, such that $0 \leq x < N$, and these methods are applied on the example

$$x \equiv 0 \pmod{3}$$
$$x \equiv 3 \pmod{4}$$
$$x \equiv 4 \pmod{5}.$$

Systematic Search

It is easy to check whether a value of x is a solution: it suffices to compute the remainder of the Euclidean division of x by each n_i. Thus, to find the solution, it suffices to check successively the integers from 0 to N until finding the solution.

Although very simple this method is very inefficient: for the simple example considered here, 40 integers (including 0 have to be checked for finding the solution 40. This is an exponential time algorithm, as the size of the input is, up to a constant factor, the number of digits of N, and the average number of operations is of the order of N.

Therefore, this method is rarely used, for hand-written computation as well on computers.

Search by Sieving

The search of the solution may be made dramatically faster by sieving. For this method, we suppose, without loss of generality, that $0 \leq a_i < n_i$ (if it were not the case, it would suffice to replace each a_i by the remainder of its division by n_i). This implies that the solution belongs to the arithmetic progression

$$a_1, a_1 + n_1, a_1 + 2n_1, \ldots$$

By testing the values of these numbers modulo n_2, one eventually finds a solution x_2 of the two first congruences. Then the solution belong to the arithmetic progression

$$x_2, x_2 + n_1 n_2, x_2 + 2n_1 n_2, \ldots$$

Testing the values of these numbers modulo n_3, , and continuing until every modulus has been tested gives eventually the solution.

This method is faster if the moduli have been ordered by decreasing value, that is if $n_1 > n_2 > \cdots > n_k$. For the example, this gives the following computation. We consider first the numbers that are congruent to 4 modulo 5 (the largest modulus), which are 4, 9 = 4 + 5, 14 = 9 + 5, ... For each of them, compute the remainder by 4 (the second largest modulus) until getting a number congruent to 3 modulo 4. Then one can proceed by adding 20 = 5×4 at each step, and computing only the remainders by 3. This gives

4 mod 4 → 0. Continue

4 + 5 = 9 mod 4 →1. Continue

9 + 5 = 14 mod 4 → 2. Continue

14 + 5 = 19 mod 4 → 3. OK, continue by considering remainders modulo 3 and adding 5×4

= 20 each time

19 mod 3 → 1. Continue

19 + 20 = 39 mod 3 → 0. OK, this is the result.

This method works well for hand-written computation with a product of moduli that is not too big. However it is much slower than other methods, for very large products of moduli. Although dramatically faster than the systematic search, this method has also an exponential time complexity, and is therefore not used on computers.

Using the Existence Construction

The constructive existence proof shows that, in the case of two moduli, the solution may be obtained by the computation of the Bézout coefficients of the moduli, followed by a few multiplications, additions and reductions modulo n_1n_2 (for getting a result in the interval $(0, n_1n_2 - 1)$. As the Bézout's coefficients may be computed with the extended Euclidean algorithm, the whole computation, at most, has a quadratic time complexity of $O((s_1 + s_2)^2)$, where s_i denotes the number of digits of n_i.

For more than two moduli, the method for two moduli allows the replacement of any two congruences by a single congruence modulo the product of the moduli. Iterating this process provides eventually the solution with a complexity, which is quadratic in the number of digits of the product of all moduli. This quadratic time complexity does not depend on the order in which the moduli are regrouped. One may regroup the two first moduli, then regrouping the resulting modulus with the next one, and so on. This is the strategy which is the easiest to implement, but it needs more computation involving large numbers.

Another strategy consists in partitioning the moduli in pairs whose product have comparable sizes (as much as possible), applying parallely the method of two moduli to each pair, and iterating with a number of moduli approximatively divided by two. This method allows an easy parallelization of the algorithm. Also, if fast algorithms (that is algorithms working in quasilinear time) are used for the basic operations, this method provides an algorithm for the whole computation that works in quasilinear time.

On the current example (which has only three moduli), both strategies are identical, and works as follows.

Bézout's identity for 3 and 4 is

$$1 \times 4 + (-1) \times 3 = 1.$$

Putting this in the formula given for proving the existence gives

$$0 \times 4 + -3 \times 3 = -9$$

for a solution of the two first congruences, the other solutions being obtained by adding to −9 any multiple of 3×4 = 12. One may continue with any of these solutions, but the solution 3 = −9 +12 is smaller (in absolute value) and thus leads probably to an easier computation

Bézout identity for 5 and 3×4 = 12 is

$$5\times 5+(-2)\times 12=1.$$

Applying the same formula again, we get a solution of the problem:

$$25\times 3-24\times 4=-21.$$

The other solutions are obtained by adding any multiple of 3×4×5 = 60, and the smallest positive solution is −21 + 60 = 39.

As a Linear Diophantine System

The system of congruences solved by the Chinese remainder theorem may be rewritten as a system of simultaneous linear Diophantine equations

$$\begin{aligned} x &= a_1 + x_1 n_1 \\ &\vdots \\ x &= a_k + x_k n_k, \end{aligned}$$

where the unknown integers are x and the x_i. Therefore, every general method for solving such systems may be used for finding the solution of Chinese remainder theorem, such as the reduction of the matrix of the system to Smith normal form or Hermite normal form. However, as usual when using a general algorithm for a more specific problem, this approach is less efficient than the method of the preceding section, based on a direct use of Bézout's identity.

Over Principal Ideal Domains

In § Theorem statement, the Chinese remainder theorem has been stated in three different ways: in terms of remainders, of congruences and of a ring isomorphism. The statement in terms of remainders does not apply, in general to principal ideal domains, as remainders are not defined in such rings. However, the two other version make sense over a principal ideal domain *R*: it suffices to replace "integer" by "element of the domain", and $\mathbb{Z}$ by *R*. These two versions of the theorem are true in this context, because the proofs (except for the first existence proof), are based on Euclid's lemma and Bézout's identity, which are true over every principal domain.

However, in general, the theorem is only an existence theorem, and does not provide any way for computing the solution, unless if one has an algorithm for computing the coefficients of Bézout's identity.

Over Univariate Polynomial Rings and Euclidean Domains

The statement in terms of remainders given in § Theorem statement cannot be generalized to any principal ideal domain, but its generalization to Euclidean domains is straightforward. The univariate polynomials over a field is the typical example of a Euclidean domain, which is not the integers. therefore, we state the theorem for the case of a ring of univariate domain $R=K[X]$ over a field K. For getting the theorem for a general Euclidean domain, it suffices to replace the degree by the Euclidean function of the Euclidean domain.

The Chinese remainder theorem for polynomials is thus: Let $P_i(X)$ (the moduli) be, for i=1, ..., k, pairwise coprime polynomials in $R = K[X]$. Let $d_i = \deg P_i$ be the degree of $P_i(X)$, and D be the sum of the d_i. If $A_i(X), \ldots, A_k(X)$ are polynomials such that $A_i(X) = 0$ or $\deg A_i < d_i$ for every i, then, there is one and only one polynomial $P(X)$, such that $\deg P < D$ and the remainder of the Euclidean division of $P(X)$ by $P_i(X)$ is $A_i(X)$ for every i.

The construction of the solution may be done as in § Existence (constructive proof) or § Existence (direct proof). However, the latter construction may be simplified by using, as follows, partial fraction decomposition instead of extended Euclidean algorithm.

Thus, we want to find a polynomial $P(X)$, which satisfies the congruences

$$P(X) \equiv A_i(X) \pmod{P_i(X)},$$

for $i = 1, \ldots, k$.

Let us consider the polynomials

$$Q(X) = \prod_{i=1}^{k} P_i(X)$$

$$Q_i(X) = \frac{Q(X)}{P_i(X)}.$$

The partial fraction decomposition of $1/Q(X)$ gives k polynomials $S_i(X)$ with degrees $\deg S_i(X) < d_i$, such that

$$\frac{1}{Q(X)} = \sum_{i=1}^{k} \frac{S_i(X)}{P_i(X)},$$

and thus

$$1 = \sum_{i=1}^{k} S_i(X) Q_i(X).$$

Then a solution of the simultaneous congruence system is given by the polynomial

$$\sum_{i=1}^{k} A_i(X) S_i(X) Q_i(X).$$

In fact, we have

$$\sum_{i=1}^{k} A_i(X) S_i(X) Q_i(X) = A_i(X) + \sum_{j=1}^{k} (A_j(X) - A_i(X)) S_j(X) Q_j(X) \equiv A_i(X) \pmod{P_i(X)},$$

for $1 \le i \le k$.

This solution may have a degree larger that vThe unique solution of degree less than D may be deduced by considering the remainder $B_i(X)$ of the Euclidean division of $A_i(X)S_i(X)$ by $P_i(X)$. This solution is

$$P(X) = \sum_{i=1}^{k} B_i(X)Q_i(X).$$

Lagrange Interpolation

A special case of Chinese remainder theorem for polynomials is Lagrange interpolation. For this, let us consider k monic polynomials of degree one:

$$P_i(X) = X - x_i.$$

They are pairwise coprime if the x_i are all different. The remainder of the division by $P_i(X)$ of a polynomial $P(X)$ is $P(x_i)$.

Now, let $A_1, \ldots, A_k$ be constants (polynomials of degree 0) in K. Both Lagrange interpolation and Chinese remainder theorem assert the existence of a unique polynomial $P(X)$, of degree less than k such that

$$P(x_i) = A_i,$$

for every i.

Lagrange interpolation formula is exactly the result, in this case, of the above construction of the solution. More precisely, let

$$Q(X) = \prod_{i=1}^{k} (X - x_i)$$

$$Q_i(X) = \frac{Q(X)}{X - x_i}.$$

The partial fraction decomposition of $1/Q(X)$ is

$$\frac{1}{Q(X)} = \sum_{i=1}^{k} \frac{1}{Q_i(x_i)(X - X_i)}.$$

In fact, reducing the right-hand side to a common denominator one gets $\sum_{i=1}^{k} \frac{1}{Q_i(x_i)(X - X_i)} = \frac{\sum_{i=1}^{k} Q_i(X)/Q_i(x_i)}{Q(X)}$, and the numerator is equal to one, as being a polynomial of degree less than k, which takes the value one for k different values of X.

Using the above general formula, we get the Lagrange interpolation formula

$$P(X)=\sum_{i=1}^{k} A_i \frac{Q_i(X)}{Q_i(x_i)}.$$

Hermite Interpolation

Hermite interpolation is an application of Chinese remainder theorem for univariate polynomials, which may involve moduli of arbitrary degrees (Lagrange interpolation involves only moduli of degree one).

The problem consists of finding a polynomial of the least possible degree, such that the polynomial and its first derivatives take given values at some fixed points.

More precisely, let $x_1,\ldots,x_k$ be k elements of the ground field K, and, for $i=1,\ldots,k$, let $a_{i,0},a_{i,1},\ldots,a_{i,r_i-1}$ be the values of the first r_i derivatives of the sought polynomial at x_i (including the 0th derivative, which is the value of the polynomial itself). The problem is to find a polynomial $P(X)$ such that its *j*th derivative takes the value $a_{i,j}$ at x_i, for $i=1,\ldots,k$ and $j=0,\ldots,r_j$.

Let us consider the polynomial

$$P_i(X)=\sum_{j=0}^{r_i-1} \frac{a_{i,j}}{j!}(X-x_i)^j.$$

This is the Taylor polynomial of order r_i-1 at x_i, , of the unknown polynomial $P(X)$. Therefore, we must have

$$P(X)\equiv P_i(X) \pmod{(X-x_i)^{r_i}}.$$

Conversely, any polynomial $P(X)$ that satisfies these k congruences, in particular verifies, for any $i=1,\ldots,k$

$$P(X)=P_i(X)+o(X-x_i)^{r_i-1}$$

therefore $P_i(X)$ is its Taylor polynomial of order r_i-1 at x_i, that is, $P(X)$ solves the initial Hermite interpolation problem. The Chinese remainder theorem asserts that there exists exactly one polynomial of degree less than the sum of the r_i, which satisfies these k congruences.

There are several ways for computing the solution $P(X)$. One may use the method described at the beginning of § Over univariate polynomial rings and Euclidean domains. One may also use the constructions given in § Existence (constructive proof) or § Existence (direct proof).

Generalization to Arbitrary Rings

The Chinese remainder theorem can be generalized to any ring, by using coprime ideals (also called comaximal ideals). Two ideals *I* and *J* are coprime if there are elements $i\in I$ and $j\in J$ such that $i+j=1$. This relation plays the role of Bézout's identity in the proofs related to this generalization, which, otherwise are very similar. The generalization may be stated as follows.

Let I_1, ..., I_k be two-sided ideals of a ring R that are pairwise coprime, and I be their intersection. Then we have the isomorphism

$$R/I \to R/I_1 \times \cdots \times R/I_k$$
$$x \bmod I \mapsto (x \bmod I_1, \cdots, x \bmod I_k),$$

between the quotient ring $R >$ and the direct product of the R/I_i, where "$x \bmod I$" denotes the image of the element x in the quotient ring defined by the ideal I. Moreover, if R is commutative, then the ideal intersection I is equal to the product of the ideals R/I_i :

$$I = I_1 \cap I_2 \cap \ldots \cap I_k = I_1 I_2 \cdots I_k$$

Applications

Sequence Numbering

The Chinese remainder theorem has been used to construct a Gödel numbering for sequences, which is involved in the proof of Gödel's incompleteness theorems.

Fast Fourier Transform

The prime-factor FFT algorithm (also called Good-Thomas algorithm) uses the Chinese remainder theorem for reducing the computation of a fast Fourier transform of size $n_1 n_2$ to the computation of two fast Fourier transforms of smaller sizes n_1 and n_2 (providing that n_1 and n_2 are coprime).

Encryption

Most implementations of RSA use the Chinese remainder theorem during signing of HTTPS certificates and during decryption.

The Chinese remainder theorem can also be used in secret sharing, which consists of distributing a set of shares among a group of people who, all together (but no one alone), can recover a certain secret from the given set of shares. Each of the shares is represented in a congruence, and the solution of the system of congruences using the Chinese remainder theorem is the secret to be recovered. Secret sharing using the Chinese remainder theorem uses, along with the Chinese remainder theorem, special sequences of integers that guarantee the impossibility of recovering the secret from a set of shares with less than a certain cardinality.

Range Ambiguity Resolution

The range ambiguity resolution techniques used with medium pulse repetition frequency radar can be seen as a special case of the Chinese remainder theorem.

Dedekind's Theorem

Dedekind's Theorem on the Linear Independence of Characters. Let M be a monoid and k an integral domain, viewed as a monoid by considering the multiplication on k. Then any finite family $(f_i)_{i\in I}$ of distinct monoid homomorphisms $f_i : M \to k$ is linearly independent. In other words, every family $(\alpha_i)_{i\in I}$ of elements $\alpha_i \in k$ satisfying

$$\sum_{i\in I} \alpha_i f_i = 0$$

must be equal to the family $(0)_{i\in I}$.

Proof. First assume that k is a field, otherwise, replace the integral domain k by its quotient field, and nothing will change. We can linearly extend the monoid homomorphisms $f_i : M \to k$ to k-algebra homomorphisms $F_i : k[M] \to k$, where $k[M]$ is the monoid ring of M over k. Then, by linearity, the condition

$$\sum_{i\in I} \alpha_i f_i = 0,$$

yields

$$\sum_{i\in I} \alpha_i F_i = 0.$$

Next, for $i, j \in I$; $i \neq j$ the two k-linear maps $F_i : k[M] \to k$ and $F_j : k[M] \to k$ are not proportional to each other. Otherwise f_i and f_j would also be proportional, and thus equal since as monoid homomorphisms they satisfy: $f_i(1) = 1 = f_j(1)$, which contradicts the assumption that they are distinct.

Therefore, the kernels Ker F_i and Ker F_j are distinct. Since $k[M]/\text{Ker } F_i \cong F_i(k[M]) = k$ is a field, Ker F_i is a maximal ideal of $k[M]$ for every $i \in I$. Because they are distinct and maximal the ideals Ker F_i and Ker F_j are coprime whenever $i \neq j$. The Chinese Remainder Theorem (for general rings) yields an isomorphism:

$$\phi : k[M]/K \to \prod_{i\in I} k[M]/\text{Ker}F$$
$${}_i\phi(x+K) = \left(x + \text{Ker}F_i\right)_{i\in I}$$

where

$$K = \prod_{i\in I} \text{Ker}F_i = \bigcap_{i\in I} \text{Ker}F_i.$$

Consequently, the map

$$\Phi : k[M] \to \prod_{i\in I} k[M]/\text{Ker}F_i$$
$$\Phi(x) = \left(x + \text{Ker}F_i\right)_{i\in I}$$

is surjective. Under the isomorphisms $k[M]/\text{Ker } F_i \to F_i(k[M]) = k$, the map Φ corresponds to:

$$\psi : k[M] \to \prod_{i \in I} k$$

$$\psi(x) = [F_i(x)]_{i \in I}$$

Now,

$$\sum_{i \in I} \alpha_i F_i = 0$$

yields

$$\sum_{i \in I} \alpha_i u_i = 0$$

for every vector $(u_i)_{i \in I}$ in the image of the map ψ. Since ψ is surjective, this means that

$$\sum_{i \in I} \alpha_i u_i = 0$$

for every vector

$$(u_i)_{i \in I} \in \prod_{i \in I} k.$$

Consequently, $(\alpha_i)_{i \in I} = (0)_{i \in I}$. QED.

Pentagonal Number Theorem

In mathematics, the pentagonal number theorem, originally due to Euler, relates the product and series representations of the Euler function. It states that

$$\prod_{n=1}^{\infty}\left(1-x^n\right) = \sum_{k=-\infty}^{\infty}\left(-1\right)^k x^{\frac{k}{2}(3k-1)} = 1 + \sum_{k=1}^{\infty}(-1)^k\left(x^{k(3k+1)/2} + x^{k(3k-1)/2}\right).$$

In other words,

$$(1-x)(1-x^2)(1-x^3)\cdots = 1 - x - x^2 + x^5 + x^7 - x^{12} - x^{15} + x^{22} + x^{26} - \cdots.$$

The exponents 1, 2, 5, 7, 12, ... on the right hand side are given by the formula $g_k = k(3k-1)/2$ for k = 1, −1, 2, −2, 3, ... and are called (generalized) pentagonal numbers. This holds as an identity of convergent power series for $|x|<1$, and also as an identity of formal power series.

A striking feature of this formula is the amount of cancellation in the expansion of the product.

Relation with Partitions

The identity implies a marvelous recurrence for calculating , the number of partitions of n:

$$p(n) = p(n-1) + p(n-2) - p(n-5) - p(n-7) + \cdots$$

or more formally,

$$p(n) = \sum_{k} (-1)^{k-1} p(n - g_k)$$

where the summation is over all nonzero integers k (positive and negative) and g_k is the kth pentagonal number.

Bijective Proof

The theorem can be given a combinatorial interpretation in terms of partitions. In particular, the left hand side is a generating function for the number of partitions of n into an even number of distinct parts minus the number of partitions of n into an odd number of distinct parts. Each partition of n into an even number of distinct parts contributes +1 to the coefficient of x^n; each partition into an odd number of distinct parts contributes −1. (The article on unrestricted partition functions discusses this type of generating function.)

For example, the coefficient of x^5 is +1 because there are two ways to split 5 into an even number of distinct parts (4+1 and 3+2), but only one way to do so for an odd number of distinct parts (the one-part partition 5). However, the coefficient of x^{12} is −1 because there are seven ways to partition 12 into an even number of distinct parts, but there are eight ways to partition 12 into an odd number of distinct parts.

This interpretation leads to a proof of the identity via involution (i.e. a bijection which is its own inverse). Consider the Ferrers diagram of any partition of n into distinct parts. For example, the diagram below shows $n = 20$ and the partition $20 = 7 + 6 + 4 + 3$.

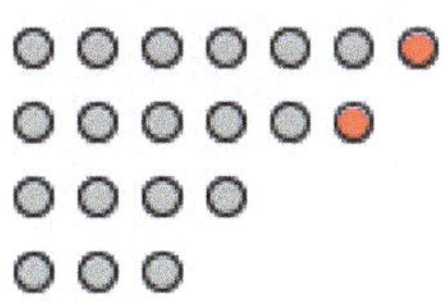

Let m be the number of elements in the smallest row of the diagram ($m = 3$ in the above example). Let s be the number of elements in the rightmost 45 degree line of the diagram ($s = 2$ dots in red above, since $7-1 = 6$, but $6-1 > 4$). If $m > s$, take the rightmost 45-degree line and move it to form a new row, as in the diagram below.

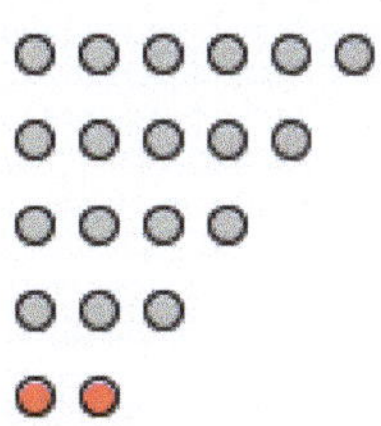

If $m \le s$ (as in our newly formed diagram where $m = 2$, $s = 5$) we may reverse the process by moving the bottom row to form a new 45 degree line (adding 1 element to each of the first m rows), taking us back to the first diagram.

A bit of thought shows that this process always changes the parity of the number of rows, and applying the process twice brings us back to the original diagram. This enables us to pair off Ferrers diagrams contributing 1 and −1 to the x^n term of the series, resulting in a net coefficient of 0. This holds for every term *except* when the process cannot be performed on every Ferrers diagram with n dots. There are two such cases:

1) $m = s$ and the rightmost diagonal and bottom row meet. For example,

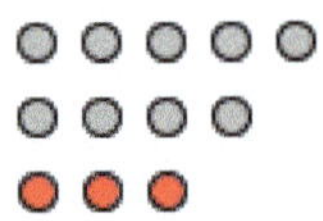

Attempting to perform the operation would lead us to:

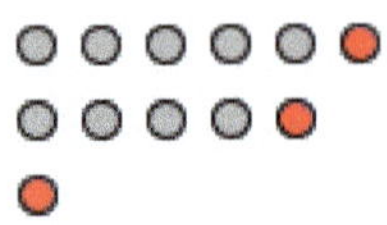

which fails to change the parity of the number of rows, and is not reversible in the sense that performing the operation again does *not* take us back to the original diagram. If there are m elements in the last row of the original diagram, then

$$n = m + (m+1) + (m+2) + \cdots + (2m-1) = \frac{m(3m-1)}{2} = \frac{k(3k-1)}{2}$$

where the new index k is taken to equal m. Note that the sign associated with this partition is $(-1)^s$, which by construction equals $(-1)^m$ and $(-1)^k$.

2) $m = s+1$ and the rightmost diagonal and bottom row meet. For example,

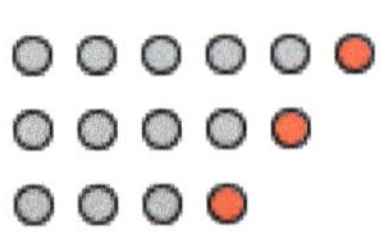

Our operation requires us to move the right diagonal to the bottom row, but that would lead to two rows of three elements, forbidden since we're counting partitions into distinct parts. This is the previous case but with one fewer row, so

$$n = m + (m+1) + (m+2) + \cdots + (2m-2) = \frac{(m-1)(3m-2)}{2} = \frac{k(3k-1)}{2},$$

where we take $k = 1-m$ (a negative integer). Here the associated sign is $(-1)^s$ with $s = m-1 = -k$, therefore the sign is again $(-1)^k$.

In summary, it has been shown that partitions into an even number of distinct parts and an odd number of distinct parts exactly cancel each other, except if n is a generalized pentagonal number $g_k = k(3k-1) >$, in which case there is exactly one Ferrers diagram left over. But this is precisely what the right side of the identity says should happen, so we are finished.

Partition Recurrence

We can rephrase the above proof, using partitions, which we denote as: $n = \lambda_1 + \lambda_2 + \cdots + \lambda_\ell$, where $\lambda_1 \geq \lambda_2 \geq \ldots \geq \lambda_\ell > 0$. The number of partitions of n is the partition function $p(n)$ having generating function:

$$\sum_{n \in \mathbb{N}_0} p(n)x^n = \prod_{k \in \mathbb{N}} (1 - x^k)^{-1}$$

Note that is the reciprocal of the product on the left hand side of our identity:

$$\left(\sum_{n=0}^{\infty} p(n)x^n\right) \cdot \left(\prod_{n=1}^{\infty} (1 - x^n)\right) = 1$$

Let us denote the expansion of our product by $\prod_{n=1}^{\infty} (1 - x^n) = \sum_{n=0}^{\infty} a_n x^n$, so that

$$\left(\sum_{n=0}^{\infty} p(n)x^n\right) \cdot \left(\sum_{n=0}^{\infty} a_n x^n\right) = 1.$$

Multiplying out the left hand side and equating coefficients on the two sides, we obtain $a_o p(0) = 1$ and $\sum_{i=0}^{n} p(n-i)a_i = 0$ for all $n \geq 1$. This gives a recurrence relation defining $p(n)$ in terms of a_n, and vice versa a recurrence for a_n in terms of $p(n)$. Thus, our desired result:

$$a_i := \begin{cases} 1 & \text{if } i = \frac{1}{2}(3k^2 \pm k) \text{ and } k \text{ is even} \\ -1 & \text{if } i = \frac{1}{2}(3k^2 \pm k) \text{ and } k \text{ is odd} \\ 0 & \text{otherwise} \end{cases}$$

for $i \geq 1$ is equivalent to the identity $\sum_i (-1)^i p(n - g_i) = 0$, where $g_i := \frac{1}{2}(3i^2 - i)$ and i ranges over all integers such that $g_i := \frac{1}{2}(3i_2 - i)$ (this range includes both positive and negative i, so as to use both kinds of generalized pentagonal numbers). This in turn means:

$$\sum_{i \text{ even}} p(n - g_i) = \sum_{i \text{ odd}} p(n - g_i),.$$

In terms of sets of partitions, this is equivalent to saying that the following sets are of equal cardinality:

$$\mathcal{X} := \bigcup_{i \text{ even}} \mathcal{P}(n - g_i) \quad \text{and} \quad \mathcal{Y} := \bigcup_{i \text{ odd}} \mathcal{P}(n - g_i),$$

where $\mathcal{P}(n)$ denotes the set of all partitions of n. All that remains is to give a bijection from one set to the other, which is accomplished by the function φ from X to Y which maps the partition $\mathcal{P}(n-g_i) \ni \lambda : n-g_i = \lambda_1 + \lambda_2 + \cdots + \lambda_\ell$ to the partition $\lambda' = \varphi(\lambda)$ defined by:

$$\varphi(\lambda) := \begin{cases} \lambda' : n - g_{i-1} = (\ell + 3i - 2) + (\lambda_1 - 1) + \cdots + (\lambda_\ell - 1) & \text{if } \ell + 3i > \lambda_1 \\ \lambda' : n - g_{i+1} = (\lambda_2 + 1) + \cdots + (\lambda_\ell + 1) + \underbrace{1 + \cdots + 1}_{\lambda_1 - \ell - 3i} & \text{if } \ell + 3i \le \lambda_1 \end{cases}.$$

This is an involution (a self-inverse mapping), and thus in particular a bijection, which proves our claim and the identity.

Six Exponentials Theorem

In mathematics, specifically transcendental number theory, the six exponentials theorem is a result that, given the right conditions on the exponents, guarantees the transcendence of at least one of a set of exponentials.

Statement

If $x_1, x_2, \ldots, x_d$ are d complex numbers that are linearly independent over the rational numbers, and $y_1, y_2, \ldots, y_l$ are l complex numbers that are also linearly independent over the rational numbers, and if $dl > d + l$, then at least one of the following dl numbers is transcendental:

$$\exp(x_i y_j), \quad (1 \le i \le d, 1 \le j \le l).$$

The most interesting case is when $d = 3$ and $l = 2$, in which case there are six exponentials, hence the name of the result. The theorem is weaker than the related but thus far unproved four exponentials conjecture, whereby the strict inequality $dl > d + l$ is replaced with $dl \ge d + l$, thus allowing $d = l = 2$.

The theorem can be stated in terms of logarithms by introducing the set L of logarithms of algebraic numbers:

$$\mathcal{L} = \{\lambda \in \mathbb{C} : e^\lambda \in \overline{\mathbb{Q}}\}.$$

The theorem then says that if λ_{ij} are elements of L for $i = 1, 2$ and $j = 1, 2, 3$, such that $\lambda_{11}, \lambda_{12}$, and λ_{13} are linearly independent over the rational numbers, and λ_{11} and λ_{21} are also linearly independent over the rational numbers, then the matrix

$$M = \begin{pmatrix} \lambda_{11} & \lambda_{12} & \lambda_{13} \\ \lambda_{21} & \lambda_{22} & \lambda_{23} \end{pmatrix}$$

has rank 2.

History

A special case of the result where x_1, x_2, and x_3 are logarithms of positive integers, $y_1 = 1$, and y_2 is real, was first mentioned in a paper by Leonidas Alaoglu and Paul Erdős from 1944 in which they try to prove that the ratio of consecutive colossally abundant numbers is always prime. They claimed that Carl Ludwig Siegel knew of a proof of this special case, but it is not recorded. Using the special case they manage to prove that the ratio of consecutive colossally abundant numbers is always either a prime or a semiprime.

The theorem was first explicitly stated and proved in its complete form independently by Serge Lang and Kanakanahalli Ramachandra in the 1960s.

Five Exponentials Theorem

A stronger, related result is the five exponentials theorem, which is as follows. Let x_1, x_2 and y_1, y_2 be two pairs of complex numbers, with each pair being linearly independent over the rational numbers, and let γ be a non-zero algebraic number. Then at least one of the following five numbers is transcendental:

$$e^{x_1 y_1}, e^{x_1 y_2}, e^{x_2 y_1}, e^{x_2 y_2}, e^{\gamma x_2 / x_1}.$$

This theorem implies the six exponentials theorem and in turn is implied by the as yet unproven four exponentials conjecture, which says that in fact one of the first four numbers on this list must be transcendental.

Sharp Six Exponentials Theorem

Another related result that implies both the six exponentials theorem and the five exponentials theorem is the sharp six exponentials theorem. This theorem is as follows. Let x_1, x_2, and x_3 be complex numbers that are linearly independent over the rational numbers, and let y_1 and y_2 be a pair of complex numbers that are linearly independent over the rational numbers, and suppose that β_{ij} are six algebraic numbers for $1 \le i \le 3$ and $1 \le j \le 2$ such that the following six numbers are algebraic:

$$e^{x_1 y_1 - \beta_{11}}, e^{x_1 y_2 - \beta_{12}}, e^{x_2 y_1 - \beta_{21}}, e^{x_2 y_2 - \beta_{22}}, e^{x_3 y_1 - \beta_{31}}, e^{x_3 y_2 - \beta_{32}}.$$

Then $x_i y_j = \beta_{ij}$ for $1 \le i \le 3$ and $1 \le j \le 2$. The six exponentials theorem then follows by setting $\beta_{ij} = 0$

for every i and j, while the five exponentials theorem follows by setting $x_3 = \gamma/x_1$ and using Baker's theorem to ensure that the x_i are linearly independent.

There is a sharp version of the five exponentials theorem as well, although it as yet unproven so is known as the sharp five exponentials conjecture. This conjecture implies both the sharp six exponentials theorem and the five exponentials theorem, and is stated as follows. Let x_1, x_2 and y_1, y_2 be two pairs of complex numbers, with each pair being linearly independent over the rational numbers, and let α, β_{11}, β_{12}, β_{21}, β_{22}, and γ be six algebraic numbers with $\gamma \neq 0$ such that the following five numbers are algebraic:

$$e^{x_1y_1-\beta_{11}}, e^{x_1y_2-\beta_{12}}, e^{x_2y_1-\beta_{21}}, e^{x_2y_2-\beta_{22}}, e^{(\gamma x_2/x_1)-\alpha}.$$

Then $x_i y_j = \beta_{ij}$ for $1 \leq i, j \leq 2$ and $\gamma x_2 = \alpha x_1$.

A consequence of this conjecture that isn't currently known would be the transcendence of e^{π^2}, by setting $x_1 = y_1 = \beta_{11} = 1$, $x_2 = y_2 = i\pi$, and all the other values in the statement to be zero.

Strong Six Exponentials Theorem

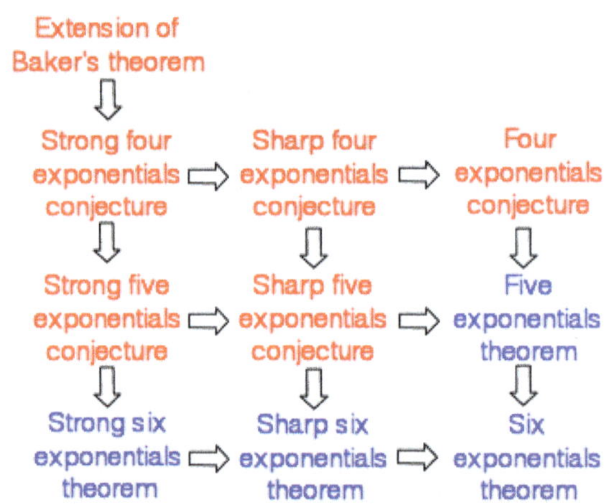

The logical implications between the various problems in this circle. Those in red are as yet unproven while those in blue are known results. The top most result refers to that discussed at Baker's theorem, while the four exponentials conjectures are detailed at the four exponentials conjecture article.

A further strengthening of the theorems and conjectures in this area are the strong versions. The strong six exponentials theorem is a result proved by Damien Roy that implies the sharp six exponentials theorem. This result concerns the vector space over the algebraic numbers generated by 1 and all logarithms of algebraic numbers, denoted here as $L^{\square}$. So $L^{\square}$ is the set of all complex numbers of the form

$$\beta_0 + \sum_{i=1}^{n} \beta_i \log \alpha_i,$$

for some $n \geq 0$, where all the β_i and α_i are algebraic and every branch of the logarithm is considered. The strong six exponentials theorem then says that if x_1, x_2, and x_3 are complex numbers that are linearly independent over the algebraic numbers, and if y_1 and y_2 are a pair of complex numbers that are also linearly independent over the algebraic numbers then at least one of the six numbers $x_i y_j$ for $1 \leq i \leq 3$ and $1 \leq j \leq 2$ is not in L^*. This is stronger than the standard six exponentials theorem which says that one of these six numbers is not simply the logarithm of an algebraic number.

There is also a strong five exponentials conjecture formulated by Michel Waldschmidt It would imply both, the strong six exponentials theorem and the sharp five exponentials conjecture. This conjecture claims that if x_1, x_2 and y_1, y_2 are two pairs of complex numbers, with each pair being linearly independent over the algebraic numbers, then at least one of the following five numbers is not in L^*:

$$x_1y_1, x_1y_2, x_2y_1, x_2y_2, x_1 >$$

All the above conjectures and theorems are consequences of the unproven extension of Baker's theorem, that logarithms of algebraic numbers that are linearly independent over the rational numbers are automatically algebraically independent too. The diagram on the right shows the logical implications between all these results.

Generalization to Commutative Group Varieties

The exponential function e^z uniformizes the exponential map of the multiplicative group G_m. Therefore, we can reformulate the six exponential theorem more abstractly as follows:

> Let $G = G_m \times G_m$ and take $u : C \to G(C)$ to be a non-zero complex-analytic group homomorphism. Define L to be the set of complex numbers l for which $u(l)$ is an algebraic point of G. If a minimal generating set of L over Q has more than two elements then the image $u(C)$ is an algebraic subgroup of $G(C)$.

(In order to derive the classical statement, set $u(z) = (e^{y_1 z}; e^{y_2 z})$ and note that $Qx_1 + Qx_2 + Qx_3$ is a subset of L).

In this way, the statement of the six exponentials theorem can be generalized to an arbitrary commutative group variety G over the field of algebraic numbers. This generalized six exponential conjecture, however, seems out of scope at the current state of transcendental number theory.

For the special, but interesting cases $G = G_m \times E$ and $G = E \times E'$, where E, E' are elliptic curves over the field of algebraic numbers, results towards the generalized six exponential conjecture were proven by Aleksander Momot. These results involve the exponential function e^z and a Weierstrass function $\wp$ resp. two Weierstrass functions $\wp, \wp'$ with algebraic invariants $g_2, g_3, g_{2'}, g_{3'}$, instead of the two exponential functions $e^{y_1 z}, e^{y_2 z}$ in the classical statement.

Let $G = G_m \times E$ and suppose E is not isogenous to a curve over a real field and that $u(C)$ is not an algebraic subgroup of $G(C)$. Then L is generated over Q either by two elements x_1, x_2, or three elements x_1, x_2, x_3 which are not all contained in a real line Rc, where c is a non-zero complex number. A similar result is shown for $G = E \times E'$.

References

- "The Unreasonable Effectiveness of Number Theory", Stefan Andrus Burr, George E. Andrews, American Mathematical Soc., 1992, ISBN 978-0-8218-5501-0
- "Applications of number theory to numerical analysis", Lo-keng Hua, Luogeng Hua, Yuan Wang, Springer-Verlag, 1981, ISBN 978-3-540-10382-0
- Apostol, Tom M. (1976), Introduction to analytic number theory, Undergraduate Texts in Mathematics, New York-Heidelberg: Springer-Verlag, ISBN 978-0-387-90163-3, MR 0434929, Zbl 0335.10001
- Davenport, Harold (2000), Multiplicative number theory, Graduate Texts in Mathematics, 74 (3rd revised ed.), New York: Springer-Verlag, ISBN 978-0-387-95097-6, MR 1790423
- Tenenbaum, Gérald (1995), Introduction to Analytic and Probabilistic Number Theory, Cambridge studies in advanced mathematics, 46, Cambridge University Press, ISBN 0-521-41261-7
- Kanemitsu, Shigeru; Chaohua Jia (2002). Number theoretic methods: future trends. Springer. pp. 271–274. ISBN 978-1-4020-1080-4.
- Michel, Anthony N.; Herget, Charles J. (1993), Applied Algebra and Functional Analysis, New York: Dover Publications, ISBN 978-0-486-67598-5
- Burris, Stanley N.; Sankappanavar, H. P. (1981), A Course in Universal Algebra, Berlin, New York: Springer-Verlag, ISBN 978-3-540-90578-3
- Baker, Alan (1990), Transcendental number theory, Cambridge Mathematical Library (2nd ed.), Cambridge University Press, ISBN 978-0-521-39791-9, MR 0422171
- Gel'fond, A. O. (1960) [1952], Transcendental and algebraic numbers, Dover Phoenix editions, New York: Dover Publications, ISBN 978-0-486-49526-2.
- Gauss, Carl Friedrich; Clarke, Arthur A. (translator into English) (1986), Disquisitiones Arithemeticae (Second, corrected ed.), New York: Springer, ISBN 978-0-387-96254-2
- Ireland, Kenneth; Rosen, Michael (1990), A Classical Introduction to Modern Number Theory (2nd ed.), Springer-Verlag, ISBN 0-387-97329-X

6

Graph Theory : An Overview

The study of graphs is known as the graph theory. A graph is made up of points, nodes and lines. They are the main objects in discrete mathematics. The following text is an overview of the subject matter incorporating all the major aspects of graph theory.

Graph Theory

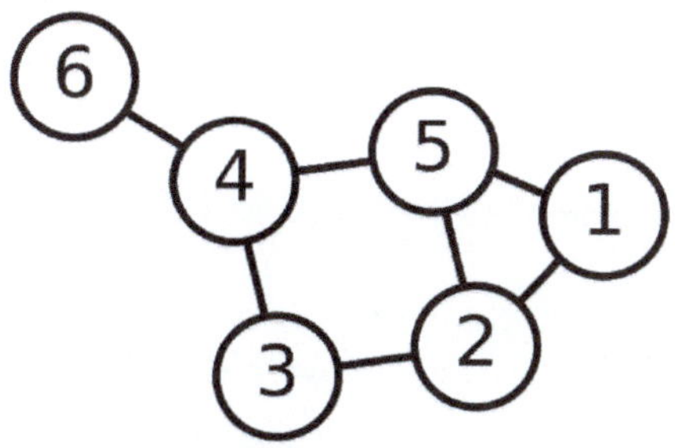

A drawing of a graph

In mathematics graph theory is the study of *graphs*, which are mathematical structures used to model pairwise relations between objects. A graph in this context is made up of *vertices*, *nodes*, or *points* which are connected by *edges*, *arcs*, or *lines*. A graph may be *undirected*, meaning that there is no distinction between the two vertices associated with each edge, or its edges may be *directed* from one vertex to another; see Graph (discrete mathematics) for more detailed definitions and for other variations in the types of graph that are commonly considered. Graphs are one of the prime objects of study in discrete mathematics.

Refer to the glossary of graph theory for basic definitions in graph theory.

Definitions

Definitions in graph theory vary. The following are some of the more basic ways of defining graphs and related mathematical structures.

Graph

In the most common sense of the term, a graph is an ordered pair $G = (V, E)$ comprising a set V of *vertices* or *nodes* or *points* together with a set E of *edges* or *arcs* or *lines*, which are 2-element subsets of V (i.e. an edge is associated with two vertices, and that association takes the form of the unordered pair comprising those two vertices). To avoid ambiguity, this type of graph may be described precisely as undirected and simple.

Other senses of *graph* stem from different conceptions of the edge set. In one more generalized notion, V is a set together with a relation of *incidence* that associates with each edge two vertices. In another generalized notion, E is a multiset of unordered pairs of (not necessarily distinct) vertices. Many authors call this type of object a multigraph or pseudograph.

All of these variants and others are described more fully below.

The vertices belonging to an edge are called the *ends* or *end vertices* of the edge. A vertex may exist in a graph and not belong to an edge.

V and E are usually taken to be finite, and many of the well-known results are not true (or are rather different) for infinite graphs because many of the arguments fail in the infinite case. The *order* of a graph is $|V|$, its number of vertices. The *size* of a graph is $|E|$, its number of edges. The *degree* or *valency* of a vertex is the number of edges that connect to it, where an edge that connects a vertex to itself (a loop) is counted twice.

For an edge $\{x, y\}$, graph theorists usually use the somewhat shorter notation xy.

Applications

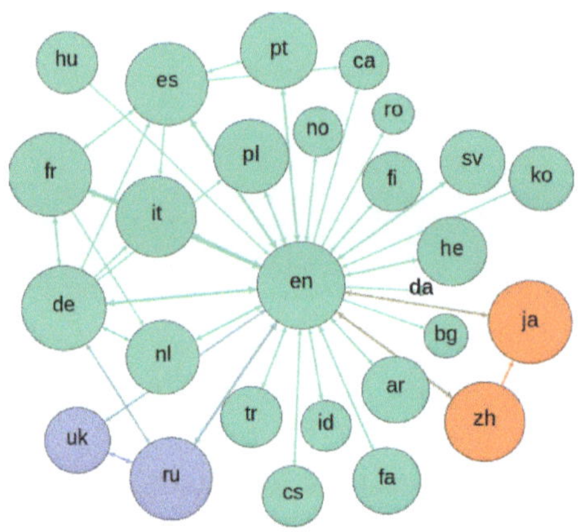

The network graph formed by Wikipedia editors (edges) contributing to different Wikipedia language versions (vertices) during one month in summer 2013

Graphs can be used to model many types of relations and processes in physical, biological, social and information systems. Many practical problems can be represented by graphs. Emphasizing their application to real-world systems, the term *network* is sometimes defined to mean a graph in which attributes (e.g. names) are associated with the nodes and/or edges.

In computer science, graphs are used to represent networks of communication, data organization, computational devices, the flow of computation, etc. For instance, the link structure of a website can be represented by a directed graph, in which the vertices represent web pages and directed edges represent links from one page to another. A similar approach can be taken to problems in social media, travel, biology, computer chip design, and many other fields. The development of algorithms to handle graphs is therefore of major interest in computer science. The transformation of graphs is often formalized and represented by graph rewrite systems. Complementary to graph transformation systems focusing on rule-based in-memory manipulation of graphs are graph databases geared towards transaction-safe, persistent storing and querying of graph-structured data.

Graph-theoretic methods, in various forms, have proven particularly useful in linguistics, since natural language often lends itself well to discrete structure. Traditionally, syntax and compositional semantics follow tree-based structures, whose expressive power lies in the principle of compositionality, modeled in a hierarchical graph. More contemporary approaches such as head-driven phrase structure grammar model the syntax of natural language using typed feature structures, which are directed acyclic graphs. Within lexical semantics, especially as applied to computers, modeling word meaning is easier when a given word is understood in terms of related words; semantic networks are therefore important in computational linguistics. Still other methods in phonology (e.g. optimality theory, which uses lattice graphs) and morphology (e.g. finite-state morphology, using finite-state transducers) are common in the analysis of language as a graph. Indeed, the usefulness of this area of mathematics to linguistics has borne organizations such as TextGraphs, as well as various 'Net' projects, such as WordNet, VerbNet, and others.

Graph theory is also used to study molecules in chemistry and physics. In condensed matter physics, the three-dimensional structure of complicated simulated atomic structures can be studied quantitatively by gathering statistics on graph-theoretic properties related to the topology of the atoms. In chemistry a graph makes a natural model for a molecule, where vertices represent atoms and edges bonds. This approach is especially used in computer processing of molecular structures, ranging from chemical editors to database searching. In statistical physics, graphs can represent local connections between interacting parts of a system, as well as the dynamics of a physical process on such systems. Similarly, in computational neuroscience graphs can be used to represent functional connections between brain areas that interact to give rise to various cognitive processes, where the vertices represent different areas of the brain and the edges represent the connections between those areas. Graphs are also used to represent the micro-scale channels of porous media, in which the vertices represent the pores and the edges represent the smaller channels connecting the pores.

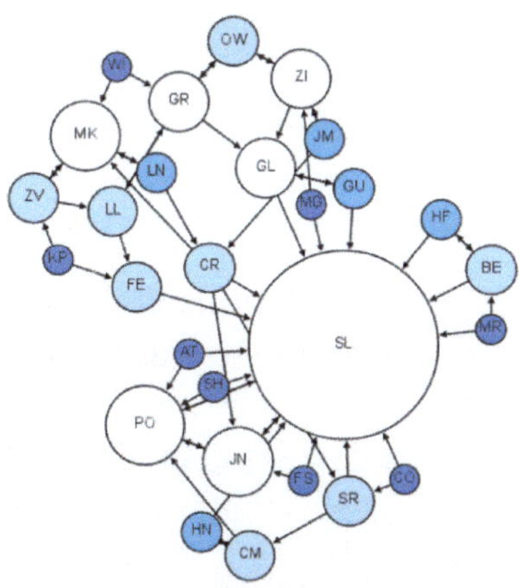

Graph theory in sociology: Moreno Sociogram (1953)

Graph theory is also widely used in sociology as a way, for example, to measure actors' prestige or to explore rumor spreading, notably through the use of social network analysis software. Under the umbrella of social networks are many different types of graphs. Acquaintanceship and friendship graphs describe whether people know each other. Influence graphs model whether certain people can influence the behavior of others. Finally, collaboration graphs model whether two people work together in a particular way, such as acting in a movie together.

Likewise, graph theory is useful in biology and conservation efforts where a vertex can represent regions where certain species exist (or inhabit) and the edges represent migration paths, or movement between the regions. This information is important when looking at breeding patterns or tracking the spread of disease, parasites or how changes to the movement can affect other species.

In mathematics, graphs are useful in geometry and certain parts of topology such as knot theory. Algebraic graph theory has close links with group theory.

A graph structure can be extended by assigning a weight to each edge of the graph. Graphs with weights, or weighted graphs, are used to represent structures in which pairwise connections have some numerical values. For example, if a graph represents a road network, the weights could represent the length of each road.

History

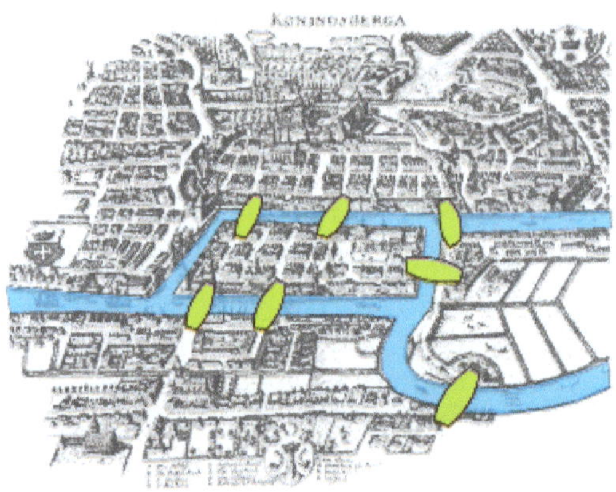

The Königsberg Bridge problem

The paper written by Leonhard Euler on the *Seven Bridges of Königsberg* and published in 1736 is regarded as the first paper in the history of graph theory. This paper, as well as the one written by Vandermonde on the *knight problem,* carried on with the *analysis situs* initiated by Leibniz. Euler's formula relating the number of edges, vertices, and faces of a convex polyhedron was studied and generalized by Cauchy and L'Huillier, and represents the beginning of the branch of mathematics known as topology.

More than one century after Euler's paper on the bridges of Königsberg and while Listing was introducing the concept of topology, Cayley was led by an interest in particular analytical forms arising from differential calculus to study a particular class of graphs, the *trees*. This study had many implications for theoretical chemistry. The techniques he used mainly concern the enumeration of graphs with particular properties. Enumerative graph theory then arose from the results of Cayley and the fundamental results published by Pólya between 1935 and 1937. These were generalized by De Bruijn in 1959. Cayley linked his results on trees with contemporary studies of chemical composition. The fusion of ideas from mathematics with those from chemistry began what has become part of the standard terminology of graph theory.

In particular, the term "graph" was introduced by Sylvester in a paper published in 1878 in *Nature*, where he draws an analogy between "quantic invariants" and "co-variants" of algebra and molecular diagrams:

> "[...] Every invariant and co-variant thus becomes expressible by a *graph* precisely identical with a Kekuléan diagram or chemicograph. [...] I give a rule for the geometrical multiplication of graphs, *i.e.* for constructing a *graph* to the product of in- or co-variants whose separate graphs are given. [...]" (italics as in the original).

The first textbook on graph theory was written by Dénes Kőnig, and published in 1936. Another book by Frank Harary, published in 1969, was "considered the world over to be the definitive textbook on the subject", and enabled mathematicians, chemists, electrical engineers and social scientists to talk to each other. Harary donated all of the royalties to fund the Pólya Prize.

One of the most famous and stimulating problems in graph theory is the four color problem: "Is it true that any map drawn in the plane may have its regions colored with four colors, in such a way that any two regions having a common border have different colors?" This problem was first posed by Francis Guthrie in 1852 and its first written record is in a letter of De Morgan addressed to Hamilton the same year. Many incorrect proofs have been proposed, including those by Cayley, Kempe, and others. The study and the generalization of this problem by Tait, Heawood, Ramsey and Hadwiger led to the study of the colorings of the graphs embedded on surfaces with arbitrary genus. Tait's reformulation generated a new class of problems, the *factorization problems*, particularly studied by Petersen and Kőnig. The works of Ramsey on colorations and more specially the results obtained by Turán in 1941 was at the origin of another branch of graph theory, *extremal graph theory*.

The four color problem remained unsolved for more than a century. In 1969 Heinrich Heesch published a method for solving the problem using computers. A computer-aided proof produced in 1976 by Kenneth Appel and Wolfgang Haken makes fundamental use of the notion of "discharging" developed by Heesch. The proof involved checking the properties of 1,936 configurations by computer, and was not fully accepted at the time due to its complexity. A simpler proof considering only 633 configurations was given twenty years later by Robertson, Seymour, Sanders and Thomas.

The autonomous development of topology from 1860 and 1930 fertilized graph theory back through the works of Jordan, Kuratowski and Whitney. Another important factor of common development of graph theory and topology came from the use of the techniques of modern algebra. The first example of such a use comes from the work of the physicist Gustav Kirchhoff, who published in 1845 his Kirchhoff's circuit laws for calculating the voltage and current in electric circuits.

The introduction of probabilistic methods in graph theory, especially in the study of Erdős and Rényi of the asymptotic probability of graph connectivity, gave rise to yet another branch, known as *random graph theory*, which has been a fruitful source of graph-theoretic results.

Graph Drawing

Graphs are represented visually by drawing a dot or circle for every vertex, and drawing an arc between two vertices if they are connected by an edge. If the graph is directed, the direction is indicated by drawing an arrow.

A graph drawing should not be confused with the graph itself (the abstract, non-visual structure) as there are several ways to structure the graph drawing. All that matters is which vertices are connected to which others by how many edges and not the exact layout. In practice it is often difficult to decide if two drawings represent the same graph. Depending on the problem domain some layouts may be better suited and easier to understand than others.

The pioneering work of W. T. Tutte was very influential in the subject of graph drawing. Among other achievements, he introduced the use of linear algebraic methods to obtain graph drawings.

Graph drawing also can be said to encompass problems that deal with the crossing number and its various generalizations. The crossing number of a graph is the minimum number of intersections between edges that a drawing of the graph in the plane must contain. For a planar graph, the crossing number is zero by definition.

Drawings on surfaces other than the plane are also studied.

Graph-theoretic Data Structures

There are different ways to store graphs in a computer system. The data structure used depends on both the graph structure and the algorithm used for manipulating the graph. Theoretically one can distinguish between list and matrix structures but in concrete applications the best structure is often a combination of both. List structures are often preferred for sparse graphs as they have smaller memory requirements. Matrix structures on the other hand provide faster access for some applications but can consume huge amounts of memory.

List structures include the incidence list, an array of pairs of vertices, and the adjacency list, which separately lists the neighbors of each vertex: Much like the incidence list, each vertex has a list of which vertices it is adjacent to.

Matrix structures include the incidence matrix, a matrix of 0's and 1's whose rows represent vertices and whose columns represent edges, and the adjacency matrix, in which both the rows and columns are indexed by vertices. In both cases a 1 indicates two adjacent objects and a 0 indicates two non-adjacent objects. The Laplacian matrix is a modified form of the adjacency matrix that incorporates information about the degrees of the vertices, and is useful in some calculations such as Kirchhoff's theorem on the number of spanning trees of a graph. The distance matrix, like the adjacency matrix, has both its rows and columns indexed by vertices, but rather than containing a 0 or a 1 in each cell it contains the length of a shortest path between two vertices.

Problems in Graph Theory

Enumeration

There is a large literature on graphical enumeration: the problem of counting graphs meeting specified conditions. Some of this work is found in Harary and Palmer (1973).

Subgraphs, Induced Subgraphs, and Minors

A common problem, called the subgraph isomorphism problem, is finding a fixed graph as a subgraph in a given graph. One reason to be interested in such a question is that many graph properties are *hereditary* for subgraphs, which means that a graph has the property if and only if all subgraphs have it too. Unfortunately, finding maximal subgraphs of a certain kind is often an NP-complete problem. For example:

- Finding the largest complete subgraph is called the clique problem (NP-complete).

A similar problem is finding induced subgraphs in a given graph. Again, some important graph properties are hereditary with respect to induced subgraphs, which means that a graph has a property if and only if all induced subgraphs also have it. Finding maximal induced subgraphs of a certain kind is also often NP-complete. For example:

- Finding the largest edgeless induced subgraph or independent set is called the independent set problem (NP-complete).

Still another such problem, the minor containment problem, is to find a fixed graph as a minor of a given graph. A minor or subcontraction of a graph is any graph obtained by taking a subgraph and contracting some (or no) edges. Many graph properties are hereditary for minors, which means that a graph has a property if and only if all minors have it too. For example, Wagner's Theorem states:

- A graph is planar if it contains as a minor neither the complete bipartite graph $K_{3,3}$ nor the complete graph K_5.

A similar problem, the subdivision containment problem, is to find a fixed graph as a subdivision of a given graph. A subdivision or homeomorphism of a graph is any graph obtained by subdividing some (or no) edges. Subdivision containment is related to graph properties such as planarity. For example, Kuratowski's Theorem states:

- A graph is planar if it contains as a subdivision neither the complete bipartite graph $K_{3,3}$ nor the complete graph K_5.

Another problem in subdivision containment is Kelmans-Seymour conjecture:

- Every 5-vertex-connected graph that is not planar contains a subdivision of the 5-vertex complete graph K_5.

Another class of problems has to do with the extent to which various species and generalizations of graphs are determined by their *point-deleted subgraphs*. For example:

- The reconstruction conjecture

Graph Coloring

Many problems have to do with various ways of coloring graphs, for example:

- Four-color theorem
- Strong perfect graph theorem
- Erdős–Faber–Lovász conjecture (unsolved)
- Total coloring conjecture, also called Behzad's conjecture (unsolved)
- List coloring conjecture (unsolved)

- Hadwiger conjecture (graph theory) (unsolved)

Subsumption and Unification

Constraint modeling theories concern families of directed graphs related by a partial order. In these applications, graphs are ordered by specificity, meaning that more constrained graphs—which are more specific and thus contain a greater amount of information—are subsumed by those that are more general. Operations between graphs include evaluating the direction of a subsumption relationship between two graphs, if any, and computing graph unification. The unification of two argument graphs is defined as the most general graph (or the computation thereof) that is consistent with (i.e. contains all of the information in) the inputs, if such a graph exists; efficient unification algorithms are known.

For constraint frameworks which are strictly compositional, graph unification is the sufficient satisfiability and combination function. Well-known applications include automatic theorem proving and modeling the elaboration of linguistic structure.

Route Problems

- Hamiltonian path problem
- Minimum spanning tree
- Route inspection problem (also called the "Chinese postman problem")
- Seven bridges of Königsberg
- Shortest path problem
- Steiner tree
- Three-cottage problem
- Traveling salesman problem (NP-hard)

Network Flow

There are numerous problems arising especially from applications that have to do with various notions of flows in networks, for example:

- Max flow min cut theorem

Visibility Problems

- Museum guard problem

Covering Problems

Covering problems in graphs are specific instances of subgraph-finding problems, and they tend to be closely related to the clique problem or the independent set problem.

- Set cover problem

- Vertex cover problem

Decomposition Problems

Decomposition, defined as partitioning the edge set of a graph (with as many vertices as necessary accompanying the edges of each part of the partition), has a wide variety of question. Often, it is required to decompose a graph into subgraphs isomorphic to a fixed graph; for instance, decomposing a complete graph into Hamiltonian cycles. Other problems specify a family of graphs into which a given graph should be decomposed, for instance, a family of cycles, or decomposing a complete graph K_n into $n - 1$ specified trees having, respectively, 1, 2, 3, ..., $n - 1$ edges.

Some specific decomposition problems that have been studied include:

- Arboricity, a decomposition into as few forests as possible
- Cycle double cover, a decomposition into a collection of cycles covering each edge exactly twice
- Edge coloring, a decomposition into as few matchings as possible
- Graph factorization, a decomposition of a regular graph into regular subgraphs of given degrees

Graph Classes

Many problems involve characterizing the members of various classes of graphs. Some examples of such questions are below:

- Enumerating the members of a class
- Characterizing a class in terms of forbidden substructures
- Ascertaining relationships among classes (e.g. does one property of graphs imply another)
- Finding efficient algorithms to decide membership in a class
- Finding representations for members of a class

Graph (Discrete Mathematics)

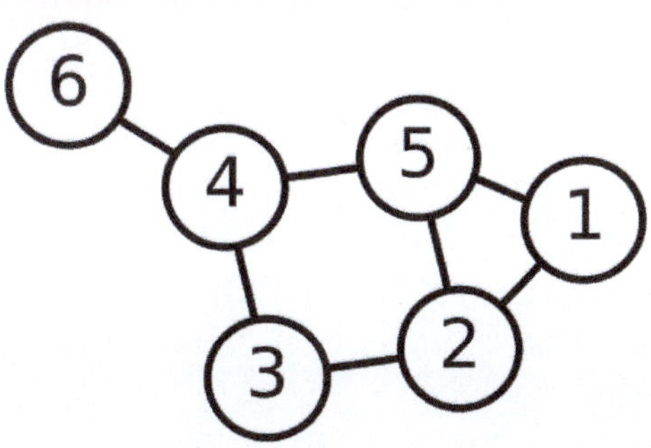

A drawing of a labeled graph on 6 vertices and 7 edges.

In mathematics, and more specifically in graph theory, a graph is a structure amounting to a set of objects in which some pairs of the objects are in some sense "related." The objects correspond to mathematical abstractions called *vertices* (also called *nodes* or *points*) and each of the related pairs of vertices is called an *edge* (also called an *arc* or *line*). Typically, a graph is depicted in diagrammatic form as a set of dots for the vertices, joined by lines or curves for the edges. Graphs are one of the objects of study in discrete mathematics.

The edges may be directed or undirected. For example, if the vertices represent people at a party, and there is an edge between two people if they shake hands, then this graph is undirected because any person A can shake hands with a person B only if B also shakes hands with A. In contrast, if any edge from a person A to a person B corresponds to A's admiring B, then this graph is directed, because admiration is not necessarily reciprocated. The former type of graph is called an *undirected graph* and the edges are called *undirected edges* while the latter type of graph is called a *directed graph* and the edges are called *directed edges*.

Graphs are the basic subject studied by graph theory. The word "graph" was first used in this sense by J. J. Sylvester in 1878.

Definitions

Definitions in graph theory vary. The following are some of the more basic ways of defining graphs and related mathematical structures.

Graph

In the most common sense of the term, a *graph* is an ordered pair $G = (V, E)$ comprising a set V of *vertices*, *nodes* or *points* together with a set E of *edges*, *arcs* or *lines*, which are 2-element subsets of V (i.e., an edge is associated with two vertices, and the association takes the form of the unordered pair of the vertices). To avoid ambiguity, this type of graph may be described precisely as *undirected* and *simple*.

Other senses of *graph* stem from different conceptions of the edge set. In one more generalized notion, E is a set together with a relation of *incidence* that associates with each edge two vertices. In another generalized notion, E is a multiset of unordered pairs of (not necessarily distinct) vertices. Many authors call this type of object a multigraph or pseudograph.

All of these variants and others are described more fully below.

The vertices belonging to an edge are called the *ends* or *end vertices* of the edge. A vertex may exist in a graph and not belong to an edge.

V and E are usually taken to be finite, and many of the well-known results are not true (or are rather different) for *infinite graphs* because many of the arguments fail in the infinite case. Moreover, V is often assumed to be non-empty, but E is allowed to be the empty set. The *order* of a graph is $|V|$, its number of vertices. The *size* of a graph is $|E|$, its number of edges. The *degree* or *valency* of a vertex is the number of edges that connect to it, where an edge that connects to the vertex at both ends (a loop) is counted twice.

For an edge $\{x, y\}$, graph theorists usually use the somewhat shorter notation xy.

Adjacency Relation

The edges E of an undirected graph G induce a symmetric binary relation $\sim$ on V that is called the *adjacency relation* of G. Specifically, for each edge $\{x, y\}$, the vertices x and y are said to be *adjacent* to one another, which is denoted $x \sim y$.

Types of Graphs

Distinction in Terms of the Main Definition

As stated above, in different contexts it may be useful to refine the term *graph* with different degrees of generality. Whenever it is necessary to draw a strict distinction, the following terms are used. Most commonly, in modern texts in graph theory, unless stated otherwise, *graph* means "undirected simple finite graph".

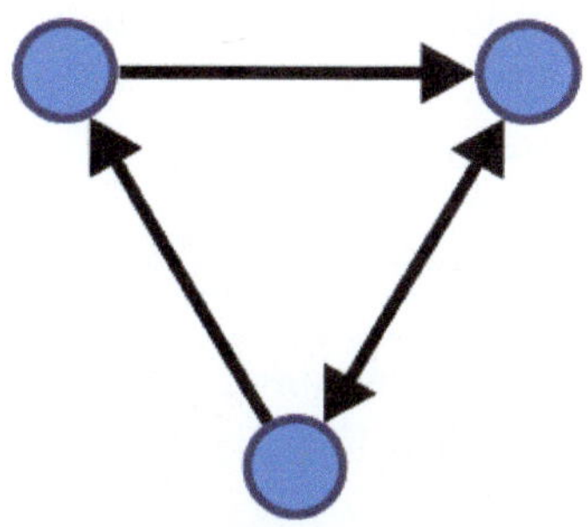

A directed graph.

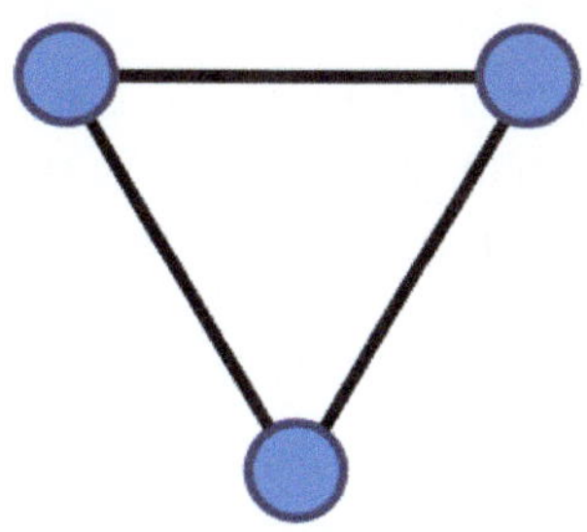

A simple undirected graph with three vertices and three edges. Each vertex has degree two, so this is also a regular graph.

Undirected Graph

An *undirected graph* is a graph in which edges have no orientation. The edge (x, y) is identical to the edge (y, x), i.e., they are not ordered pairs, but sets $\{x, y\}$ (or 2-multisets) of vertices. The maximum number of edges in an undirected graph without a loop is $n(n - 1)/2$.

Directed Graph

A *directed graph* or *digraph* is a graph in which edges have orientations. It is written as an ordered pair $G = (V, A)$ (sometimes $G = (V, E)$) with

V a set whose elements are called *vertices*, *nodes*, or *points*;

A a set of ordered pairs of vertices, called *arrows, directed edges* (sometimes simply *edges* with the corresponding set named E instead of A), *directed arcs*, or *directed lines*.

An arrow (x, y) is considered to be directed *from x to y*; y is called the *head* and x is called the *tail* of the arrow; y is said to be a *direct successor* of x and x is said to be a *direct predecessor* of y. If a path leads from x to y, then y is said to be a *successor* of x and *reachable* from x, and x is said to be a *predecessor* of y. The arrow (y, x) is called the *inverted arrow* of (x, y).

A directed graph G is called *symmetric* if, for every arrow in G, the corresponding inverted arrow also belongs to G. A symmetric loopless directed graph $G = (V, A)$ is equivalent to a simple undirected graph $G' = (V, E)$, where the pairs of inverse arrows in A correspond one-to-one with the edges in E; thus the number of edges in G' is $|E| = |A|/2$, that is half the number of arrows in G.

Oriented Graph

An *oriented graph* is a directed graph in which at most one of (x, y) and (y, x) may be arrows of the graph. That is, it is a directed graph that can be formed as an orientation of an undirected graph. However, some authors use "oriented graph" to mean the same as "directed graph".

Mixed Graph

A *mixed graph* is a graph in which some edges may be directed and some may be undirected. It is written as an ordered triple $G = (V, E, A)$ with V, E, and A defined as above. Directed and undirected graphs are special cases.

Multigraph

Multiple edges are two or more edges that connect the same two vertices. A *loop* is an edge (directed or undirected) that connects a vertex to itself; it may be permitted or not, according to the application. In this context, an edge with two different ends is called a *link*.

A *multigraph*, as opposed to a simple graph, is an undirected graph in which multiple edges (and sometimes loops) are allowed.

Where graphs are defined so as to *disallow* both multiple edges and loops, a multigraph is often defined to mean a graph which can have both multiple edges and loops, although many use the term *pseudograph* for this meaning. Where graphs are defined so as to *allow* both multiple edges and loops, a multigraph is often defined to mean a graph without loops.

Simple Graph

A *simple graph*, as opposed to a multigraph, is an undirected graph in which both multiple edges and loops are disallowed. In a simple graph the edges form a *set* (rather than a multiset) and each edge is an unordered pair of *distinct* vertices. In a simple graph with n vertices, the degree of every vertex is at most $n - 1$.

Quiver

A *quiver* or *multidigraph* is a directed multigraph. A quiver may also have directed loops in it.

Weighted Graph

A *weighted graph* is a graph in which a number (the weight) is assigned to each edge. Such weights might represent for example costs, lengths or capacities, depending on the problem at hand. Some authors call such a graph a *network*. Weighted correlation networks can be defined by soft-thresholding the pairwise correlations among variables (e.g. gene measurements). Such graphs arise in many contexts, for example in shortest path problems such as the traveling salesman problem.

Half-edges, Loose Edges

In certain situations it can be helpful to allow edges with only one end, called *half-edges*, or no ends, called *loose edges*.

Important Classes of Graph

Regular Graph

A *regular graph* is a graph in which each vertex has the same number of neighbours, i.e., every vertex has the same degree. A regular graph with vertices of degree *k* is called a *k*regular graph or regular graph of degree *k*.

Complete Graph

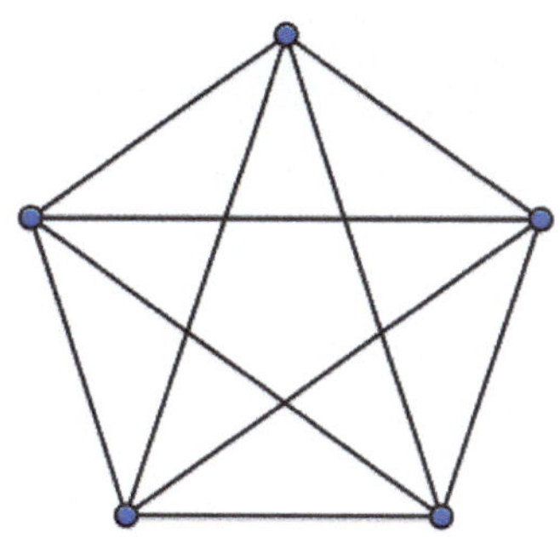

A complete graph with 5 vertices. Each vertex has an edge to every other vertex.

A *complete graph* is a graph in which each pair of vertices is joined by an edge. A complete graph contains all possible edges.

Finite Graph

A *finite graph* is a graph in which the vertex set and the edge set are finite sets. Otherwise, it is called an *infinite graph*.

Most commonly in graph theory it is implied that the graphs discussed are finite. If the graphs are infinite, that is usually specifically stated.

Connected Graph

In an undirected graph, an unordered pair of vertices $\{x, y\}$ is called *connected* if a path leads from x to y. Otherwise, the unordered pair is called *disconnected*.

A *connected graph* is an undirected graph in which every unordered pair of vertices in the graph is connected. Otherwise, it is called a *disconnected graph.*

In a directed graph, an ordered pair of vertices (x, y) is called *strongly connected* if a directed path leads from x to y. Otherwise, the ordered pair is called *weakly connected* if an undirected path leads from x to y after replacing all of its directed edges with undirected edges. Otherwise, the ordered pair is called *disconnected.*

A *strongly connected graph* is a directed graph in which every ordered pair of vertices in the graph is strongly connected. Otherwise, it is called a *weakly connected graph* if every ordered pair of vertices in the graph is weakly connected. Otherwise it is called a *disconnected graph.*

A *k-vertex-connected graph* or *k-edge-connected graph* is a graph in which no set of $k - 1$ vertices (respectively, edges) exists that, when removed, disconnects the graph. A k-vertex-connected graph is often called simply a *k-connected graph.*

Bipartite Graph

A *bipartite graph* is a graph in which the vertex set can be partitioned into two sets, W and X, so that no two vertices in W share a common edge and no two vertices in X share a common edge. Alternatively, it is a graph with a chromatic number of 2.

In a complete bipartite graph, the vertex set is the union of two disjoint sets, W and X, so that every vertex in W is adjacent to every vertex in X but there are no edges within W or X.

Path Graph

A *path graph* or *linear graph* of order $n \geq 2$ is a graph in which the vertices can be listed in an order $v_1, v_2, \ldots, v_n$ such that the edges are the $\{v_i, v_{i+1}\}$ where $i = 1, 2, \ldots, n - 1$. Path graphs can be characterized as connected graphs in which the degree of all but two vertices is 2 and the degree of the two remaining vertices is 1. If a path graph occurs as a subgraph of another graph, it is a path in that graph.

Planar Graph

A *planar graph* is a graph whose vertices and edges can be drawn in a plane such that no two of the edges intersect.

Cycle Graph

A *cycle graph* or *circular graph* of order $n \geq 3$ is a graph in which the vertices can be listed in an order $v_1, v_2, \ldots, v_n$ such that the edges are the $\{v_i, v_{i+1}\}$ where $i = 1, 2, \ldots, n - 1$, plus the edge $\{v_n, v_1\}$. Cycle graphs can be characterized as connected graphs in which the degree of all vertices is 2. If a cycle graph occurs as a subgraph of another graph, it is a cycle or circuit in that graph.

Tree

A *tree* is a connected graph with no cycles.

A *forest* is a graph with no cycles, i.e. the disjoint union of one or more trees.

Advanced Classes

More advanced kinds of graphs are:

- Petersen graph and its generalizations;
- perfect graphs;
- cographs;
- chordal graphs;
- other graphs with large automorphism groups: vertex-transitive, arc-transitive, and distance-transitive graphs;
- strongly regular graphs and their generalizations distance-regular graphs.

Properties of Graphs

Two edges of a graph are called *adjacent* if they share a common vertex. Two arrows of a directed graph are called *consecutive* if the head of the first one is the tail of the second one. Similarly, two vertices are called *adjacent* if they share a common edge (*consecutive* if the first one is the tail and the second one is the head of an arrow), in which case the common edge is said to *join* the two vertices. An edge and a vertex on that edge are called *incident*.

The graph with only one vertex and no edges is called the *trivial graph*. A graph with only vertices and no edges is known as an *edgeless graph*. The graph with no vertices and no edges is sometimes called the *null graph* or *empty graph*, but the terminology is not consistent and not all mathematicians allow this object.

Normally, the vertices of a graph, by their nature as elements of a set, are distinguishable. This kind of graph may be called *vertex-labeled*. However, for many questions it is better to treat vertices as indistinguishable. (Of course, the vertices may be still distinguishable by the properties of the graph itself, e.g., by the numbers of incident edges.) The same remarks apply to edges, so graphs with labeled edges are called *edge-labeled*. Graphs with labels attached to edges or vertices are more generally designated as *labeled*. Consequently, graphs in which vertices are indistinguishable and edges are indistinguishable are called *unlabeled*. (Note that in the literature, the term *labeled* may apply to other kinds of labeling, besides that which serves only to distinguish different vertices or edges.)

The category of all graphs is the slice category Set $\downarrow$ D where D: Set $\rightarrow$ Set is the functor taking a set s to $s \times s$.

Examples

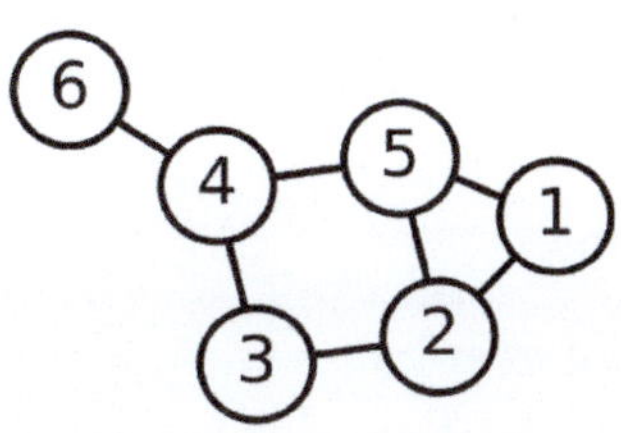

A graph with six nodes.

- The diagram at right is a graphic representation of the following graph:

  $V = \{1, 2, 3, 4, 5, 6\}$;

  $E = \{\{1, 2\}, \{1, 5\}, \{2, 3\}, \{2, 5\}, \{3, 4\}, \{4, 5\}, \{4, 6\}\}$.

- In category theory, a small category can be represented by a directed multigraph in which the objects of the category are represented as vertices and the morphisms as directed edges. Then, the functors between categories induce some, but not necessarily all, of the digraph morphisms of the graph.
- In computer science, directed graphs are used to represent knowledge (e.g., conceptual graph), finite state machines, and many other discrete structures.
- A binary relation R on a set X defines a directed graph. An element x of X is a direct predecessor of an element y of X if and only if xRy.
- A directed graph can model information networks such as Twitter, with one user following another.
- Particularly regular examples of directed graphs are given by the Cayley graphs of finitely-generated groups, as well as Schreier coset graphs

Graph Operations

There are several operations that produce new graphs from initial ones, which might be classified into the following categories:

- *unary operations*, which create a new graph from an initial one, such as:
    - edge contraction,
    - line graph,
    - dual graph,
    - complement graph,
    - graph rewriting;
- *binary operations*, which create a new graph from two initial ones, such as:
    - disjoint union of graphs,
    - cartesian product of graphs,
    - tensor product of graphs,
    - strong product of graphs,
    - lexicographic product of graphs,
    - series-parallel graphs.

Generalizations

In a hypergraph, an edge can join more than two vertices.

An undirected graph can be seen as a simplicial complex consisting of 1-simplices (the edges) and 0-simplices (the vertices). As such, complexes are generalizations of graphs since they allow for higher-dimensional simplices.

Every graph gives rise to a matroid.

In model theory, a graph is just a structure. But in that case, there is no limitation on the number of edges: it can be any cardinal number.

In computational biology, power graph analysis introduces power graphs as an alternative representation of undirected graphs.

In geographic information systems, geometric networks are closely modeled after graphs, and borrow many concepts from graph theory to perform spatial analysis on road networks or utility grids.

Algebraic Graph Theory

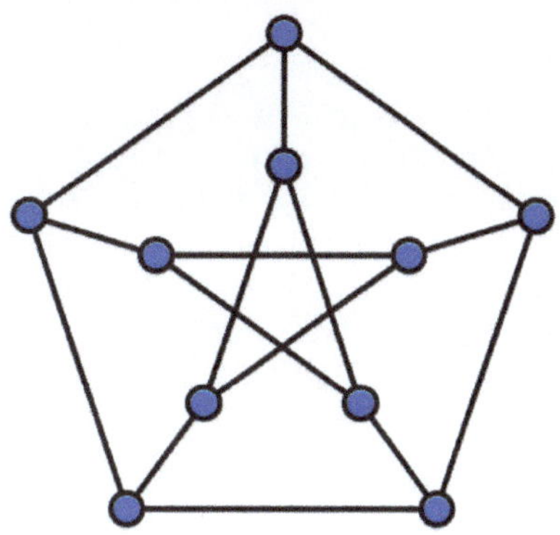

A highly symmetrical graph, the Petersen graph, which is vertex-transitive, symmetric, distance-transitive, and distance-regular. It has diameter 2. Its automorphism group has 120 elements, and is in fact the symmetric group .

Algebraic graph theory is a branch of mathematics in which algebraic methods are applied to problems about graphs. This is in contrast to geometric, combinatoric, or algorithmic approaches. There are three main branches of algebraic graph theory, involving the use of linear algebra, the use of group theory, and the study of graph invariants.

Branches of Algebraic Graph Theory

Using Linear Algebra

The first branch of algebraic graph theory involves the study of graphs in connection with linear algebra. Especially, it studies the spectrum of the adjacency matrix, or the Laplacian matrix of a graph (this part of algebraic graph theory is also called spectral graph theory). For the Petersen graph, for example, the spectrum of the adjacency matrix is (−2, −2, −2, −2, 1, 1, 1, 1, 1, 3). Several theorems relate properties of the spectrum to other graph properties. As a simple example, a connected graph with diameter D will have at least $D+1$ distinct values in its spectrum. Aspects of graph spectra have been used in analysing the synchronizability of networks.

Using Group Theory

The second branch of algebraic graph theory involves the study of graphs in connection to group theory, particularly automorphism groups and geometric group theory. The focus is placed on various families of graphs based on symmetry (such as symmetric graphs, vertex-transitive graphs, edge-transitive graphs, distance-transitive graphs, distance-regular graphs, and strongly regular graphs), and on the inclusion relationships between these families. Certain of such categories of graphs are sparse enough that lists of graphs can be drawn up. By Frucht's theorem, all groups can be represented as the automorphism group of a connected graph (indeed, of a cubic graph). Another connection with group theory is that, given any group, symmetrical graphs known as Cayley graphs can be generated, and these have properties related to the structure of the group.

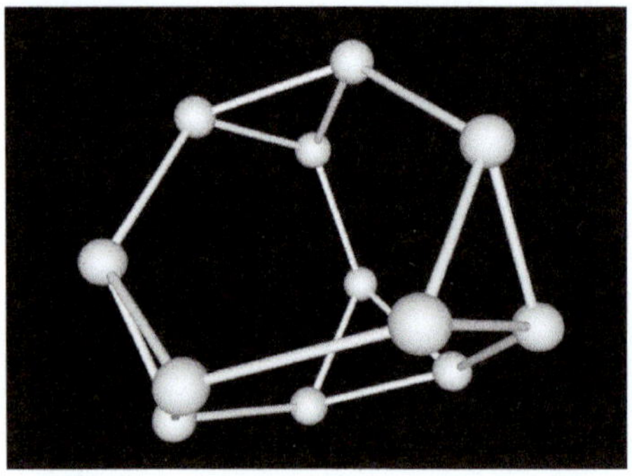

A Cayley graph for the alternating group A_4, forming a truncated tetrahedron in three dimensions. All Cayley graphs are vertex-transitive, but some vertex-transitive graphs (like the Petersen graph) are not Cayley graphs.

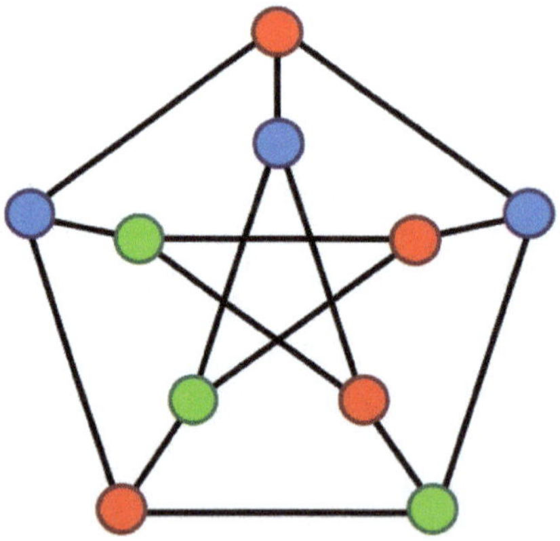

A proper vertex coloring of the Petersen graph with 3 colors, the minimum number possible. According to the chromatic polynomial, there are 120 such colorings with 3 colors.

This second branch of algebraic graph theory is related to the first, since the symmetry properties of a graph are reflected in its spectrum. In particular, the spectrum of a highly symmetrical graph, such as the Petersen graph, has few distinct values (the Petersen graph has 3, which is the minimum possible, given its diameter). For Cayley graphs, the spectrum can be related directly to the structure of the group, in particular to its irreducible characters.

Studying Graph Invariants

Finally, the third branch of algebraic graph theory concerns algebraic properties of invariants of graphs, and especially the chromatic polynomial, the Tutte polynomial and knot invariants. The chromatic polynomial of a graph, for example, counts the number of its proper vertex colorings. For the Petersen graph, this polynomial is $t(t-1)(t-2)(t^7-12t^6+67t^5-230t^4+529t^3-814t^2+775t-352)$. In particular, this means that the Petersen graph cannot be properly colored with one or two colors, but can be colored in 120 different ways with 3 colors. Much work in this area of algebraic graph theory was motivated by attempts to prove the four color theorem. However, there are still many open problems, such as characterizing graphs which have the same chromatic polynomial, and determining which polynomials are chromatic.

Extremal Graph Theory

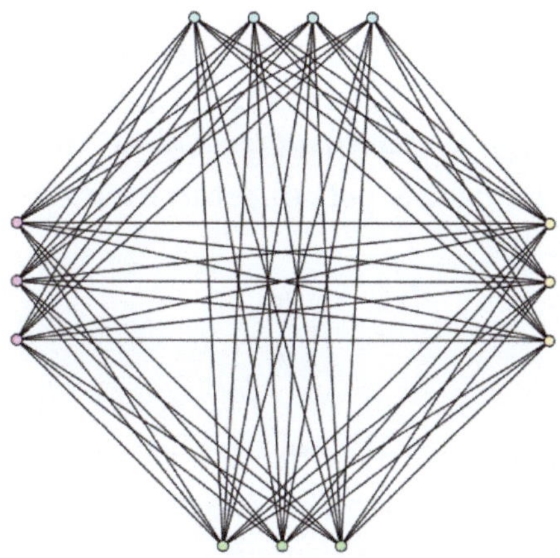

The Turán graph T(n,r) is an example of an extremal graph. It has the maximum possible number of edges for a graph on n vertices without (r+1)-cliques. This is T(13,4).

Extremal graph theory is a branch of the mathematical field of graph theory. Extremal graph theory studies extremal (maximal or minimal) graphs which satisfy a certain property. Extremality can be taken with respect to different graph invariants, such as order, size or girth. More abstractly, it studies how global properties of a graph influence local substructures of the graph.

Examples

For example, a simple extremal graph theory question is "which acyclic graphs on n vertices have the maximum number of edges?" The extremal graphs for this question are trees on n vertices, which have $n - 1$ edges. More generally, a typical question is the following: given a graph property P, an invariant u, and a set of graphs H, we wish to find the minimum value of m such that every graph in H which has u larger than m possess property P. In the example above, H was the set of n-vertex graphs, P was the property of being cyclic, and u was the number of edges in the graph. Thus every graph on n vertices with more than $n - 1$ edges must contain a cycle.

Several foundational results in extremal graph theory are questions of the above-mentioned form. For instance, the question of how many edges an n-vertex graph can have before it must contain as subgraph a clique of size k is answered by Turán's theorem. Instead of cliques, if the same question is asked for complete multi-partite graphs, the answer is given by the Erdős–Stone theorem.

History

Extremal graph theory, in its strictest sense, is a branch of graph theory developed and loved by Hungarians.

Bollobás (2004)

Extremal graph theory started in 1941 when Turán proved his theorem determining those graphs of order n, not containing the complete graph K_k of order k, and extremal with respect to size (that is, with as many edges as possible). Another crucial year for the subject was 1975 when Szemerédi proved his result a vital tool in attacking extremal problems.

Density Results

A typical result in extremal graph theory is Turán's theorem. It answers the following question. What is the maximum possible number of edges in an undirected graph G with n vertices which does not contain K_3 (three vertices A, B, C with edges AB, AC, BC; i.e. a triangle) as a subgraph? The complete bipartite graph where the partite sets differ in their size by at most 1, is the only extremal graph with this property. It contains

$$\left\lfloor \frac{n^2}{4} \right\rfloor$$

edges. Similar questions have been studied with various other subgraphs H instead of K_3; for instance, the Zarankiewicz problem concerns the largest graph that does not contain a fixed complete bipartite graph as a subgraph, and the even circuit theorem concerns the largest graph without a fixed-length even cycle. Turán also found the (unique) largest graph not containing K_k which is named after him, namely the Turán graph. This graph is the complete join of "k-1" independent sets (as equi-sized as possible) and has at most

$$\left\lfloor \frac{(k-2)n^2}{2(k-1)} \right\rfloor = \left\lfloor \left(1 - \frac{1}{k-1}\right) \frac{n^2}{2} \right\rfloor$$

edges. For C_4, the largest graph on n vertices not containing C_4 has

$$\left(\frac{1}{2} + o(1)\right) n^{3/2}$$

edges.

Minimum Degree Conditions

The preceding theorems give conditions for a small object to appear within a (perhaps) very large graph. At the opposite extreme, one might search for conditions which force the existence of a structure which covers every vertex. But it is possible for a graph with

$$\binom{n-1}{2}$$

edges to have an isolated vertex - even though almost every possible edge is present in the graph - which means that even a graph with very high density may have no interesting structure covering every vertex. Simple edge counting conditions, which give no indication as to how the edges in the graph are distributed, thus often tend to give uninteresting results for very large structures. Instead, we introduce the concept of minimum degree. The minimum degree of a graph *G* is defined to be

$$\delta(G) = \min_{v \in G} d(v).$$

Specifying a large minimum degree removes the objection that there may be a few 'pathological' vertices; if the minimum degree of a graph *G* is 1, for example, then there can be no isolated vertices (even though *G* may have very few edges).

A classic result is Dirac's theorem, which states that every graph *G* with *n* vertices and minimum degree at least *n*/2 contains a Hamilton cycle.

Clique (Graph Theory)

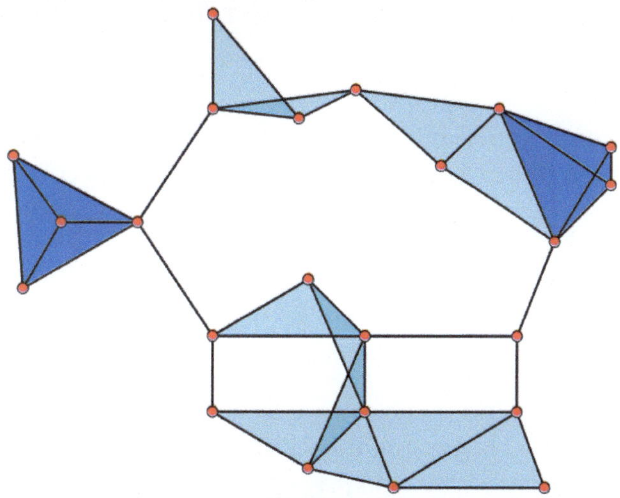

A graph with
23 × 1-vertex cliques (the vertices),
42 × 2-vertex cliques (the edges),
19 × 3-vertex cliques (light and dark blue triangles), and
2 × 4-vertex cliques (dark blue areas).
The 11 light blue triangles form maximal cliques. The two dark blue 4-cliques are both maximum and maximal, and the clique number of the graph is 4.

In the mathematical area of graph theory, a clique (/ˈkliːk/ or /ˈklɪk/) is a subset of vertices of an undirected graph such that its induced subgraph is complete; that is, every two distinct vertices in the clique are adjacent. Cliques are one of the basic concepts of graph theory and are used in many other mathematical problems and constructions on graphs. Cliques have also been studied in computer science: the task of finding whether there is a clique of a given size in a graph (the clique problem) is NP-complete, but despite this hardness result, many algorithms for finding cliques have been studied.

Although the study of complete subgraphs goes back at least to the graph-theoretic reformulation of Ramsey theory by Erdős & Szekeres (1935), the term *clique* comes from Luce & Perry (1949), who used complete subgraphs in social networks to model cliques of people; that is, groups of people all of whom know each other. Cliques have many other applications in the sciences and particularly in bioinformatics.

Definitions

A clique, C, in an undirected graph $G = (V, E)$ is a subset of the vertices, $C \subseteq V$, such that every two distinct vertices are adjacent. This is equivalent to the condition that the subgraph of G induced by C is complete. In some cases, the term clique may also refer to the subgraph directly.

A maximal clique is a clique that cannot be extended by including one more adjacent vertex, that is, a clique which does not exist exclusively within the vertex set of a larger clique.

A maximum clique of a graph, G, is a clique, such that there is no clique with more vertices.

The clique number $\omega(G)$ of a graph G is the number of vertices in a maximum clique in G.

The intersection number of G is the smallest number of cliques that together cover all edges of G.

The clique cover number of a graph G is the smallest number of cliques of G whose union covers $V(G)$.

The opposite of a clique is an independent set, in the sense that every clique corresponds to an independent set in the complement graph. The clique cover problem concerns finding as few cliques as possible that include every vertex in the graph.

A related concept is a biclique, a complete bipartite subgraph. The bipartite dimension of a graph is the minimum number of bicliques needed to cover all the edges of the graph.

Mathematics

Mathematical results concerning cliques include the following.

- Turán's theorem (Turán 1941) gives a lower bound on the size of a clique in dense graphs. If a graph has sufficiently many edges, it must contain a large clique. For instance, every graph with n vertices and more than edges must contain a three-vertex clique.
- Ramsey's theorem (Graham, Rothschild & Spencer 1990) states that every graph or its complement graph contains a clique with at least a logarithmic number of vertices.
- According to a result of Moon & Moser (1965), a graph with $3n$ vertices can have at most 3^n maximal cliques. The graphs meeting this bound are the Moon–Moser graphs $K_{3,3,\ldots}$, a special case of the Turán graphs arising as the extremal cases in Turán's theorem.

- Hadwiger's conjecture, still unproven, relates the size of the largest clique minor in a graph (its Hadwiger number) to its chromatic number.
- The Erdős–Faber–Lovász conjecture is another unproven statement relating graph coloring to cliques.

Several important classes of graphs may be defined by their cliques:

- A cluster graph is a graph whose connected components are cliques.
- A block graph is a graph whose biconnected components are cliques.
- A chordal graph is a graph whose vertices can be ordered into a perfect elimination ordering, an ordering such that the neighbors of each vertex v that come later than v in the ordering form a clique.
- A cograph is a graph all of whose induced subgraphs have the property that any maximal clique intersects any maximal independent set in a single vertex.
- An interval graph is a graph whose maximal cliques can be ordered in such a way that, for each vertex v, the cliques containing v are consecutive in the ordering.
- A line graph is a graph whose edges can be covered by edge-disjoint cliques in such a way that each vertex belongs to exactly two of the cliques in the cover.
- A perfect graph is a graph in which the clique number equals the chromatic number in every induced subgraph.
- A split graph is a graph in which some clique contains at least one endpoint of every edge.
- A triangle-free graph is a graph that has no cliques other than its vertices and edges.

Additionally, many other mathematical constructions involve cliques in graphs. Among them,

- The clique complex of a graph G is an abstract simplicial complex $X(G)$ with a simplex for every clique in G
- A simplex graph is an undirected graph $\kappa(G)$ with a vertex for every clique in a graph G and an edge connecting two cliques that differ by a single vertex. It is an example of median graph, and is associated with a median algebra on the cliques of a graph: the median $m(A,B,C)$ of three cliques A, B, and C is the clique whose vertices belong to at least two of the cliques A, B, and C.
- The clique-sum is a method for combining two graphs by merging them along a shared clique.
- Clique-width is a notion of the complexity of a graph in terms of the minimum number of distinct vertex labels needed to build up the graph from disjoint unions, relabeling operations, and operations that connect all pairs of vertices with given labels. The graphs with clique-width one are exactly the disjoint unions of cliques.
- The intersection number of a graph is the minimum number of cliques needed to cover all the graph's edges.

- The clique graph of a graph is the intersection graph of its maximal cliques.

Closely related concepts to complete subgraphs are subdivisions of complete graphs and complete graph minors. In particular, Kuratowski's theorem and Wagner's theorem characterize planar graphs by forbidden complete and complete bipartite subdivisions and minors, respectively.

Computer Science

In computer science, the clique problem is the computational problem of finding a maximum clique, or all cliques, in a given graph. It is NP-complete, one of Karp's 21 NP-complete problems (Karp 1972). It is also fixed-parameter intractable, and hard to approximate. Nevertheless, many algorithms for computing cliques have been developed, either running in exponential time (such as the Bron–Kerbosch algorithm) or specialized to graph families such as planar graphs or perfect graphs for which the problem can be solved in polynomial time.

Applications

The word "clique", in its graph-theoretic usage, arose from the work of Luce & Perry (1949), who used complete subgraphs to model cliques (groups of people who all know each other) in social networks. For continued efforts to model social cliques graph-theoretically.

Many different problems from bioinformatics have been modeled using cliques. For instance, Ben-Dor, Shamir & Yakhini (1999) model the problem of clustering gene expression data as one of finding the minimum number of changes needed to transform a graph describing the data into a graph formed as the disjoint union of cliques; Tanay, Sharan & Shamir (2002) discuss a similar biclustering problem for expression data in which the clusters are required to be cliques. Sugihara (1984) uses cliques to model ecological niches in food webs. Day & Sankoff (1986) describe the problem of inferring evolutionary trees as one of finding maximum cliques in a graph that has as its vertices characteristics of the species, where two vertices share an edge if there exists a perfect phylogeny combining those two characters. Samudrala & Moult (1998) model protein structure prediction as a problem of finding cliques in a graph whose vertices represent positions of subunits of the protein. And by searching for cliques in a protein-protein interaction network, Spirin & Mirny (2003) found clusters of proteins that interact closely with each other and have few interactions with proteins outside the cluster. Power graph analysis is a method for simplifying complex biological networks by finding cliques and related structures in these networks.

In electrical engineering, Prihar (1956) uses cliques to analyze communications networks, and Paull & Unger (1959) use them to design efficient circuits for computing partially specified Boolean functions. Cliques have also been used in automatic test pattern generation: a large clique in an incompatibility graph of possible faults provides a lower bound on the size of a test set. Cong & Smith (1993) describe an application of cliques in finding a hierarchical partition of an electronic circuit into smaller subunits.

In chemistry, Rhodes et al. (2003) use cliques to describe chemicals in a chemical database that have a high degree of similarity with a target structure. Kuhl, Crippen & Friesen (1983) use cliques to model the positions in which two chemicals will bind to each other.

Cycle (Graph Theory)

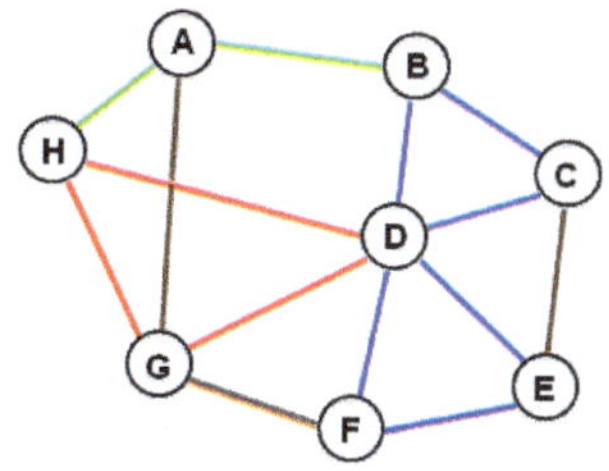

A graph with edges colored to illustrate path H-A-B (green), closed path or walk with a repeated vertex B-D-E-F-D-C-B (blue) and a cycle with no repeated edge or vertex H-D-G-H (red)

In graph theory, there are several different types of objects called cycles, principally a closed walk and a simple cycle; also, e.g., an element of the cycle space of the graph.

Definitions

A closed walk consists of a sequence of vertices starting and ending at the same vertex, with each two consecutive vertices in the sequence adjacent to each other in the graph. In a directed graph, each edge must be traversed by the walk consistently with its direction: the edge must be oriented from the earlier of two consecutive vertices to the later of the two vertices in the sequence. The choice of starting vertex is not important: traversing the same cyclic sequence of edges from different starting vertices produces the same closed walk.

A simple cycle may be defined either as a closed walk with no repetitions of vertices and edges allowed, other than the repetition of the starting and ending vertex, or as the set of edges in such a walk. The two definitions are equivalent in directed graphs, where simple cycles are also called directed cycles: the cyclic sequence of vertices and edges in a walk is completely determined by the set of edges that it uses. In undirected graphs the set of edges of a cycle can be traversed by a walk in either of two directions, giving two possible directed cycles for every undirected cycle. (For closed walks more generally, in directed or undirected graphs, the multiset of edges does not unambiguously determine the vertex ordering.) A circuit can be a closed walk allowing repetitions of vertices but not edges; however, it can also be a simple cycle, so explicit definition is recommended when it is used.

In order to maintain a consistent terminology, for the rest of this article, "cycle" means a simple cycle, except where otherwise stated.

Chordless Cycles

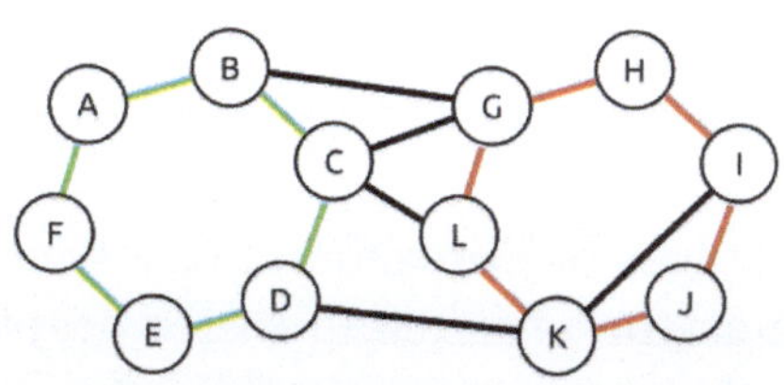

In this graph the green cycle (A-B-C-D-E-F-A) is chordless whereas the red cycle (G-H-I-J-K-L-G) is not. This is because the edge K-I is a chord.

A chordless cycle in a graph, also called a hole or an induced cycle, is a cycle such that no two vertices of the cycle are connected by an edge that does not itself belong to the cycle. An antihole is the complement of a graph hole. Chordless cycles may be used to characterize perfect graphs: by the strong perfect graph theorem, a graph is perfect if and only if none of its holes or antiholes have an odd number of vertices that is greater than three. A chordal graph, a special type of perfect graph, has no holes of any size greater than three.

The girth of a graph is the length of its shortest cycle; this cycle is necessarily chordless. Cages are defined as the smallest regular graphs with given combinations of degree and girth.

A peripheral cycle is a cycle in a graph with the property that every two edges not on the cycle can be connected by a path whose interior vertices avoid the cycle. In a graph that is not formed by adding one edge to a cycle, a peripheral cycle must be an induced cycle.

Cycle Space

The term *cycle* may also refer to an element of the cycle space of a graph. This consists of the edge sets that have even degree at every vertex; it forms a vector space over the two-element finite field. Using methods from algebraic topology, it may be generalized to vector spaces or modules over other rings such as the integers, real numbers, etc. By Veblen's theorem, every element of the cycle space may be formed by as an edge-disjoint union of simple cycles. A cycle basis of the graph is a set of simple cycles that forms a basis of the cycle space.

Cycle Detection

The existence of a cycle in directed and undirected graphs can be determined by whether depth-first search (DFS) finds an edge that points to an ancestor of the current vertex (it contains a back edge). In an undirected graph, finding any already visited vertex will indicate a back edge. All the back edges which DFS skips over are part of cycles. In the case of undirected graphs, only $O(n)$ time is required to find a cycle in an n-vertex graph, since at most $n - 1$ edges can be tree edges.

Many topological sorting algorithms will detect cycles too, since those are obstacles for topological order to exist. Also, if a directed graph has been divided into strongly connected components, cycles only exist within the components and not between them, since cycles are strongly connected.

For directed graphs, Rocha–Thatte Algorithm is a distributed cycle detection algorithm. Distributed cycle detection algorithms are useful for processing large-scale graphs using a distributed graph processing system on a computer cluster (or supercomputer).

Applications of cycle detection include the use of wait-for graphs to detect deadlocks in concurrent systems.

Covering Graphs by Cycles

In his 1736 paper on the Seven Bridges of Königsberg, widely considered to be the birth of graph theory, Leonhard Euler proved that, for a finite undirected graph to have a closed walk that visits each edge exactly once, it is necessary and sufficient that it be connected except for isolated vertices (that is, all edges are contained in one component) and have even degree at each vertex. The corresponding characterization for the existence of a closed walk visiting each edge exactly once in a directed graph is that the graph be strongly connected and have equal numbers of incoming and outgoing edges at each vertex. In either case, the resulting walk is known as an Euler cycle or Euler tour. If a finite undirected graph has even degree at each of its vertices, regardless of whether it is connected, then it is possible to find a set of simple cycles that together cover each edge exactly once: this is Veblen's theorem. When a connected graph does not meet the conditions of Euler's theorem, a closed walk of minimum length covering each edge at least once can nevertheless be found in polynomial time by solving the route inspection problem.

The problem of finding a single simple cycle that covers each vertex exactly once, rather than covering the edges, is much harder. Such a cycle is known as a Hamiltonian cycle, and determining whether it exists is NP-complete. Much research has been published concerning classes of graphs that can be guaranteed to contain Hamiltonian cycles; one example is Ore's theorem that a Hamiltonian cycle can always be found in a graph for which every non-adjacent pair of vertices have degrees summing to at least the total number of vertices in the graph.

The cycle double cover conjecture states that, for every bridgeless graph, there exists a multiset of simple cycles that covers each edge of the graph exactly twice. Proving that this is true (or finding a counterexample) remains an open problem.

Graph Classes Defined by Cycles

Several important classes of graphs can be defined by or characterized by their cycles. These include:

- Bipartite graph, a graph without odd cycles.
- Cactus graph, a graph in which every nontrivial biconnected component is a cycle
- Cycle graph, a graph that consists of a single cycle.
- Chordal graph, a graph with no induced cycles of length greater than three
- Directed acyclic graph, a directed graph with no cycles
- Perfect graph, a graph with no induced cycles or their complements of odd length greater than three
- Pseudoforest, a graph in which each connected component has at most one cycle
- Strongly connected graph, a directed graph in which every edge is part of a cycle
- Triangle-free graph, a graph without three-vertex cycles

Split (Graph Theory)

In graph theory, a split of an undirected graph is a cut whose cut-set forms a complete bipartite graph. A graph is prime if it has no splits. The splits of a graph can be collected into a tree-like structure called the split decomposition or join decomposition, which can be constructed in linear time. This decomposition has been used for fast recognition of circle graphs and distance-hereditary graphs, as well as for other problems in graph algorithms.

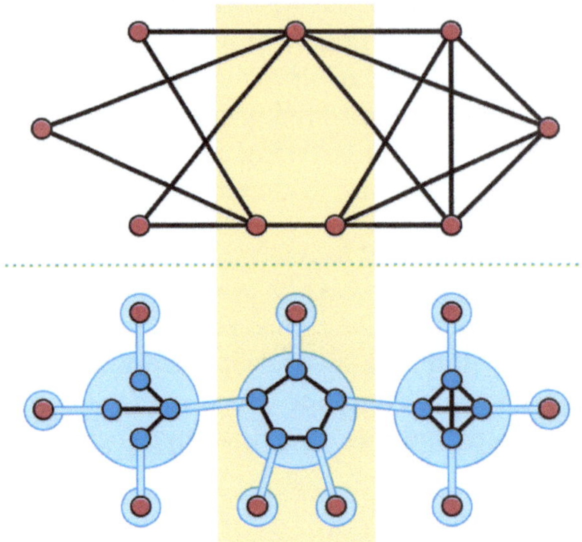

A graph with two nontrivial strong splits (top) and its split decomposition (bottom). The three quotient graphs are a star (left), a prime graph (center), and a complete graph (right).

Splits and split decompositions were first introduced by Cunningham (1982), who also studied variants of the same notions for directed graphs.

Definitions

A cut of an undirected graph is a partition of the vertices into two nonempty subsets, the sides of the cut. The subset of edges that have one endpoint in each side are called a cut-set. When a cut-set forms a complete bipartite graph, its cut is called a split. Thus, a split can be described as a partition of the vertices of the graph into two subsets X and Y, such that every neighbor of X in Y is adjacent to every neighbor of Y in X.

A cut or split is trivial when one of its two sides has only one vertex in it; every trivial cut is a split. A graph is said to be prime (with respect to splits) if it has no nontrivial splits.

Two splits are said to cross if each side of one split has a non-empty intersection with each side of the other split. A split is called strong when it is not crossed by any other split. As a special case, every trivial split is strong. The strong splits of a graph give rise to a structure called the split decomposition or join decomposition of the graph. This decomposition can be represented by a tree whose leaves correspond one-to-one with the given graph, and whose edges correspond one-to-one with the strong splits of the graph, such that the partition of leaves formed by removing any edge from the tree is the same as the partition of vertices given by the associated strong split.

Each internal node i of the split decomposition tree of a graph G is associated with a graph G_i, called the quotient graph for node i. The quotient graph can be formed by deleting i from the tree, forming subsets of vertices in G corresponding to the leaves in each of the resulting subtrees, and collapsing each of these vertex sets into a single vertex. Every quotient graph has one of three forms: it may be a prime graph, a complete graph, or a star.

A graph may have exponentially different splits, but they are all represented in the split decomposition tree, either as an edge of the tree (for a strong split) or as an arbitrary partition of a complete or star quotient graph (for a split that is not strong).

Examples

In a complete graph or complete bipartite graph, every cut is a split.

In a cycle graph of length four, the partition of the vertices given by 2-coloring the cycle is a nontrivial split, but for cycles of any longer length there are no nontrivial splits.

A bridge of a graph that is not 2-edge-connected corresponds to a split, with each side of the split formed by the vertices on one side of the bridge. The cut-set of the split is just the single bridge edge, which is a special case of a complete bipartite subgraph. Similarly, if v is an articulation point of a graph that is not 2-vertex-connected, then the graph has multiple splits in which v and some but not all of the components formed by its deletion are on one side, and the remaining components are on the other side. In these examples, the cut-set of the split forms a star.

Algorithms

Cunningham (1982) already showed that it is possible to find the split decomposition in polynomial time. After subsequent improvements to the algorithm, linear time algorithms were discovered by Dahlhaus (2000) and Charbit, de Montgolfier & Raffinot (2012).

Applications

Split decomposition has been applied in the recognition of several important graph classes:

- A distance-hereditary graph is a graph whose split decomposition contains no prime quotients. Based on this characterization, it is possible to use the split decomposition to recognize distance-hereditary graphs in linear time.
- Parity graphs can be recognized in linear time as the graphs in which each split quotient is either complete or bipartite.
- A circle graph is the intersection graph of a family of chords of a circle. A given graph is a circle graph if and only if each of the quotients of its split decomposition is a circle graph, so testing whether a graph is a circle graph can be reduced to the same problem on the prime quotient graphs of the graph. More, when a circle graph is prime, the combinatorial structure of the set of chords representing it is uniquely determined, which simplifies the task of recognizing this structure. Based on these ideas, it is possible to recognize circle graphs in polynomial time.

Split decomposition has also been used to simplify the solution of some problems that are NP-hard on arbitrary graphs:

- As Cunningham (1982) already observed, the maximum independent set of any graph may be found by a dynamic programming algorithm on a bottom-up traversal of its split decomposition tree. At each node we choose the maximum weight independent set on its quotient graph, weighted by the sizes of the independent sets already computed at child nodes.

- Although another algorithm given by Cunningham (1982) is flawed, a similar bottom-up traversal can be used to compute the maximum clique of a graph by combining computations of weighted maximum cliques in its quotient graphs.
- Rao (2008) also presents algorithms for connected dominating sets, complete dominating sets, and graph coloring.

These methods can lead to polynomial time algorithms for graphs in which each quotient graph has a simple structure that allows its subproblem to be computed efficiently. For instance, this is true of the graphs in which each quotient graph has constant size.

Graph Factorization

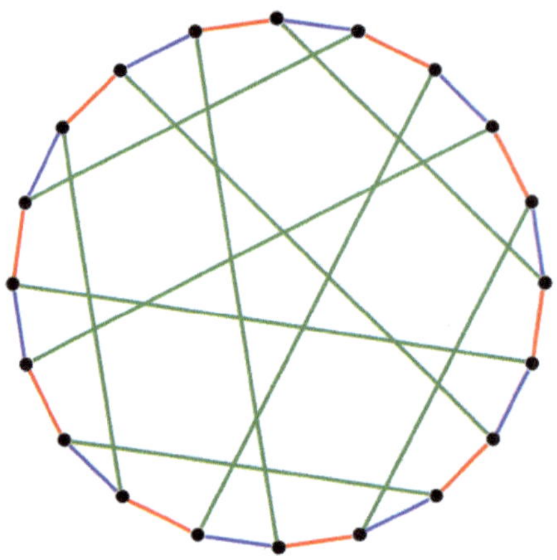

1-factorization of Desargues graph: each color class is a 1-factor.

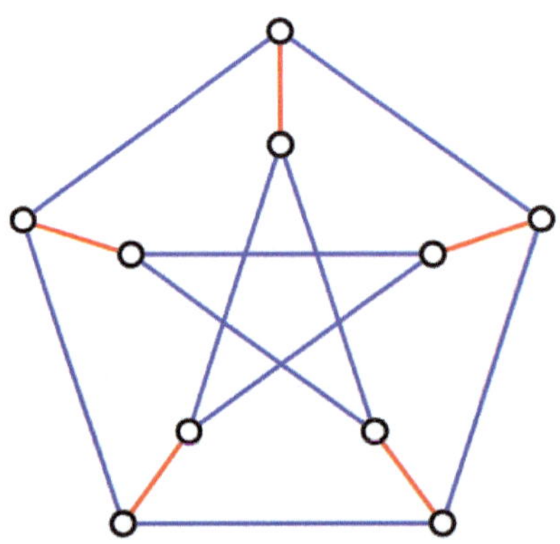

Petersen graph can be partitioned into a 1-factor (red) and a 2-factor (blue). However, the graph is not 1-factorable.

In graph theory, a factor of a graph G is a spanning subgraph, i.e., a subgraph that has the same vertex set as G. A ***k*-factor** of a graph is a spanning k-regular subgraph, and a ***k*-factorization** partitions the edges of the graph into disjoint k-factors. A graph G is said to be ***k*-factorable** if it admits a k-factorization. In particular, a 1-factor is a perfect matching, and a 1-factorization of a k-regular graph is an edge coloring with k colors. A 2-factor is a collection of cycles that spans all vertices of the graph.

1-factorization

If a graph is 1-factorable, then it has to be a regular graph. However, not all regular graphs are 1-factorable. A k-regular graph is 1-factorable if it has chromatic index k; examples of such graphs include:

- Any regular bipartite graph. Hall's marriage theorem can be used to show that a k-regular bipartite graph contains a perfect matching. One can then remove the perfect matching to obtain a $(k - 1)$-regular bipartite graph, and apply the same reasoning repeatedly.
- Any complete graph with an even number of nodes.

However, there are also k-regular graphs that have chromatic index $k + 1$, and these graphs are not 1-factorable; examples of such graphs include:

- Any regular graph with an odd number of nodes.
- The Petersen graph.

Complete Graphs

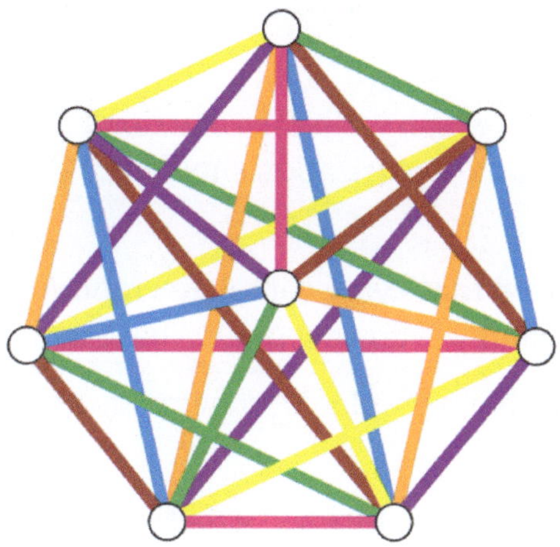

1-factorization of K_8 in which each 1-factor consists of an edge from the center to a vertex of a heptagon together with all possible perpendicular edges

A 1-factorization of a complete graph corresponds to pairings in a round-robin tournament. The 1-factorization of complete graphs is a special case of Baranyai's theorem concerning the 1-factorization of complete hypergraphs.

One method for constructing a 1-factorization of a complete graph involves placing all but one of the vertices on a circle, forming a regular polygon, with the remaining vertex at the center of the circle. With this arrangement of vertices, one way of constructing a 1-factor of the graph is to choose an edge e from the center to a single polygon vertex together with all possible edges that lie on lines perpendicular to e. The 1-factors that can be constructed in this way form a 1-factorization of the graph.

The number of distinct 1-factorizations of K_2, K_4, K_6, K_8, ... is 1, 1, 6, 6240, 1225566720, 252282619805368320, 98758655816833727741338583040, ... OEIS A000438.

1-Factorization Conjecture

Let G be a k-regular graph with $2n$ nodes. If k is sufficiently large, it is known that G has to be 1-factorable:

- If $k = 2n - 1$, then G is the complete graph K_{2n}, and hence 1-factorable.
- If $k = 2n - 2$, then G can be constructed by removing a perfect matching from K_{2n}. Again, G is 1-factorable.
- Chetwynd & Hilton (1985) show that if $k \geq 12n/7$, then G is 1-factorable.

The 1-factorization conjecture is a long-standing conjecture that states that $k \approx n$ is sufficient. In precise terms, the conjecture is:

- If n is odd and $k \geq n$, then G is 1-factorable. If n is even and $k \geq n - 1$ then G is 1-factorable.

The overfull conjecture implies the 1-factorization conjecture.

Perfect 1-factorization

A perfect pair from a 1-factorization is a pair of 1-factors whose union induces a Hamiltonian cycle.

A perfect 1-factorization (P1F) of a graph is a 1-factorization having the property that every pair of 1-factors is a perfect pair. A perfect 1-factorization should not be confused with a perfect matching (also called a 1-factor).

In 1964, Anton Kotzig conjectured that every complete graph K_{2n} where $n \geq 2$ has a perfect 1-factorization. So far, it is known that the following graphs have a perfect 1-factorization:

- the infinite family of complete graphs K_{2p} where p is an odd prime (by Anderson and also Nakamura, independently),
- the infinite family of complete graphs K_{p+1} where p is an odd prime,
- and sporadic additional results, including K_{2n} where $2n \in \{16, 28, 36, 40, 50, 126, 170, 244, 344, 730, 1332, 1370, 1850, 2198, 3126, 6860, 12168, 16808, 29792\}$. Some newer results are collected here.

If the complete graph K_{n+1} has a perfect 1-factorization, then the complete bipartite graph $K_{n,n}$ also has a perfect 1-factorization.

2-factorization

If a graph is 2-factorable, then it has to be $2k$-regular for some integer k. Julius Petersen showed in 1891 that this necessary condition is also sufficient: any $2k$-regular graph is 2-factorable.

If a connected graph is $2k$-regular and has an even number of edges it may also be k-factored, by choosing each of the two factors to be an alternating subset of the edges of an Euler tour. This applies only to connected graphs; disconnected counterexamples include disjoint unions of odd cycles, or of copies of K_{2k+1}.

References

- Zwillinger, Daniel (2002-11-27). CRC Standard Mathematical Tables and Formulae (31st ed.). Chapman & Hall/CRC. ISBN 1-58488-291-3.
- Balakrishnan, V.K. (2005). Schaum's outline of theory and problems of graph theory ([Nachdr.]. ed.). McGraw–Hill. ISBN 978-0070054899.
- Gross, Jonathan L.; Yellen, Jay (2005), "4.6 Graphs and Vector Spaces", Graph Theory and Its Applications (2nd ed.), CRC Press, pp. 197–207, ISBN 9781584885054.
- Tucker, Alan (2006). "Chapter 2: Covering Circuits and Graph Colorings". Applied Combinatorics (5th ed.). Hoboken: John Wiley & sons. p. 49. ISBN 978-0-471-73507-6.
- Silberschatz, Abraham; Peter Galvin; Greg Gagne (2003). Operating System Concepts. John Wiley & Sons, INC. p. 260. ISBN 0-471-25060-0.
- Diestel, Reinhard (2005), Graph Theory (3rd ed.), Springer, ISBN 3-540-26182-6, Chapter 2: "Matching, covering and packing". Electronic edition.
- Bondy, John Adrian; Murty, U. S. R. (1976), Graph Theory with Applications, North-Holland, ISBN 0-444-19451-7, Section 5.1: "Matchings".
- West, Douglas B. "1-Factorization Conjecture (1985?)". Open Problems – Graph Theory and Combinatorics. Retrieved 2010-01-09.

7

Topology : An Essential Aspect

Topology studies a collection of subsets that turn the given set into a topological space. Topology has a number of properties; some of these are connectedness and compactness. The essential aspects of topology that are dealt within this chapter are general topology, algebraic topology, differential topology and geometric topology.

Topology

In mathematics, topology (from the Greek τόπος, *place*, and λόγος, *study*) is concerned with the properties of space that are preserved under continuous deformations, such as stretching and bending, but not tearing or gluing. This can be studied by considering a collection of subsets, called open sets, that satisfy certain properties, turning the given set into what is known as a topological space. Important topological properties include connectedness and compactness.

Möbius strips, which have only one surface and one edge, are a kind of object studied in topology.

Topology developed as a field of study out of geometry and set theory, through analysis of concepts such as space, dimension, and transformation. Such ideas go back to Gottfried Leibniz, who in the 17th century envisioned the *geometria situs* (Greek-Latin for "geometry of place") and *analysis situs* (Greek-Latin for "picking apart of place"). Leonhard Euler's Seven Bridges of Königsberg Problem and Polyhedron Formula are arguably the field's first theorems. The term *topology* was introduced by Johann Benedict Listing in the 19th century, although it was not until the first decades of the 20th century that the idea of a topological space was developed. By the middle of the 20th century, topology had become a major branch of mathematics.

Topology has many subfields:

- General topology, also called point-set topology, establishes the foundational aspects of topology and investigates properties of topological spaces and concepts inherent to topological spaces. It defines the basic notions used in all other branches of topology (including concepts like compactness and connectedness).

- Algebraic topology tries to measure degrees of connectivity using algebraic constructs such as homology and homotopy groups.
- Differential topology is the field dealing with differentiable functions on differentiable manifolds. It is closely related to differential geometry and together they make up the geometric theory of differentiable manifolds.
- Geometric topology primarily studies manifolds and their embeddings (placements) in other manifolds. A particularly active area is low-dimensional topology, which studies manifolds of four or fewer dimensions. This includes knot theory, the study of mathematical knots.

A three-dimensional depiction of a thickened trefoil knot, the simplest non-trivial knot

History

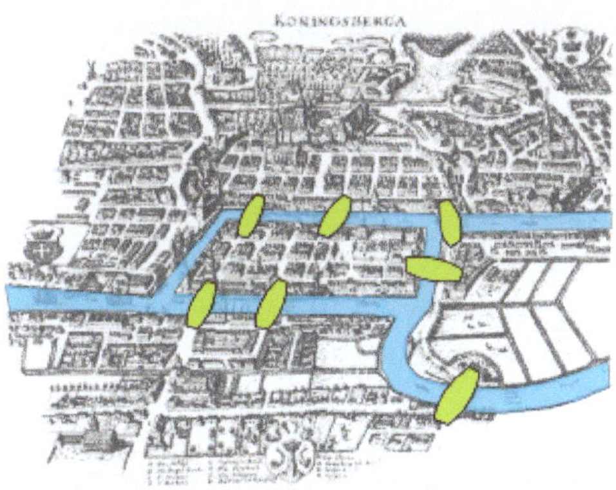

The Seven Bridges of Königsberg was a problem solved by Euler.

Topology, as a well-defined mathematical discipline, originates in the early part of the twentieth century, but some isolated results can be traced back several centuries. Among these are certain questions in geometry investigated by Leonhard Euler. His 1736 paper on the Seven Bridges of Königsberg is regarded as one of the first practical applications of topology. On 14 November 1750 Euler wrote to a friend that he had realised the importance of the *edges* of a polyhedron. This led to his polyhedron formula, $V - E + F = 2$ (where V, E and F respectively indicate the number of vertices, edges and faces of the polyhedron). Some authorities regard this analysis as the first theorem, signalling the birth of topology.

Further contributions were made by Augustin-Louis Cauchy, Ludwig Schläfli, Johann Benedict Listing, Bernhard Riemann and Enrico Betti. Listing introduced the term "Topologie" in *Vorstu-*

dien zur Topologie, written in his native German, in 1847, having used the word for ten years in correspondence before its first appearance in print. The English form "topology" was used in 1883 in Listing's obituary in the journal *Nature* to distinguish "...qualitative geometry from the ordinary geometry in which quantitative relations chiefly are treated." The term "topologist" in the sense of a specialist in topology was used in 1905 in the magazine *Spectator*.

Their work was corrected, consolidated and greatly extended by Henri Poincaré. In 1895 he published his ground-breaking paper on *Analysis Situs*, which introduced the concepts now known as homotopy and homology, which are now considered part of algebraic topology.

Unifying the work on function spaces of Georg Cantor, Vito Volterra, Cesare Arzelà, Jacques Hadamard, Giulio Ascoli and others, Maurice Fréchet introduced the metric space in 1906. A metric space is now considered a special case of a general topological space, with any given topological space potentially giving rise to many distinct metric spaces. In 1914, Felix Hausdorff coined the term "topological space" and gave the definition for what is now called a Hausdorff space. Currently, a topological space is a slight generalization of Hausdorff spaces, given in 1922 by Kazimierz Kuratowski.

Modern topology depends strongly on the ideas of set theory, developed by Georg Cantor in the later part of the 19th century. In addition to establishing the basic ideas of set theory, Cantor considered point sets in Euclidean space as part of his study of Fourier series.

Introduction

Topology can be formally defined as "the study of qualitative properties of certain objects (called topological spaces) that are invariant under a certain kind of transformation (called a continuous map), especially those properties that are invariant under a certain kind of invertible transformation (called homeomorphism)."

Topology is also used to refer to a structure imposed upon a set X, a structure that essentially 'characterizes' the set X as a topological space by taking proper care of properties such as convergence, connectedness and continuity, upon transformation.

Topological spaces show up naturally in almost every branch of mathematics. This has made topology one of the great unifying ideas of mathematics.

The motivating insight behind topology is that some geometric problems depend not on the exact shape of the objects involved, but rather on the way they are put together. For example, the square and the circle have many properties in common: they are both one dimensional objects (from a topological point of view) and both separate the plane into two parts, the part inside and the part outside.

In one of the first papers in topology, Leonhard Euler demonstrated that it was impossible to find a route through the town of Königsberg (now Kaliningrad) that would cross each of its seven bridges exactly once. This result did not depend on the lengths of the bridges, nor on their distance from one another, but only on connectivity properties: which bridges connect to which islands or riverbanks. This problem in introductory mathematics called *Seven Bridges of Königsberg* led to the branch of mathematics known as graph theory.

A continuous deformation (a type of homeomorphism) of a mug into a doughnut (torus) and back

Similarly, the hairy ball theorem of algebraic topology says that "one cannot comb the hair flat on a hairy ball without creating a cowlick." This fact is immediately convincing to most people, even though they might not recognize the more formal statement of the theorem, that there is no non-vanishing continuous tangent vector field on the sphere. As with the *Bridges of Königsberg*, the result does not depend on the shape of the sphere; it applies to any kind of smooth blob, as long as it has no holes.

To deal with these problems that do not rely on the exact shape of the objects, one must be clear about just what properties these problems *do* rely on. From this need arises the notion of homeomorphism. The impossibility of crossing each bridge just once applies to any arrangement of bridges homeomorphic to those in Königsberg, and the hairy ball theorem applies to any space homeomorphic to a sphere.

Intuitively, two spaces are homeomorphic if one can be deformed into the other without cutting or gluing. A traditional joke is that a topologist cannot distinguish a coffee mug from a doughnut, since a sufficiently pliable doughnut could be reshaped to a coffee cup by creating a dimple and progressively enlarging it, while shrinking the hole into a handle.

Homeomorphism can be considered the most basic *topological equivalence*. Another is homotopy equivalence. This is harder to describe without getting technical, but the essential notion is that two objects are homotopy equivalent if they both result from "squishing" some larger object.

An introductory exercise is to classify the uppercase letters of the English alphabet according to homeomorphism and homotopy equivalence. The result depends partially on the font used. The figures use the sans-serif Myriad font. Homotopy equivalence is a rougher relationship than homeomorphism; a homotopy equivalence class can contain several homeomorphism classes. The simple case of homotopy equivalence described above can be used here to show two letters are homotopy equivalent. For example, O fits inside P and the tail of the P can be squished to the "hole" part.

Homeomorphism classes are:

- no holes,
- no holes three tails,
- no holes four tails,

- one hole no tail,
- one hole one tail,
- one hole two tails,
- two holes no tail, and
- a bar with four tails (the "bar" on the *K* is almost too short to see).

Homotopy classes are larger, because the tails can be squished down to a point. They are:

- one hole,
- two holes, and
- no holes.

To classify the letters correctly, we must show that two letters in the same class are equivalent and two letters in different classes are not equivalent. In the case of homeomorphism, this can be done by selecting points and showing their removal disconnects the letters differently. For example, X and Y are not homeomorphic because removing the center point of the X leaves four pieces; whatever point in Y corresponds to this point, its removal can leave at most three pieces. The case of homotopy equivalence is harder and requires a more elaborate argument showing an algebraic invariant, such as the fundamental group, is different on the supposedly differing classes.

Letter topology has practical relevance in stencil typography. For instance, Braggadocio font stencils are made of one connected piece of material.

Concepts

Topologies on Sets

The term topology also refers to a specific mathematical idea central to the area of mathematics called topology. Informally, a topology tells how elements of a set relate spatially to each other. The same set can have different topologies. For instance, the real line, the complex plane, and the Cantor set can be thought of as the same set with different topologies.

Formally, let X be a set and let τ be a family of subsets of X. Then τ is called a *topology on X* if:

1. Both the empty set and X are elements of τ
2. Any union of elements of τ is an element of τ
3. Any intersection of finitely many elements of τ is an element of τ

If τ is a topology on X, then the pair (X, τ) is called a *topological space*. The notation X_τ may be used to denote a set X endowed with the particular topology τ.

The members of τ are called *open sets* in X. A subset of X is said to be closed if its complement is in τ (i.e., its complement is open). A subset of X may be open, closed, both (clopen set), or neither. The empty set and X itself are always both closed and open. An open set containing a point x is called a 'neighborhood' of x.

A set with a topology is called a topological space.

Continuous Functions and Homeomorphisms

A function or map from one topological space to another is called *continuous* if the inverse image of any open set is open. If the function maps the real numbers to the real numbers (both spaces with the Standard Topology), then this definition of continuous is equivalent to the definition of continuous in calculus. If a continuous function is one-to-one and onto, and if the inverse of the function is also continuous, then the function is called a homeomorphism and the domain of the function is said to be homeomorphic to the range. Another way of saying this is that the function has a natural extension to the topology. If two spaces are homeomorphic, they have identical topological properties, and are considered topologically the same. The cube and the sphere are homeomorphic, as are the coffee cup and the doughnut. But the circle is not homeomorphic to the doughnut.

Manifolds

While topological spaces can be extremely varied and exotic, many areas of topology focus on the more familiar class of spaces known as manifolds. A manifold is a topological space that resembles Euclidean space near each point. More precisely, each point of an n-dimensional manifold has a neighbourhood that is homeomorphic to the Euclidean space of dimension n. Lines and circles, but not figure eights, are one-dimensional manifolds. Two-dimensional manifolds are also called surfaces. Examples include the plane, the sphere, and the torus, which can all be realized without self-intersection in three dimensions, but also the Klein bottle and real projective plane, which cannot.

Topics

General Topology

General topology is the branch of topology dealing with the basic set-theoretic definitions and constructions used in topology. It is the foundation of most other branches of topology, including differential topology, geometric topology, and algebraic topology. Another name for general topology is point-set topology.

The fundamental concepts in point-set topology are *continuity*, *compactness*, and *connectedness*. Intuitively, continuous functions take nearby points to nearby points. Compact sets are those that can be covered by finitely many sets of arbitrarily small size. Connected sets are sets that cannot be divided into two pieces that are far apart. The words *nearby*, *arbitrarily small*, and *far apart* can all be made precise by using open sets. If we change the definition of *open set*, we change what continuous functions, compact sets, and connected sets are. Each choice of definition for *open set* is called a *topology*. A set with a topology is called a *topological space*.

Metric spaces are an important class of topological spaces where distances can be assigned a number called a *metric*. Having a metric simplifies many proofs, and many of the most common topological spaces are metric spaces.

Algebraic Topology

Algebraic topology is a branch of mathematics that uses tools from abstract algebra to study topological spaces. The basic goal is to find algebraic invariants that classify topological spaces up to homeomorphism, though usually most classify up to homotopy equivalence.

The most important of these invariants are homotopy groups, homology, and cohomology.

Although algebraic topology primarily uses algebra to study topological problems, using topology to solve algebraic problems is sometimes also possible. Algebraic topology, for example, allows for a convenient proof that any subgroup of a free group is again a free group.

Differential Topology

Differential topology is the field dealing with differentiable functions on differentiable manifolds. It is closely related to differential geometry and together they make up the geometric theory of differentiable manifolds.

More specifically, differential topology considers the properties and structures that require only a smooth structure on a manifold to be defined. Smooth manifolds are 'softer' than manifolds with extra geometric structures, which can act as obstructions to certain types of equivalences and deformations that exist in differential topology. For instance, volume and Riemannian curvature are invariants that can distinguish different geometric structures on the same smooth manifold—that is, one can smoothly "flatten out" certain manifolds, but it might require distorting the space and affecting the curvature or volume.

Geometric Topology

Geometric topology is a branch of topology that primarily focuses on low-dimensional manifolds (i.e. dimensions 2,3 and 4) and their interaction with geometry, but it also includes some higher-dimensional topology. Some examples of topics in geometric topology are orientability, handle decompositions, local flatness, and the planar and higher-dimensional Schönflies theorem.

In high-dimensional topology, characteristic classes are a basic invariant, and surgery theory is a key theory.

Low-dimensional topology is strongly geometric, as reflected in the uniformization theorem in 2 dimensions – every surface admits a constant curvature metric; geometrically, it has one of 3 possible geometries: positive curvature/spherical, zero curvature/flat, negative curvature/hyperbolic – and the geometrization conjecture (now theorem) in 3 dimensions – every 3-manifold can be cut into pieces, each of which has one of eight possible geometries.

2-dimensional topology can be studied as complex geometry in one variable (Riemann surfaces are complex curves) – by the uniformization theorem every conformal class of metrics is equivalent to a unique complex one, and 4-dimensional topology can be studied from the point of view of complex geometry in two variables (complex surfaces), though not every 4-manifold admits a complex structure.

Generalizations

Occasionally, one needs to use the tools of topology but a "set of points" is not available. In pointless topology one considers instead the lattice of open sets as the basic notion of the theory, while Grothendieck topologies are structures defined on arbitrary categories that allow the definition of sheaves on those categories, and with that the definition of general cohomology theories.

Applications

Biology

Knot theory, a branch of topology, is used in biology to study the effects of certain enzymes on DNA. These enzymes cut, twist, and reconnect the DNA, causing knotting with observable effects such as slower electrophoresis. Topology is also used in evolutionary biology to represent the relationship between phenotype and genotype. Phenotypic forms that appear quite different can be separated by only a few mutations depending on how genetic changes map to phenotypic changes during development.

Computer Science

Topological data analysis uses techniques from algebraic topology to determine the large scale structure of a set (for instance, determining if a cloud of points is spherical or toroidal). The main method used by topological data analysis is:

1. Replace a set of data points with a family of simplicial complexes, indexed by a proximity parameter.
2. Analyse these topological complexes via algebraic topology — specifically, via the theory of persistent homology.
3. Encode the persistent homology of a data set in the form of a parameterized version of a Betti number, which is called a barcode.

Physics

In physics, topology is used in several areas such as condensed matter physics, quantum field theory and cosmology.

A topological quantum field theory (or topological field theory or TQFT) is a quantum field theory that computes topological invariants.

Although TQFTs were invented by physicists, they are also of mathematical interest, being related to, among other things, knot theory and the theory of four-manifolds in algebraic topology, and to the theory of moduli spaces in algebraic geometry. Donaldson, Jones, Witten, and Kontsevich have all won Fields Medals for work related to topological field theory.

The topological classification of Calabi-Yau manifolds has important implications in string theory, as different manifolds can sustain different kinds of strings.

In cosmology, topology can be used to describe the overall shape of the universe. This area is known as spacetime topology.

Robotics

The various possible positions of a robot can be described by a manifold called configuration space. In the area of motion planning, one finds paths between two points in configuration space. These paths represent a motion of the robot's joints and other parts into the desired location and pose.

General Topology

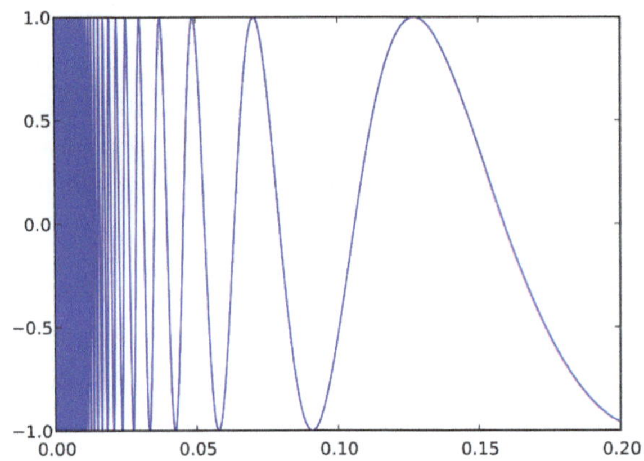

The Topologist's sine curve, a useful example in point-set topology. It is connected but not path-connected.

In mathematics, general topology is the branch of topology that deals with the basic set-theoretic definitions and constructions used in topology. It is the foundation of most other branches of topology, including differential topology, geometric topology, and algebraic topology. Another name for general topology is point-set topology.

The fundamental concepts in point-set topology are *continuity*, *compactness*, and *connectedness*:

- Continuous functions, intuitively, take nearby points to nearby points.
- Compact sets are those that can be covered by finitely many sets of arbitrarily small size.
- Connected sets are sets that cannot be divided into two pieces that are far apart.

The words 'nearby', 'arbitrarily small', and 'far apart' can all be made precise by using open sets. If we change the definition of 'open set', we change what continuous functions, compact sets, and connected sets are. Each choice of definition for 'open set' is called a *topology*. A set with a topology is called a *topological space*.

Metric spaces are an important class of topological spaces where distances can be assigned a number called a *metric*. Having a metric simplifies many proofs, and many of the most common topological spaces are metric spaces.

History

General topology grew out of a number of areas, most importantly the following:

- the detailed study of subsets of the real line (once known as the *topology of point sets*, this usage is now obsolete)
- the introduction of the manifold concept
- the study of metric spaces, especially normed linear spaces, in the early days of functional analysis.

General topology assumed its present form around 1940. It captures, one might say, almost everything in the intuition of continuity, in a technically adequate form that can be applied in any area of mathematics.

A Topology on a Set

Let X be a set and let τ be a family of subsets of X. Then τ is called a *topology on* X if:

1. Both the empty set and X are elements of τ
2. Any union of elements of τ is an element of τ
3. Any intersection of finitely many elements of τ is an element of τ

If τ is a topology on X, then the pair (X, τ) is called a *topological space*. The notation X_τ may be used to denote a set X endowed with the particular topology τ.

The members of τ are called *open sets* in X. A subset of X is said to be closed if its complement is in τ (i.e., its complement is open). A subset of X may be open, closed, both (clopen set), or neither. The empty set and X itself are always both closed and open.

Basis for a Topology

A base (or basis) B for a topological space X with topology T is a collection of open sets in T such that every open set in T can be written as a union of elements of B. We say that the base *generates* the topology T. Bases are useful because many properties of topologies can be reduced to statements about a base that generates that topology—and because many topologies are most easily defined in terms of a base that generates them.

Subspace and Quotient

Every subset of a topological space can be given the subspace topology in which the open sets are the intersections of the open sets of the larger space with the subset. For any indexed family of topological spaces, the product can be given the product topology, which is generated by the inverse images of open sets of the factors under the projection mappings. For example, in finite products, a basis for the product topology consists of all products of open sets. For infinite products, there is the additional requirement that in a basic open set, all but finitely many of its projections are the entire space.

A quotient space is defined as follows: if X is a topological space and Y is a set, and if $f : X \to Y$ is a surjective function, then the quotient topology on Y is the collection of subsets of Y that have open inverse images under f. In other words, the quotient topology is the finest topology on Y for which f is continuous. A common example of a quotient topology is when an equivalence relation is defined on the topological space X. The map f is then the natural projection onto the set of equivalence classes.

Examples of Topological Spaces

A given set may have many different topologies. If a set is given a different topology, it is viewed as a different topological space. Any set can be given the discrete topology in which every subset is open. The only convergent sequences or nets in this topology are those that are eventually constant. Also, any set can be given the trivial topology (also called the indiscrete topology), in which only the empty set and the whole space are open. Every sequence and net in this topology converges to every point of the space. This example shows that in general topological spaces, limits of sequences need not be unique. However, often topological spaces must be Hausdorff spaces where limit points are unique.

There are many ways to define a topology on R, the set of real numbers. The standard topology on R is generated by the open intervals. The set of all open intervals forms a base or basis for the topology, meaning that every open set is a union of some collection of sets from the base. In particular, this means that a set is open if there exists an open interval of non zero radius about every point in the set. More generally, the Euclidean spaces R^n can be given a topology. In the usual topology on R^n the basic open sets are the open balls. Similarly, C, the set of complex numbers, and C^n have a standard topology in which the basic open sets are open balls.

Every metric space can be given a metric topology, in which the basic open sets are open balls defined by the metric. This is the standard topology on any normed vector space. On a finite-dimensional vector space this topology is the same for all norms.

Many sets of linear operators in functional analysis are endowed with topologies that are defined by specifying when a particular sequence of functions converges to the zero function.

Any local field has a topology native to it, and this can be extended to vector spaces over that field.

Every manifold has a natural topology since it is locally Euclidean. Similarly, every simplex and every simplicial complex inherits a natural topology from R^n.

The Zariski topology is defined algebraically on the spectrum of a ring or an algebraic variety. On R^n or C^n, the closed sets of the Zariski topology are the solution sets of systems of polynomial equations.

A linear graph has a natural topology that generalises many of the geometric aspects of graphs with vertices and edges.

The Sierpiński space is the simplest non-discrete topological space. It has important relations to the theory of computation and semantics.

There exist numerous topologies on any given finite set. Such spaces are called finite topological spaces. Finite spaces are sometimes used to provide examples or counterexamples to conjectures about topological spaces in general.

Any set can be given the cofinite topology in which the open sets are the empty set and the sets whose complement is finite. This is the smallest T_1 topology on any infinite set.

Any set can be given the cocountable topology, in which a set is defined as open if it is either empty or its complement is countable. When the set is uncountable, this topology serves as a counterexample in many situations.

The real line can also be given the lower limit topology. Here, the basic open sets are the half open intervals $[a, b)$. This topology on R is strictly finer than the Euclidean topology defined above; a sequence converges to a point in this topology if and only if it converges from above in the Euclidean topology. This example shows that a set may have many distinct topologies defined on it.

If Γ is an ordinal number, then the set $\Gamma = [0, \Gamma)$ may be endowed with the order topology generated by the intervals (a, b), $[0, b)$ and (a, Γ) where a and b are elements of Γ.

Continuous Functions

Continuity is expressed in terms of neighborhoods: f is continuous at some point $x \in X$ if and only if for any neighborhood V of $f(x)$, there is a neighborhood U of x such that $f(U) \subseteq V$. Intuitively, continuity means no matter how "small" V becomes, there is always a U containing x that maps inside V and whose image under f contains $f(x)$. This is equivalent to the condition that the preimages of the open (closed) sets in Y are open (closed) in X. In metric spaces, this definition is equivalent to the ε–δ-definition that is often used in analysis.

An extreme example: if a set X is given the discrete topology, all functions

$$f: X \to T$$

to any topological space T are continuous. On the other hand, if X is equipped with the indiscrete topology and the space T set is at least T_0, then the only continuous functions are the constant functions. Conversely, any function whose range is indiscrete is continuous.

Alternative Definitions

Several equivalent definitions for a topological structure exist and thus there are several equivalent ways to define a continuous function.

Neighborhood Definition

Definitions based on preimages are often difficult to use directly. The following criterion expresses continuity in terms of neighborhoods: f is continuous at some point $x \in X$ if and only if for any neighborhood V of $f(x)$, there is a neighborhood U of x such that $f(U) \subseteq V$. Intuitively, continuity means no matter how "small" V becomes, there is always a U containing x that maps inside V.

If X and Y are metric spaces, it is equivalent to consider the neighborhood system of open balls centered at x and $f(x)$ instead of all neighborhoods. This gives back the above δ-ε definition of continuity in the context of metric spaces. However, in general topological spaces, there is no notion of nearness or distance.

Note, however, that if the target space is Hausdorff, it is still true that f is continuous at a if and only if the limit of f as x approaches a is $f(a)$. At an isolated point, every function is continuous.

Sequences and Nets

In several contexts, the topology of a space is conveniently specified in terms of limit points. In many instances, this is accomplished by specifying when a point is the limit of a sequence, but for some spaces that are too large in some sense, one specifies also when a point is the limit of more general sets of points indexed by a directed set, known as nets. A function is continuous only if it takes limits of sequences to limits of sequences. In the former case, preservation of limits is also sufficient; in the latter, a function may preserve all limits of sequences yet still fail to be continuous, and preservation of nets is a necessary and sufficient condition.

In detail, a function $f: X \rightarrow Y$ is sequentially continuous if whenever a sequence (x_n) in X converges to a limit x, the sequence $(f(x_n))$ converges to $f(x)$. Thus sequentially continuous functions "preserve sequential limits". Every continuous function is sequentially continuous. If X is a first-countable space and countable choice holds, then the converse also holds: any function preserving sequential limits is continuous. In particular, if X is a metric space, sequential continuity and continuity are equivalent. For non first-countable spaces, sequential continuity might be strictly weaker than continuity. (The spaces for which the two properties are equivalent are called sequential spaces.) This motivates the consideration of nets instead of sequences in general topological spaces. Continuous functions preserve limits of nets, and in fact this property characterizes continuous functions.

Closure Operator Definition

Instead of specifying the open subsets of a topological space, the topology can also be determined by a closure operator (denoted cl), which assigns to any subset $A \subseteq X$ its closure, or an interior operator (denoted int), which assigns to any subset A of X its interior. In these terms, a function

$$f:(X,\mathrm{cl}) \rightarrow (X',\mathrm{cl}')$$

between topological spaces is continuous in the sense above if and only if for all subsets A of X

$$f(\mathrm{cl}(A)) \subseteq \mathrm{cl}'(f(A)).$$

That is to say, given any element x of X that is in the closure of any subset A, $f(x)$ belongs to the closure of $f(A)$. This is equivalent to the requirement that for all subsets A' of X'

$$f^{-1}(\mathrm{cl}'(A')) \supseteq \mathrm{cl}(f^{-1}(A')).$$

Moreover,

$$f:(X,\mathrm{int}) \rightarrow (X',\mathrm{int}')$$

is continuous if and only if

$$f^{-1}(\mathrm{int}'(A)) \subseteq \mathrm{int}(f^{-1}(A))$$

for any subset A of X.

Properties

If $f: X \to Y$ and $g: Y \to Z$ are continuous, then so is the composition $g \circ f: X \to Z$. If $f: X \to Y$ is continuous and

- X is compact, then $f(X)$ is compact.
- X is connected, then $f(X)$ is connected.
- X is path-connected, then $f(X)$ is path-connected.
- X is Lindelöf, then $f(X)$ is Lindelöf.
- X is separable, then $f(X)$ is separable.

The possible topologies on a fixed set X are partially ordered: a topology τ_1 is said to be coarser than another topology τ_2 (notation: $\tau_1 \subseteq \tau_2$) if every open subset with respect to τ_1 is also open with respect to τ_2. Then, the identity map

$$\mathrm{id}_X: (X, \tau_2) \to (X, \tau_1)$$

is continuous if and only if $\tau_1 \subseteq \tau_2$. More generally, a continuous function

$$(X, \tau_X) \to (Y, \tau_Y)$$

stays continuous if the topology τ_Y is replaced by a coarser topology and/or τ_X is replaced by a finer topology.

Homeomorphisms

Symmetric to the concept of a continuous map is an open map, for which *images* of open sets are open. In fact, if an open map f has an inverse function, that inverse is continuous, and if a continuous map g has an inverse, that inverse is open. Given a bijective function f between two topological spaces, the inverse function f^{-1} need not be continuous. A bijective continuous function with continuous inverse function is called a *homeomorphism*.

If a continuous bijection has as its domain a compact space and its codomain is Hausdorff, then it is a homeomorphism.

Defining Topologies Via Continuous Functions

Given a function

$$f: X \to S,$$

where X is a topological space and S is a set (without a specified topology), the final topology on S is defined by letting the open sets of S be those subsets A of S for which $f^{-1}(A)$ is open in X. If S has an existing topology, f is continuous with respect to this topology if and only if the existing topology is coarser than the final topology on S. Thus the final topology can be characterized as the finest topology on S that makes f continuous. If f is surjective, this topology is canonically identified with the quotient topology under the equivalence relation defined by f.

Dually, for a function f from a set S to a topological space, the initial topology on S has as open subsets A of S those subsets for which $f(A)$ is open in X. If S has an existing topology, f is continuous with respect to this topology if and only if the existing topology is finer than the initial topology on S. Thus the initial topology can be characterized as the coarsest topology on S that makes f continuous. If f is injective, this topology is canonically identified with the subspace topology of S, viewed as a subset of X.

More generally, given a set S, specifying the set of continuous functions

$$S \to X$$

into all topological spaces X defines a topology. Dually, a similar idea can be applied to maps

$$X \to S.$$

This is an instance of a universal property.

Compact Sets

Formally, a topological space X is called *compact* if each of its open covers has a finite subcover. Otherwise it is called *non-compact*. Explicitly, this means that for every arbitrary collection

$$\{U_\alpha\}_{\alpha \in A}$$

of open subsets of X such that

$$X = \bigcup_{\alpha \in A} U_\alpha,$$

there is a finite subset J of A such that

$$X = \bigcup_{i \in J} U_i.$$

Some branches of mathematics such as algebraic geometry, typically influenced by the French school of Bourbaki, use the term *quasi-compact* for the general notion, and reserve the term *compact* for topological spaces that are both Hausdorff and *quasi-compact*. A compact set is sometimes referred to as a *compactum*, plural *compacta*.

Every closed interval in R of finite length is compact. More is true: In R^n, a set is compact if and only if it is closed and bounded.

Every continuous image of a compact space is compact.

A compact subset of a Hausdorff space is closed.

Every continuous bijection from a compact space to a Hausdorff space is necessarily a homeomorphism.

Every sequence of points in a compact metric space has a convergent subsequence.

Every compact finite-dimensional manifold can be embedded in some Euclidean space R^n.

Connected Sets

A topological space X is said to be disconnected if it is the union of two disjoint nonempty open sets. Otherwise, X is said to be connected. A subset of a topological space is said to be connected if it is connected under its subspace topology. Some authors exclude the empty set (with its unique topology) as a connected space, but this article does not follow that practice.

For a topological space X the following conditions are equivalent:

1. X is connected.
2. X cannot be divided into two disjoint nonempty closed sets.
3. The only subsets of X that are both open and closed (clopen sets) are X and the empty set.
4. The only subsets of X with empty boundary are X and the empty set.
5. X cannot be written as the union of two nonempty separated sets.
6. The only continuous functions from X to {0,1}, the two-point space endowed with the discrete topology, are constant.

Every interval in R is connected.

The continuous image of a connected space is connected.

Connected Components

The maximal connected subsets (ordered by inclusion) of a nonempty topological space are called the connected components of the space. The components of any topological space X form a partition of X: they are disjoint, nonempty, and their union is the whole space. Every component is a closed subset of the original space. It follows that, in the case where their number is finite, each component is also an open subset. However, if their number is infinite, this might not be the case; for instance, the connected components of the set of the rational numbers are the one-point sets, which are not open.

Let Γ_x be the connected component of x in a topological space X, and $\Gamma_{x'}$ be the intersection of all open-closed sets containing x (called quasi-component of x.) Then $\Gamma_x \subset \Gamma'_x$ where the equality holds if X is compact Hausdorff or locally connected.

Disconnected Spaces

A space in which all components are one-point sets is called totally disconnected. Related to this property, a space X is called totally separated if, for any two distinct elements x and y of X, there exist disjoint open neighborhoods U of x and V of y such that X is the union of U and V. Clearly any totally separated space is totally disconnected, but the converse does not hold. For example, take two copies of the rational numbers Q, and identify them at every point except zero. The resulting space, with the quotient topology, is totally disconnected. However, by considering the two copies of zero, one sees that the space is not totally separated. In fact, it is not even Hausdorff, and the condition of being totally separated is strictly stronger than the condition of being Hausdorff.

Path-connected Sets

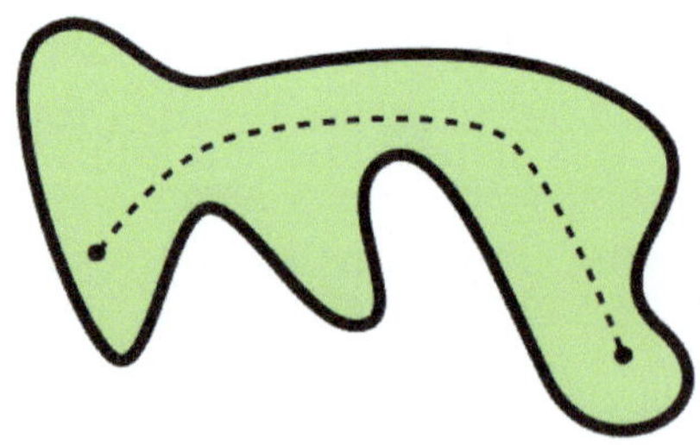

This subspace of $\mathbf{R}^2$ is path-connected, because a path can be drawn between any two points in the space.

A path from a point x to a point y in a topological space X is a continuous function f from the unit interval [0,1] to X with $f(0) = x$ and $f(1) = y$. A path-component of X is an equivalence class of X under the equivalence relation, which makes x equivalent to y if there is a path from x to y. The space X is said to be path-connected (or pathwise connected or 0-connected) if there is at most one path-component, i.e. if there is a path joining any two points in X. Again, many authors exclude the empty space.

Every path-connected space is connected. The converse is not always true: examples of connected spaces that are not path-connected include the extended long line L^* and the *topologist's sine curve*.

However, subsets of the real line R are connected if and only if they are path-connected; these subsets are the intervals of R. Also, open subsets of R^n or C^n are connected if and only if they are path-connected. Additionally, connectedness and path-connectedness are the same for finite topological spaces.

Products of Spaces

Given X such that

$$X := \prod_{i \in I} X_i,$$

is the Cartesian product of the topological spaces X_i, indexed by $i \in I$, and the canonical projections $p_i : X \to X_i$, the product topology on X is defined as the coarsest topology (i.e. the topology with the fewest open sets) for which all the projections p_i are continuous. The product topology is sometimes called the Tychonoff topology.

The open sets in the product topology are unions (finite or infinite) of sets of the form $\prod_{i \in I} U_i$, where each U_i is open in X_i and $U_i \neq X_i$ only finitely many times. In particular, for a finite product (in particular, for the product of two topological spaces), the products of base elements of the X_i gives a basis for the product $\prod_{i \in I} X_i$.

The product topology on X is the topology generated by sets of the form $p_i^{-1}(U)$, where i is in I and U is an open subset of X_i. In other words, the sets $\{p_i^{-1}(U)\}$ form a subbase for the topology on X. A subset of X is open if and only if it is a (possibly infinite) union of intersections of finitely many sets of the form $p_i^{-1}(U)$. The $p_i^{-1}(U)$ are sometimes called open cylinders, and their intersections are cylinder sets.

In general, the product of the topologies of each X_i forms a basis for what is called the box topology on X. In general, the box topology is finer than the product topology, but for finite products they coincide.

Related to compactness is Tychonoff's theorem: the (arbitrary) product of compact spaces is compact.

Separation Axioms

Many of these names have alternative meanings in some of mathematical literature, as explained on History of the separation axioms; for example, the meanings of "normal" and "T_4" are sometimes interchanged, similarly "regular" and "T_3", etc. Many of the concepts also have several names; however, the one listed first is always least likely to be ambiguous.

Most of these axioms have alternative definitions with the same meaning; the definitions given here fall into a consistent pattern that relates the various notions of separation defined in the previous section. Other possible definitions can be found in the individual articles.

In all of the following definitions, X is again a topological space.

- X is T_0, or *Kolmogorov*, if any two distinct points in X are topologically distinguishable. (It is a common theme among the separation axioms to have one version of an axiom that requires T_0 and one version that doesn't.)
- X is T_1, or *accessible* or *Fréchet*, if any two distinct points in X are separated. Thus, X is T_1 if and only if it is both T_0 and R_0. (Though you may say such things as *T_1 space, Fréchet topology*, and *Suppose that the topological space* X *is Fréchet*, avoid saying *Fréchet space* in this context, since there is another entirely different notion of Fréchet space in functional analysis.)
- X is *Hausdorff*, or T_2 or *separated*, if any two distinct points in X are separated by neighbourhoods. Thus, X is Hausdorff if and only if it is both T_0 and R_1. A Hausdorff space must also be T_1.
- X is $T_{2½}$, or *Urysohn*, if any two distinct points in X are separated by closed neighbourhoods. A $T_{2½}$ space must also be Hausdorff.
- X is *regular*, or T_3, if it is T_0 and if given any point x and closed set F in X such that x does not belong to F, they are separated by neighbourhoods. (In fact, in a regular space, any such x and F is also separated by closed neighbourhoods.)

- *X* is *Tychonoff*, or $T_{3½}$, *completely* T_3, or *completely regular*, if it is T_0 and if f, given any point *x* and closed set *F* in *X* such that *x* does not belong to *F*, they are separated by a continuous function.
- *X* is *normal*, or T_4, if it is Hausdorff and if any two disjoint closed subsets of *X* are separated by neighbourhoods. (In fact, a space is normal if and only if any two disjoint closed sets can be separated by a continuous function; this is Urysohn's lemma.)
- *X* is *completely normal*, or T_5 or *completely* T_4, if it is T_1 and if any two separated sets are separated by neighbourhoods. A completely normal space must also be normal.
- *X* is *perfectly normal*, or T_6 or *perfectly* T_4, if it is T_1 and if any two disjoint closed sets are precisely separated by a continuous function. A perfectly normal Hausdorff space must also be completely normal Hausdorff.

The Tietze extension theorem: In a normal space, every continuous real-valued function defined on a closed subspace can be extended to a continuous map defined on the whole space.

Countability Axioms

An axiom of countability is a property of certain mathematical objects (usually in a category) that requires the existence of a countable set with certain properties, while without it such sets might not exist.

Important countability axioms for topological spaces:

- sequential space: a set is open if every sequence convergent to a point in the set is eventually in the set
- first-countable space: every point has a countable neighbourhood basis (local base)
- second-countable space: the topology has a countable base
- separable space: there exists a countable dense subspace
- Lindelöf space: every open cover has a countable subcover
- σ-compact space: there exists a countable cover by compact spaces

Relations:

- Every first countable space is sequential.
- Every second-countable space is first-countable, separable, and Lindelöf.
- Every σ-compact space is Lindelöf.
- A metric space is first-countable.
- For metric spaces second-countability, separability, and the Lindelöf property are all equivalent.

Metric Spaces

A metric space is an ordered pair (M,d) where M is a set and d is a metric on M , i.e., a function

$$d : M \times M \to \mathbb{R}$$

such that for any $x, y, z \in M$, the following holds:

1. $d(x, y) \geq 0$ (*non-negative*),
2. $d(x, y) = 0$ iff $x = y$ (*identity of indiscernibles*),
3. $d(x, y) = d(y, x)$ (*symmetry*) and
4. $d(x, z) \leq d(x, y) + d(y, z)$ (*triangle inequality*).

The function d is also called *distance function* or simply *distance*. Often, d is omitted and one just writes M for a metric space if it is clear from the context what metric is used.

Every metric space is paracompact and Hausdorff, and thus normal.

The metrization theorems provide necessary and sufficient conditions for a topology to come from a metric.

Baire Category Theory

The Baire category theorem says: If *X* is a complete metric space or a locally compact Hausdorff space, then the interior of every union of countably many nowhere dense sets is empty.

Any open subspace of a Baire space is itself a Baire space.

Main Areas of Research

Three iterations of a Peano curve construction, whose limit is a space-filling curve. The Peano curve is studied in continuum theory, a branch of **general topology**.

Continuum Theory

A continuum (pl *continua*) is a nonempty compact connected metric space, or less frequently, a compact connected Hausdorff space. Continuum theory is the branch of topology devoted to the study of continua.

Pointless Topology

Pointless topology (also called point-free or pointfree topology) is an approach to topology that avoids mentioning points. The name 'pointless topology' is due to John von Neumann. The ideas of pointless topology are closely related to mereotopologies, in which regions (sets) are treated as foundational without explicit reference to underlying point sets.

Dimension Theory

Dimension theory is a branch of general topology dealing with dimensional invariants of topological spaces.

Topological Algebras

A topological algebra A over a topological field K is a topological vector space together with a continuous multiplication

$$\cdot : A \times A \to A$$
$$(a,b) \mapsto a \cdot b$$

that makes it an algebra over K. A unital associative topological algebra is a topological ring.

The term was coined by David van Dantzig; it appears in the title of his doctoral dissertation (1931).

Metrizability Theory

In topology and related areas of mathematics, a metrizable space is a topological space that is homeomorphic to a metric space. That is, a topological space (X,τ) is said to be metrizable if there is a metric

$$d : X \times X \to [0,\infty)$$

such that the topology induced by d is τ. Metrization theorems are theorems that give sufficient conditions for a topological space to be metrizable.

Set-theoretic Topology

Set-theoretic topology is a subject that combines set theory and general topology. It focuses on topological questions that are independent of Zermelo–Fraenkel set theory(ZFC). A famous problem is the normal Moore space question, a question in general topology that was the subject of intense research. The answer to the normal Moore space question was eventually proved to be independent of ZFC.

Algebraic Topology

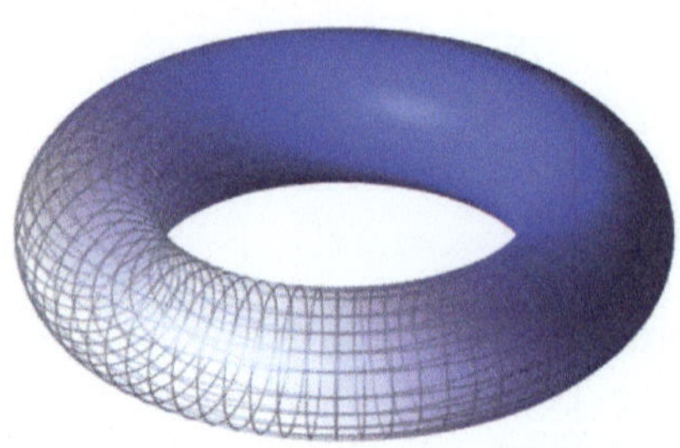

A torus, one of the most frequently studied objects in algebraic topology

Algebraic topology is a branch of mathematics that uses tools from abstract algebra to study topological spaces. The basic goal is to find algebraic invariants that classify topological spaces up to homeomorphism, though usually most classify up to homotopy equivalence.

Although algebraic topology primarily uses algebra to study topological problems, using topology to solve algebraic problems is sometimes also possible. Algebraic topology, for example, allows for a convenient proof that any subgroup of a free group is again a free group.

Main Branches of Algebraic Topology

Below are some of the main areas studied in algebraic topology:

Homotopy Groups

In mathematics, homotopy groups are used in algebraic topology to classify topological spaces. The first and simplest homotopy group is the fundamental group, which records information about loops in a space. Intuitively, homotopy groups record information about the basic shape, or holes, of a topological space.

Homology

In algebraic topology and abstract algebra, homology (in part from Greek ὁμός *homos* "identical") is a certain general procedure to associate a sequence of abelian groups or modules with a given mathematical object such as a topological space or a group.

Cohomology

In homology theory and algebraic topology, cohomology is a general term for a sequence of abelian groups defined from a co-chain complex. That is, cohomology is defined as the abstract study of cochains, cocycles, and coboundaries. Cohomology can be viewed as a method of assigning algebraic invariants to a topological space that has a more refined algebraic structure than does homology. Cohomology arises from the algebraic dualization of the construction of homology. In less abstract language, cochains in the fundamental sense should assign 'quantities' to the *chains* of homology theory.

Manifolds

A manifold is a topological space that near each point resembles Euclidean space. Examples include the plane, the sphere, and the torus, which can all be realized in three dimensions, but also the Klein bottle and real projective plane which cannot be realized in three dimensions, but can be realized in four dimensions. Typically, results in algebraic topology focus on global, non-differentiable aspects of manifolds; for example Poincaré duality.

Knot Theory

Knot theory is the study of mathematical knots. While inspired by knots that appear in daily life in shoelaces and rope, a mathematician's knot differs in that the ends are joined together so that it cannot be undone. In precise mathematical language, a knot is an embedding of a circle in 3-dimensional Euclidean space, R^3. Two mathematical knots are equivalent if one can be transformed into the other via a deformation of R^3 upon itself (known as an ambient isotopy); these transformations correspond to manipulations of a knotted string that do not involve cutting the string or passing the string through itself.

Complexes

A simplicial complex is a topological space of a certain kind, constructed by "gluing together" points, line segments, triangles, and their *n*-dimensional counterparts. Simplicial complexes should not be confused with the more abstract notion of a simplicial set appearing in modern simplicial homotopy theory. The purely combinatorial counterpart to a simplicial complex is an abstract simplicial complex.

A CW complex is a type of topological space introduced by J. H. C. Whitehead to meet the needs of homotopy theory. This class of spaces is broader and has some better categorical properties than simplicial complexes, but still retains a combinatorial nature that allows for computation (often with a much smaller complex).

Method of Algebraic Invariants

An older name for the subject was combinatorial topology, implying an emphasis on how a space X was constructed from simpler ones (the modern standard tool for such construction is the CW-complex). In the 1920s and 1930s, there was growing emphasis on investigating topological spaces by finding correspondences from them to algebraic groups, which led to the change of name to algebraic topology. The combinatorial topology name is still sometimes used to emphasize an algorithmic approach based on decomposition of spaces.

In the algebraic approach, one finds a correspondence between spaces and groups that respects the relation of homeomorphism (or more general homotopy) of spaces. This allows one to recast statements about topological spaces into statements about groups, which have a great deal of manageable structure, often making these statement easier to prove. Two major ways in which this can be done are through fundamental groups, or more generally homotopy theory, and through homology and cohomology groups. The fundamental groups give us basic information about the structure of a topological space, but they are often nonabelian and can be difficult to work with. The fundamental group of a (finite) simplicial complex does have a finite presentation.

Homology and cohomology groups, on the other hand, are abelian and in many important cases finitely generated. Finitely generated abelian groups are completely classified and are particularly easy to work with.

Setting in Category Theory

In general, all constructions of algebraic topology are functorial; the notions of category, functor and natural transformation originated here. Fundamental groups and homology and cohomology groups are not only *invariants* of the underlying topological space, in the sense that two topological spaces which are homeomorphic have the same associated groups, but their associated morphisms also correspond — a continuous mapping of spaces induces a group homomorphism on the associated groups, and these homomorphisms can be used to show non-existence (or, much more deeply, existence) of mappings.

One of the first mathematicians to work with different types of cohomology was Georges de Rham. One can use the differential structure of smooth manifolds via de Rham cohomology, or Čech or sheaf cohomology to investigate the solvability of differential equations defined on the manifold in question. De Rham showed that all of these approaches were interrelated and that, for a closed, oriented manifold, the Betti numbers derived through simplicial homology were the same Betti numbers as those derived through de Rham cohomology. This was extended in the 1950s, when Eilenberg and Steenrod generalized this approach. They defined homology and cohomology as functors equipped with natural transformations subject to certain axioms (e.g., a weak equivalence of spaces passes to an isomorphism of homology groups), verified that all existing (co)homology theories satisfied these axioms, and then proved that such an axiomatization uniquely characterized the theory.

Applications of Algebraic Topology

Classic applications of algebraic topology include:

- The Brouwer fixed point theorem: every continuous map from the unit n-disk to itself has a fixed point.
- The free rank of the nth homology group of a simplicial complex is the n-th Betti number, which allows one to calculate the Euler-Poincaré characteristic.
- One can use the differential structure of smooth manifolds via de Rham cohomology, or Čech or sheaf cohomology to investigate the solvability of differential equations defined on the manifold in question.
- A manifold is orientable when the top-dimensional integral homology group is the integers, and is non-orientable when it is 0.
- The n-sphere admits a nowhere-vanishing continuous unit vector field if and only if n is odd. (For $n = 2$, this is sometimes called the «hairy ball theorem".)
- The Borsuk–Ulam theorem: any continuous map from the n-sphere to Euclidean n-space identifies at least one pair of antipodal points.
- Any subgroup of a free group is free. This result is quite interesting, because the statement is purely algebraic yet the simplest known proof is topological. Namely, any free group G may be realized as the fundamental group of a graph X. The main theorem on covering spaces tells us that every subgroup H of G is the fundamental group of some covering space Y of X; but every such Y is again a graph. Therefore, its fundamental group H is free. On the other hand, this type of application is also handled more simply by the use of covering morphisms of groupoids, and that technique has yielded subgroup theorems not yet proved by methods of algebraic topology.
- Topological combinatorics

Differential Topology

In mathematics, differential topology is the field dealing with differentiable functions on differentiable manifolds. It is closely related to differential geometry and together they make up the geometric theory of differentiable manifolds.

Description

Differential topology considers the properties and structures that require only a smooth structure on a manifold to be defined. Smooth manifolds are 'softer' than manifolds with extra geometric structures, which can act as obstructions to certain types of equivalences and deformations that exist in differential topology. For instance, volume and Riemannian curvature are invariants that can distinguish different geometric structures on the same smooth manifold—that is, one can smoothly "flatten out" certain manifolds, but it might require distorting the space and affecting the curvature or volume.

On the other hand, smooth manifolds are more rigid than the topological manifolds. John Milnor discovered that some spheres have more than one smooth structure—see exotic sphere and Donaldson's theorem. Kervaire exhibited topological manifolds with no smooth structure at all. Some constructions of smooth manifold theory, such as the existence of tangent bundles, can be done in the topological setting with much more work, and others cannot.

One of the main topics in differential topology is the study of special kinds of smooth mappings between manifolds, namely immersions and submersions, and the intersections of submanifolds via transversality. More generally one is interested in properties and invariants of smooth manifolds which are carried over by diffeomorphisms, another special kind of smooth mapping. Morse theory is another branch of differential topology, in which topological information about a manifold is deduced from changes in the rank of the Jacobian of a function.

Differential Topology Versus Differential Geometry

Differential topology and differential geometry are first characterized by their *similarity*. They both study primarily the properties of differentiable manifolds, sometimes with a variety of structures imposed on them.

One major difference lies in the nature of the problems that each subject tries to address. In one view, differential topology distinguishes itself from differential geometry by studying primarily those problems which are *inherently global*. Consider the example of a coffee cup and a donut. From the point of view of differential topology, the donut and the coffee cup are *the same* (in a sense). This is an inherently global view, though, because there is no way for the differential topologist to tell whether the two objects are the same (in this sense) by looking at just a tiny (*local*) piece of either of them. He or she must have access to each entire (*global*) object.

From the point of view of differential geometry, the coffee cup and the donut are *different* because it is impossible to rotate the coffee cup in such a way that its configuration matches that of the donut. This is also a global way of thinking about the problem. But an important distinction is that the geometer does not need the entire object to decide this. By looking, for instance, at just a tiny piece of the handle, he can decide that the coffee cup is different from the donut because the handle is thinner (or more curved) than any piece of the donut.

To put it succinctly, differential topology studies structures on manifolds which, in a sense, have no interesting local structure. Differential geometry studies structures on manifolds which do have an interesting local (or sometimes even infinitesimal) structure.

More mathematically, for example, the problem of constructing a diffeomorphism between two manifolds of the same dimension is inherently global since *locally* two such manifolds are always diffeomorphic. Likewise, the problem of computing a quantity on a manifold which is invariant under differentiable mappings is inherently global, since any local invariant will be *trivial* in the sense that it is already exhibited in the topology of R^n. Moreover, differential topology does not restrict itself necessarily to the study of diffeomorphism. For example, symplectic topology—a subbranch of differential topology—studies global properties of symplectic manifolds. Differential geometry concerns itself with problems—which may be local *or* global—that always have some non-trivial local properties. Thus differential geometry may study differentiable manifolds equipped with a *connection*, a *metric* (which may be Riemannian, pseudo-Riemannian, or Finsler), a special sort of *distribution* (such as a CR structure), and so on.

This distinction between differential geometry and differential topology is blurred, however, in questions specifically pertaining to local diffeomorphism invariants such as the tangent space at a point. Differential topology also deals with questions like these, which specifically pertain to the properties of differentiable mappings on R^n (for example the tangent bundle, jet bundles, the Whitney extension theorem, and so forth).

The distinction is concise in abstract terms. Differential topology is the study of the (infinitesimal, local, and global) properties of structures on manifolds that have only trivial local moduli; differential geometry, at least one non-trivial local moduli.

Geometric Topology

In mathematics, geometric topology is the study of manifolds and maps between them, particularly embeddings of one manifold into another.

A Seifert surface bounded by a set of Borromean rings. Seifert surfaces for links are a useful tool in geometric topology.

History

Geometric topology as an area distinct from algebraic topology may be said to have originated in the 1935 classification of lens spaces by Reidemeister torsion, which required distinguishing spaces that are homotopy equivalent but not homeomorphic. This was the origin of *simple* homotopy theory.

Differences between Low-dimensional and High-dimensional Topology

Manifolds differ radically in behavior in high and low dimension.

High-dimensional topology refers to manifolds of dimension 5 and above, or in relative terms, embeddings in codimension 3 and above. Low-dimensional topology is concerned with questions in dimensions up to 4, or embeddings in codimension up to 2.

Dimension 4 is special, in that in some respects (topologically), dimension 4 is high-dimensional, while in other respects (differentiably), dimension 4 is low-dimensional; this overlap yields phenomena exceptional to dimension 4, such as exotic differentiable structures on R^4. Thus the topological classification of 4-manifolds is in principle easy, and the key questions are: does a topological manifold admit a differentiable structure, and if so, how many? Notably, the smooth case of dimension 4 is the last open case of the generalized Poincaré conjecture.

The distinction is because surgery theory works in dimension 5 and above (in fact, it works topologically in dimension 4, though this is very involved to prove), and thus the behavior of manifolds in dimension 5 and above is controlled algebraically by surgery theory. In dimension 4 and below (topologically, in dimension 3 and below), surgery theory does not work, and other phenomena occur. Indeed, one approach to discussing low-dimensional manifolds is to ask "what would surgery theory predict to be true, were it to work?" – and then understand low-dimensional phenomena as deviations from this.

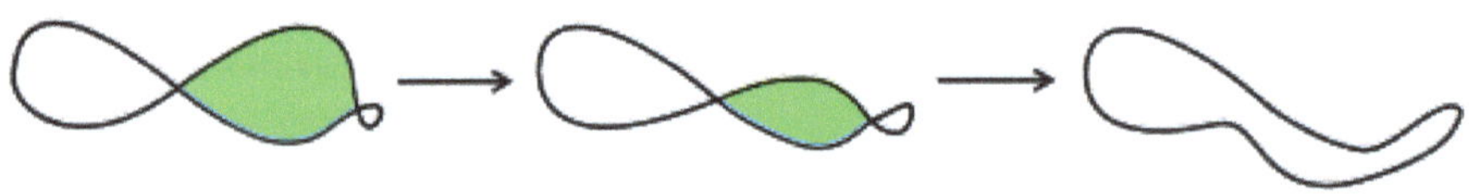

The Whitney trick requires 2+1 dimensions, hence surgery theory requires 5 dimensions.

The precise reason for the difference at dimension 5 is because the Whitney embedding theorem, the key technical trick which underlies surgery theory, requires 2+1 dimensions. Roughly, the Whitney trick allows one to "unknot" knotted spheres – more precisely, remove self-intersections of immersions; it does this via a homotopy of a disk – the disk has 2 dimensions, and the homotopy adds 1 more – and thus in codimension greater than 2, this can be done without intersecting itself; hence embeddings in codimension greater than 2 can be understood by surgery. In surgery theory, the key step is in the middle dimension, and thus when the middle dimension has codimension more than 2 (loosely, 2½ is enough, hence total dimension 5 is enough), the Whitney trick works. The key consequence of this is Smale's *h*-cobordism theorem, which works in dimension 5 and above, and forms the basis for surgery theory.

A modification of the Whitney trick can work in 4 dimensions, and is called Casson handles – because there are not enough dimensions, a Whitney disk introduces new kinks, which can be resolved by another Whitney disk, leading to a sequence ("tower") of disks. The limit of this tower yields a topological but not differentiable map, hence surgery works topologically but not differentiably in dimension 4.

Important Tools in Geometric Topology

Fundamental Group

In all dimensions, the fundamental group of a manifold is a very important invariant, and determines much of the structure; in dimensions 1, 2 and 3, the possible fundamental groups are restricted, while in every dimension 4 and above every finitely presented group is the fundamental group of a manifold (note that it is sufficient to show this for 4- and 5-dimensional manifolds, and then to take products with spheres to get higher ones).

Orientability

A manifold is orientable if it has a consistent choice of orientation, and a connected orientable manifold has exactly two different possible orientations. In this setting, various equivalent formulations of orientability can be given, depending on the desired application and level of generality. Formulations applicable to general topological manifolds often employ methods of homology theory, whereas for differentiable manifolds more structure is present, allowing a formulation in terms of differential forms. An important generalization of the notion of orientability of a space is that of orientability of a family of spaces parameterized by some other space (a fiber bundle) for which an orientation must be selected in each of the spaces which varies continuously with respect to changes in the parameter values.

Handle Decompositions

A handle decomposition of an m-manifold M is a union

$$\varnothing = M_{-1} \subset M_0 \subset M_1 \subset M_2 \subset \ldots \subset M_{m-1} \subset M_m = M$$

where each M_i is obtained from M_{i-1} by the attaching of i-handles. A handle decomposition is to a manifold what a CW-decomposition is to a topological space—in many regards the purpose of a handle decomposition is to have a language analogous to CW-complexes, but adapted to the world of smooth manifolds. Thus an i-handle is the smooth analogue of an i-cell. Handle decompositions of manifolds arise naturally via Morse theory. The modification of handle structures is closely linked to Cerf theory.

A 3-ball with three 1-handles attached.

Local Flatness

Local flatness is a property of a submanifold in a topological manifold of larger dimension. In the category of topological manifolds, locally flat submanifolds play a role similar to that of embedded submanifolds in the category of smooth manifolds.

Suppose a d dimensional manifold N is embedded into an n dimensional manifold M (where $d < n$). If $x \in N$, we say N is locally flat at x if there is a neighborhood $U \subset M$ of x such that the topological pair $(U, U \cap N)$ is homeomorphic to the pair $(\mathbb{R}^n, \mathbb{R}^d)$, with a standard inclusion of $\mathbb{R}^d$ as a subspace of $\mathbb{R}^n$. That is, there exists a homeomorphism $U \to R^n$ such that the image of $U \cap N$ coincides with $\mathbb{R}^d$.

Schönflies Theorems

The generalized Schoenflies theorem states that, if an $(n - 1)$-dimensional sphere S is embedded into the n-dimensional sphere S^n in a locally flat way (that is, the embedding extends to that of a thickened sphere), then the pair (S^n, S) is homeomorphic to the pair (S^n, S^{n-1}), where S^{n-1} is the equator of the n-sphere. Brown and Mazur received the Veblen Prize for their independent proofs of this theorem.

Branches of Geometric Topology

Low-dimensional Topology

Low-dimensional topology includes:

- Surface (topology)s (2-manifolds)
- 3-manifolds
- 4-manifolds

each have their own theory, where there are some connections.

Low-dimensional topology is strongly geometric, as reflected in the uniformization theorem in 2 dimensions – every surface admits a constant curvature metric; geometrically, it has one of 3 possible geometries: positive curvature/spherical, zero curvature/flat, negative curvature/hyperbolic – and the geometrization conjecture (now theorem) in 3 dimensions – every 3-manifold can be cut into pieces, each of which has one of 8 possible geometries.

2-dimensional topology can be studied as complex geometry in one variable (Riemann surfaces are complex curves) – by the uniformization theorem every conformal class of metrics is equivalent to a unique complex one, and 4-dimensional topology can be studied from the point of view of complex geometry in two variables (complex surfaces), though not every 4-manifold admits a complex structure.

Knot Theory

Knot theory is the study of mathematical knots. While inspired by knots which appear in daily life in shoelaces and rope, a mathematician's knot differs in that the ends are joined together so that it cannot be undone. In mathematical language, a knot is an embedding of a circle in 3-dimensional Euclidean space, R^3 (since we're using topology, a circle isn't bound to the classical geometric concept, but to all of its homeomorphisms). Two mathematical knots are equivalent if one can be transformed into the other via a deformation of R^3 upon itself (known as an ambient isotopy); these transformations correspond to manipulations of a knotted string that do not involve cutting the string or passing the string through itself.

To gain further insight, mathematicians have generalized the knot concept in several ways. Knots can be considered in other three-dimensional spaces and objects other than circles can be used; see *knot (mathematics)*. Higher-dimensional knots are n-dimensional spheres in m-dimensional Euclidean space.

High-dimensional Geometric Topology

In high-dimensional topology, characteristic classes are a basic invariant, and surgery theory is a key theory.

A characteristic class is a way of associating to each principal bundle on a topological space X a cohomology class of X. The cohomology class measures the extent to which the bundle is "twisted" — particularly, whether it possesses sections or not. In other words, characteristic classes are global invariants which measure the deviation of a local product structure from a global product structure. They are one of the unifying geometric concepts in algebraic topology, differential geometry and algebraic geometry.

Surgery theory is a collection of techniques used to produce one manifold from another in a 'controlled' way, introduced by Milnor (1961). Surgery refers to cutting out parts of the manifold and replacing it with a part of another manifold, matching up along the cut or boundary. This is closely related to, but not identical with, handlebody decompositions. It is a major tool in the study and classification of manifolds of dimension greater than 3.

More technically, the idea is to start with a well-understood manifold M and perform surgery on it to produce a manifold M' having some desired property, in such a way that the effects on the homology, homotopy groups, or other interesting invariants of the manifold are known.

The classification of exotic spheres by Kervaire and Milnor (1963) led to the emergence of surgery theory as a major tool in high-dimensional topology.

Topological Data Analysis

In applied mathematics, Topological data analysis (TDA) is an approach to the analysis of datasets using techniques from topology. Extraction of information from datasets that are high-dimensional, incomplete and noisy is generally challenging. TDA provides a general framework to analyze such data in a manner that is insensitive to the particular metric and provides dimension reduction and robustness to noise. Beyond, it inherits functorality, one of the keys to the modern mathematics, from its topological nature, which makes it adaptive to new tools from mathematics.

The initial motivation is to study the shape of data. TDA has combined algebraic topology and other tools from pure mathematics to give mathematically rigorous and quantitative study of "shape". The main tool is persistent homology, an adaptation of homology to point cloud data. Persistent homology has been applied to many types of data across many fields. Moreover, its mathematical foundation is also of theoretical importance. The unique features of TDA make it a promising bridge between topology and geometry.

Basic Theory

Intuition

The premise underlying TDA is that shape matters. Real data in high dimensions is nearly always sparse, and tends to have relevant low dimensional features. The task of TDA is to precisely characterize this observation. One illustrative example is a predator-prey system governed by a Lotka-Volterra equation. One can easily observe that data forms a closed circle. TDA provides tools to detect and quantify such recurrent motion.

Without prior domain knowledge, the correct threshold for a data set is difficult to choose. The main insight of persistent homology is that we can use the information obtained from all values of a parameter. Of course this insight alone is easy to make; the hard part is encoding this huge amount of information into an understandable and easy-to-represent way. With TDA, there is a mathematical interpretation when the information is a homology group. In general, the assumption is that features that persist for a wide range of parameters are "true" features. Features persisting for only a short period are presumed to be noise, although the theoretical justification for this is unclear.

Early History

Precursors to the full concept of persistent homology appeared gradually over time. In 1990, Patrizio Frosini introduced the size function, which is equivalent to the 0th persistent homology. Nearly a decade later, Vanessa Robins studied the images of homomorphisms induced by inclusion. Finally, shortly thereafter, Edelsbrunner et al. introduced the concept of persistent homology together with an efficient algorithm and its presentation as persistence diagram. Carlsson et al. reformulated the initial definition and gave an equivalent visualization method called persistence barcodes, interpreting persistence in the language of commutative algebra.

Concepts

Some widely used concepts are introduced below. Note that some definitions may vary from author to author.

A point cloud is often defined as a finite set of points in some Euclidean space, but may be taken to be any finite metric space.

The Čech complex of a point cloud is the *nerve* of the *cover* of balls of a fixed radius around each point in the cloud.

A persistence module $\mathbb{U}$ indexed by $\mathbb{Z}$ is a vector space U_t for each $t \in Z$, and a linear map $u_t^s : U_s \to U_t$ whenever $s \leq t$, such that $u_t^t = 1$ for all t and $u_t^s u_s^r = u_t^r$ whenever An equivalent definition is a functor from $\mathbb{Z}$ considered as a partially ordered set to the category of vector spaces.

The persistent homology group PH of a persistence module is defined as $PH_k(X) = \prod H_k(X_r)$, where X_r is the Čech complex of radius r of the point cloud X and is the homology group.

A persistence barcode is a multiset of intervals in $\mathbb{R}$, and a persistence diagram is a multiset of points in $\Delta (:= \{(u,v) \in \mathbb{R}^2 \mid u,v \geq 0, u \leq v\})$.

The Wasserstein distance between two persistence diagrams X and Y is defined as

$$W_p[L_q](X,Y) := \inf_{\phi:X\to Y} [\sum_{x \in X} \| x - \phi(x) \|_q]^{\frac{1}{p}}$$

where $1 \leq p,q \leq \infty$ and ϕ ranges over bijections between X and Y. Please refer to figure 3.1 in Munch for illustration.

The bottleneck distance between X and Y is

$$W_\infty[L_q](X,Y) := \inf_{\phi:X\to Y} \sup_{x \in X} \| x - \phi(x) \|_q .$$

This is a special case of Wasserstein distance, letting $p = \infty$.

Basic Property

Structure Theorem

The first classification theorem for persistent homology appeared in 2005: for a finitely generated persistence module C with field F coefficients,

$$H(C;F) \simeq \bigoplus_i x^{t_i} \cdot F[x] \oplus \left(\bigoplus_j x^{r_j} \cdot (F[x]/(x^{s_j} \cdot F[x])) \right).$$

Intuitively, the free parts correspond to the homology generators that appear at filtration level t_i and never disappear, while the torsion parts correspond to those that appear at filtration level r_j and last for s_j steps of the filtration (or equivalently, disappear at filtration level $s_j + r_j$).

Persistent homology is visualized through a barcode or persistence diagram. The barcode has its root in abstract mathematics, though not at first sight; essentially, the derived category of chain complexes over a field is equivalent to the graded category of vector spaces.

Stability

Stability is desirable because it provides robustness against noise. If X is any space which is homeomorphic to a simplicial complex, and $f, g : X \to \mathbb{R}$ are continuous tame functions, then the persistence vector spaces $\{H_k(f^{-1}([0,r])\}$ and $\{H_k(g^{-1}([0,r])\}$ are finitely presented, and $W_\infty(D(f), D(g)) \leq \| f - g \|_\infty$, where W_∞ refers to the bottleneck distance.

Workflow

The basic workflow in TDA is:

1. If X is a point cloud, replace X with a nested family of simplicial complexes X_r (such as the Čech or Vietoris-Rips complex). This process converts the point cloud into a filtration of simplicial complexes. Taking the homology of each complex in this filtration gives a persistence module
   $H_i(X_{r_0}) \to H_i(X_{r_1}) \to H_i(X_{r_2}) \to \cdots$
2. Apply the structure theorem to provide a parameterized version of Betti number, persistence diagram, or equivalently, barcode.

Graphically speaking,

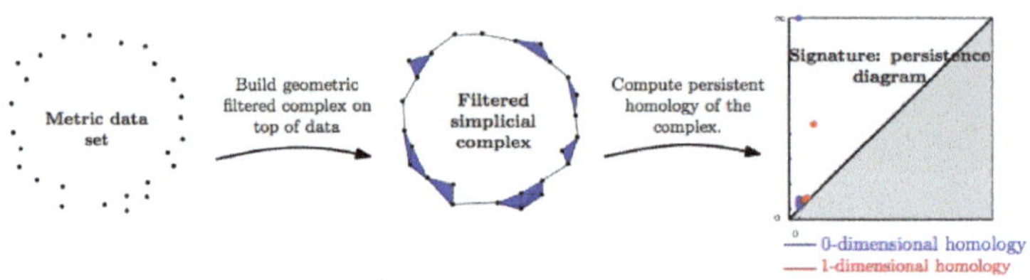

A usual use of persistence in TDA

Computation

The first algorithm for persistent homology over F_2 was given by Edelsbrunner et al. Carlsson et al. gave the first practical algorithm to compute the persistent homology over all fields. Edelsbrunner and Harer's book gives general guidance on computational topology.

One issue that arises in computation is the choice of complex. The Čech complex and Vietoris-Rips complex are most natural at first glance; however, their size grows rapidly with the number of data points. The Vietoris-Rips complex is preferred over Čech complex because its definition is simpler and the Čech complex requires extra effort to define in a general finite metric space. Efficient ways to lower the computational cost of homology have been studied. For example, the α-complex and witness complex are used to reduce the dimension and size of complexes.

Recently, discrete Morse theory has shown promise for computational homology because it can reduce a given simplicial complex to a much smaller cellular complex which is homotopic to the original one. This reduction can in fact be performed as the complex is constructed by using matroid theory, leading to further performance increases. Another recent algorithm saves time by ignoring the homology classes with low persistence.

Various software packages are available, such as javaPlex, Dionysus, Perseus, PHAT, DIPHA, and Gudhi. A comparison between these tool is done by Otter et al. Also, an R package TDA is capable of calculating recently invented concepts like landscape and the kernel distance estimator.

Visualization

High-dimensional data is impossible to visualize directly. Many methods have been invented to extract a low-dimensional structure from the data set, such as principal component analysis and multidimensional scaling. However, it is important to note that the problem itself is ill-posed, since many different topological features can be found in the same data set. Thus, the study of visualization of high-dimensional spaces is of central importance to TDA, although it does not necessarily involve the use of persistent homology. However, recent attempts have been made to use persistent homology in data visualization.

Carlsson et al. have proposed a general method called MAPPER. It inherits the idea of Serre that a covering preserves homotopy. A generalized formulation of MAPPER is as follows:

Let X and Z be topological spaces and let $f : X \to Z$ be a continuous map. Let $\mathbb{U} = \{U_\alpha\}_{\alpha \in A}$ be a finite open covering of Z. The output of MAPPER is the nerve of the pullback cover $M(\mathbb{U}, f) := N(f^{-1}(\mathbb{U}))$, where each preimage is split into its connected components. This is a very general concept, of which the Reeb graph and merge trees are special cases.

This is not quite the original definition. Carlsson et al. choose Z to be $\mathbb{R}$ or $\mathbb{R}^2$, and cover it with open sets such that at most two intersect. This restriction means that the output is in the form of a complex network. Because the topology of a finite point cloud is trivial, clustering methods (such as single linkage) are used to produce the analogue of connected sets in the preimage $f^{-1}(U)$ when MAPPER is applied to actual data.

Mathematically speaking, MAPPER is a variation of the Reeb graph. If the $M(\mathbb{U}, f)$ is at most one dimensional, then for each $i \geq 0$,

$$H_i(X) \simeq H_0(N(\mathbb{U}); \hat{F}_i) \oplus H_1(N(\mathbb{U}); \hat{F}_{i-1}).$$

The added flexibility also has disadvantages. One problem is instability, in that some change of the choice of the cover can lead to major change of the output of the algorithm. Work has been done to overcome this problem.

Three successful applications of MAPPER can be found in Carlsson et al. A comment on the applications in this paper by J. Curry is that "a common feature of interest in applications is the presence of flares or tendrils."

A free implementation of MAPPER is available online written by Daniel Müllner and Aravindakshan Babu. MAPPER also forms the basis of Ayasdi's data visualization platform.

Mathematical Foundation

Although persistent homology is a "21st century child of algebraic homology", its mathematical foundation has been vastly developed. An incomplete list of active mathematicians working on this field may include leading figures Gunnar Carlsson, Herbert Edelsbrunner, Vin De Silva, Peter Bubenik, Frédéric Chazal, Robert Ghrist, and rising scholars such as Michael Lesnick, Justin Curry, Jonathan Scott.

Multidimensional Persistence

Multidimensional persistence is important to TDA. The concept arises in both theory and practice. The first investigation of multidimensional persistence was early in the development of TDA, and is one of the founding papers of TDA. The first application to appear in the literature is a method for shape comparison, similar to the invention of TDA.

The definition of an n-dimensional persistence module in $\mathbb{R}^n$ is

- vector space V_s is assigned to each point in $s=(s_1,\dots,s_n)$
- map $\rho_s^t : V_s \to V_t$ is assigned if $s \le t\ (s_i \le t_i, i=1,\dots,n)$
- maps satisfy $\rho_r^t = \rho_s^t \circ \rho_r^s$ for all $r \le s \le t$

It might be worth noting that there are controversies on the definition of multidimensional persistence.

One of the advantages of one-dimensional persistence is its representability by a diagram or barcode. However, discrete complete invariants of multidimensional persistence modules do not exist. The main reason for this is that the structure of the collection of indecomposables is extremely complicated by Gabriel's theorem in the theory of quiver representations, although a finitely n-dim persistence module can be uniquely decomposed into a direct sum of indecomposables due to the Kull-Schmidt theorem.

Nonetheless, many results have been established. Carlsson et al. introduced the rank invariant $\rho_M(u,v)$, defined as the $\rho_M(u,v) = \operatorname{rank}(x^{u-v} : M_u \to M_v)$, in which M is a finitely generated n-graded module. In one dimension, it is equivalent to the barcode. In the literature, the rank invariant is often referred as the persistent Betti numbers (PBNs). In many theoretical works, authors have used a more restricted definition, an analogue from sublevel set persistence. Specifically, the persistence Betti numbers of a function $f : X \to \mathrm{R}^k$ are given by the function $\beta_f : \Delta^+ \to \mathrm{N}$, taking each $(u,v) \in \Delta^+$ to $\beta_f(u,v) := \operatorname{rank}(H(X(f \le u) \to H(X(f \le v)))$, where $\Delta^+ := \{(u,v) \in \mathbb{R} \times \mathbb{R} : u \le v\}$ and $X(f \le u) := \{x \in X : f(x) \le u\}$.

Some basic properties include monotonicity and diagonal jump. Persistent Betti numbers will be finite if X is a compact and locally contractible subspace of .

Using a foliation method, the k-dim PBNs can be decomposed into a family of 1-dim PBNs by dimensionality deduction. This method has also led to a proof that multi-dim PBNs are stable. The discontinuities of PBNs only occur at points $\rho(u,\star)$ where either u is a discontinuous point of $\rho_M(\star,v)$ or v is a discontinuous point of $\rho(u,\star)$ under the assumption that $f \in C^0(X,\mathrm{R}^k)$ and X is a compact, triangulable topological space.

Persistent space, a generalization of persistent diagram, is defined as the multiset of all points with multiplicity larger than 0 and the diagonal. It provides a stable and complete representation of PBNs. An ongoing work by Carlsson et al. is trying to give geometric interpretation of persistent homology, which might provide insights on how to combine machine learning theory with topological data analysis.

The first practical algorithm to compute multidimensional persistence was invented very early. After then, many other algorithms have been proposed, based on such concepts as discrete morse theory and finite sample estimating.

Other Persistences

The standard paradigm in TDA is often referred as sublevel persistence. Apart from multidimensional persistence, many works have been done to extend this special case.

Zigzag Persistence

The nonzero maps in persistence module are restricted by the preorder relationship in the category. However, mathematicians have found that the unanimousness of direction is not essential to many results. "The philosophical point is that the decomposition theory of graph representations is somewhat independent of the orientation of the graph edges". Zigzag persistence is important to the theoretical side. The examples given in Carlsson's review paper to illustrate the importance of functorality all share some of its features.

Extended Persistence and Levelset Persistence

Some attempts is to lose the stricter restriction of the function. Please refer to the Categorization and cosheaf and Impact on mathematics sections for more information.

It's natural to extend persistence homology to other basic concepts in algebraic topology, such as cohomology and relative homology/cohomology. An interesting application is the computation of circular coordinates for a data set via the first persistent cohomology group.

Circular Persistence

Normal persistence homology studies real-valued functions. The circle-valued map might be useful, "persistence theory for circle-valued maps promises to play the role for some vector fields as does the standard persistence theory for scalar fields", as commented in D. Burghelea et al. The main difference is that Jordan cells(very similar in format to the ones in linear algebra) are nontrivial in circle-valued functions, which would be zero in real-valued case, and combing with barcodes give the invariants of a tame map, under moderate conditions.

Two techniques they use are More-Novikov theory and graph representation theory. More recent results can be found in D. Burghelea et al. For example, the tameness requirement can be replaced by the much weaker condition, continuous.

Persistence with Torsion

The proof of the structure theorem relies on the base domain being field, so not many attempts have been made on persistence homology with torsion. Frosini defined a pseudometric on this specific module and proved its stability. One of its novelty is that it doesn't depend on some classification theory to define the metric.

Categorization and Cosheaves

One advantage of category theory is its ability to lift concrete results to a higher level, showing relationships between seemingly unconnected objects. Bubenik et al. offers a short introduction of category theory fitted for TDA.

Category theory is the language of modern algebra, and has been widely used in the study of algebraic geometry and topology. It has been noted that "the key observation of is that the persistence diagram produced by depends only on the algebraic structure carried by this diagram." The use of category theory in TDA has proved to be fruitful.

Following the notations made in Bubenik et al., the indexing category P is any preordered set (not necessarily $\mathbb{N}$ or $\mathbb{R}$), the target category D is any category (instead of the commonly used $\mathrm{Vect}_{\mathbb{F}}$), and functors $P \to D$ are called generalized persistence modules in D, over P.

One advantage of using category theory in TDA is a clearer understanding of concepts and the discovery of new relationships between proofs. Take two examples for illustration. The understanding of the correspondence between interleaving and matching is of huge importance, since matching has been the method used in the beginning (modified from Morse theory). A summary of works can be found in Vin de Silva et al. Many theorems can be proved much more easily in a more intuitive setting. Another example is the relationship between the construction of different complexes from point clouds. It has long been noticed that Čech and Vietoris-Rips complexes are related. Specifically, $V_r(X) \subset C_{\sqrt{2}r}(X) \subset V_{2r}(X)$. The essential relationship between Cech and Rips complexes can be seen much more clearly in categorical language.

The language of category theory also helps cast results in terms recognizable to the broader mathematical community. Bottleneck distance is widely used in TDA because of the results on stability with respect to the bottleneck distance. In fact, the interleaving distance is the terminal object in a poset category of stable metrics on multidimensional persistence modules in a prime field.

Sheaves, a central concept in modern algebraic geometry, are intrinsically related to category theory. Roughly speaking, sheaves are the mathematical tool for understanding how local information determines global information. Justin Curry regards level set persistence as the study of fibers of continuous functions. The objects that he studies are very similar to those by MAPPER, but with sheaf theory as the theoretical foundation. Although no breakthrough in the theory of TDA has yet used sheaf theory, it is promising since there are many beautiful theorems in algebraic geometry relating to sheaf theory. For example, a natural theoretical question is whether different filtration methods result in the same output.

Stability

Stability is of central importance to data analysis, since real data carry noises. By usage of category theory, Bubenik et al. have distinguished between soft and hard stability theorems, and proved that soft cases are formal.

Soft stability theorem asserts that HF is Lipschitz, and hard stability theorem asserts that J is Lipschitz.

Bottleneck distance is widely used in TDA. The isometry theorem asserts that the interleaving distance d_I is equal to the bottleneck distance. Bubenik et al. have abstracted the definition to that between functors $F, G : P \to D$ when P is equipped with a sublinear projection or superlinear family, in which still remains a pseudometric. Considering the magnificent characters of interleaving distance, here we introduce the general definition of interleaving distance(instead of the first introduced one): Let $\Gamma, K \in \mathrm{Trans}_P$ $\Gamma, K \in \mathrm{Tran}_P$ (a function from P to P which is monotone and satisfies $x \leq \Gamma(x)$ for all $x \in P$). A (Γ, K)-interleaving between F and F consists of natural transformations $\varphi : F \Rightarrow G\Gamma$ and $\psi : G \Rightarrow FK$, such that $(\psi\Gamma)\varphi = F\eta_{K\Gamma}$ and $(\varphi\Gamma) = \psi G\eta_{\Gamma K}$.

The two main results are

- Let P be a preordered set with a sublinear projection or superlinear family. Let $H : D \to E$ be a functor between arbitrary categories D, E. Then for any two functors $F, G : P \to D$, we have $d_I(HF, HG) \leq d_I(F, G)$.
- Let P be a poset of a metric space Y, X be a topological space. And let $f, g : X \to Y$ (not necessarily continuous) be functions, and F, G to be the corresponding persistence diagram. Then $d_I(F, G) \leq d_\infty(f, g) := \sup_{x \in X} d_Y(f(x), g(x))$.

These two results summarize many results on stability of different models of persistence.

For the stability theorem of multidimensional persistence, please refer to the subsection of persistence.

Structure Theorem

The structure theorem is of central importance to TDA; as commented by G. Carlsson, "what makes homology useful as a discriminator between topological spaces is the fact that there is a classification theorem for finitely generated abelian groups."

The main argument used in the proof of the original structure theorem is the standard classification theorem for finitely generated modules over a principal ideal domain. However, this argument fails in $(\mathbb{R}, \leq)$ since $\mathbb{R}$ is not a PID. Carlsson gave a detailed discussion in the most influential review paper in TDA.

In general, not every persistence module can be decomposed into intervals. Many attempts have been made loosing the assumptions. The case for pointwise dimensional persistence modules indexed by a locally finite subset of $\mathbb{R}$ is solved based on the work of Webb. The most notable result is done by Crawley-Boevey, which solved the case of $\mathbb{R}$. Crawley-Boevey's theorem states that any pointwise finite-dimensional persistence module is a direct sum of interval modules.

To understand the definition of his theorem, some concepts need introducing. An interval in $(\mathbb{R}, \leq)$

is defined as a subset $I \subset \mathbb{R}$ having the property that if $r, t \in I$ and if there is an $s \in \mathbb{R}$ such that $r \le s \le t$, then $s \in I$ as well. An interval module k_I assigns to each element $s \in I$ the vector space k and assigns the zero vector space to elements in $\mathbb{R} / I$. All maps ρ_s^t are the zero map, unless $s, t \in I$ and $s \le t$, in which case ρ_s^t is the identity map. Interval modules are indecomposable.

Although this is a very powerful theorem, it still doesn't extend to the q-tame case. A persistence module is q-tame if the rank(ρ_s^t) is finite for all $s \le t$. There are examples that q-tame persistence module fails to be pointwise finite. However, it turns out that similar structure theorem still exists if the features that exist only at one index value are removed. Actually, the infinite dimension wouldn't persist. Specifically, the observable category Ob is defined as Pers >, in which Eph denotes the full subcategory of Pers whose objects are the ephemeral modules($\rho_s^t = 0$ whenever $s < t$).

Note that all these extended results listed here don't apply to the zigzag persistence. There is some work on the stability of zigzag persistence.

Statistics

Real data is always finite, thus the study of it is stochastic in essence. To distinguish between true nature and artifacts is the power of statistics. Note that persistent homology has no mechanism to distinguish between low-probability features and high-probability features.

One way of statistics is to study statistical properties of summaries on topological features of point cloud. A reference of the works done on "the study of random abstract simplicial complexes generated from stochastic processes and non-asymptotic bounds on the convergence or consistency of topological summaries as the number of points increase" can be found in K. Turner et al.

Anotherway,andalsothemoreimportantone,istostudytheprobabilitydistributiononthepersistence

space. The persistence space B_∞ is the $\coprod_n B_n /$ 佷, where B_n are all the barcodes containing exactly n intervals and the equivalences are $\{[x_1, y_1], [x_2, y_2], \ldots, [x_n, y_n]\}$ 佷 $\{[x_1, y_1], [x_2, y_2], \ldots, [x_{n-1}, y_{n-1}]\}$ if that $x_n = y_n$. This space is fairly complicated, for example, not complete endowed with the bottleneck metric. The first attempt made on is by Y. Mileyko et al. The space of persistence diagrams D_p in their paper is defined as

$$D_p := \left\{ d \mid \sum_{x \in d} \left(2 \inf_{y \in \Delta} \| x - y \| \right)^p \right\}$$

where Δ is the diagonal line in $\mathbb{R}^2$. A very nice property is that D_p is complete and separable in the

Wasserstein metric $W_p(u, v) = \left(\inf_{\gamma \in \Gamma(u,v)} \int_{\mathbb{X} \times \mathbb{X}} \rho^p(x, y) \mathrm{d}\gamma(x, y) \right)^{\frac{1}{p}}$. Expectation and variance and conditional probability, three basic concepts in probability theory, can be defined in the Fréchet sense. Since then, many statistical tools are converted into TDA. Works on null hypothesis significance test, confidence interval, and robust estimate are notable steps.

An interesting concept, persistent landscape, invented by Peter Bubenik, has led another direction, namely the different representation of barcode. The persistent landscape over a persistent module M is defined as a function $\lambda : \mathbb{N} \times \mathbb{R} \to \overline{\mathbb{R}}$, $\lambda(k,t) := \sup(m \geq 0 \mid \beta^{t-m,t-m} \geq k)$, where $\overline{\mathbb{R}}$ denotes the extended real line and $\beta^{a,b} = \dim(\operatorname{im}(M(a \leq b)))$. While it inherits all good properties of barcode representation (stability, easy representation, etc.), its space is very nice: not only statistical inferences can be defined, some problems in Y. Mileyko et al.'s work, such as the expectation is not necessarily unique, can be overcome. Effective algorithm is available. Another approach is to use revised persistence, which is Image, kernel and cokernel persistence.

Applications

Classification of Applications

More than one way exist to classify the applications of TDA. Perhaps the most natural way is by field. A very incomplete list of successful applications includes data skeletonization, shape study, graph reconstruction, image analysis, material, progression analysis of disease, sensor network, signal analysis, cosmic web, complex network, fractal geometry, viral evolution, and the propagation of contagions on networks.

Another way is by distinguishing the techniques by G. Carlsson,

one being the study of homological invariants of data one individual data sets, and the other is the use of homological invariants in the study of databases where the data points themselves have geometric structure.

Characteristics of TDA in Applications

Ayasdi is a data analysis company relying heavily on TDA, cofounded by a number of leading researchers in the field. There are several notable interesting features of the recent applications of TDA:

1. Combining tools from all main branches of mathematics. Besides the obvious need for algebra and topology, partial differential equations, algebraic geometry, presentation theory, statistics, combinatorics, and Riemannian geometry have all found use in TDA.
2. Quantitative analysis. Topology is considered to be very soft since many concepts are invariant under homotopy. However, persistent topology is able to record the birth(appearance) and death(disappearance) of topological feature, thus extra geometric information is embedded in it. One evidence in theory is a partially positive result on the uniqueness of reconstruction of curves; two in application are on the quantitative analysis of Fullerene stability and quantitative analysis of self-similarity, separately.
3. The role of short persistence. Short persistence has also been found to be useful, despite the common belief that noise is the cause of the phenomena. This is interesting to the mathematical theory.

One of the main fields of data analysis today is machine learning. Some examples of machine learning in TDA can be found in Adcock et al. A conference is dedicated to the link between TDA and machine learning. In order to apply tools from machine leaning, the information obtained from TDA should be represented in vector form. An ongoing and promising attempt is the persistence landscape discussed above. Another attempt uses the concept of persistence images. However, one problem of this method is the loss of stability, since the hard stability theorem depends on the barcode representation.

Impact on Mathematics

Topological data analysis and persistent homology have had impacts on Morse theory. Morse theory has played a very important role in the theory of TDA, including on computation. Some work in persistent homology has extended results about Morse functions to tame functions or, even to continuous functions. A forgotten result of R. Deheuvels long before the invention of persistent homology extends Morse theory to all continuous functions.

One recent result is that the category of Reeb graphs is equivalent to a particular class of cosheaf. This is motivated by theoretical work in TDA, since the Reeb graph is related to Morse theory and MAPPER is derived from it. The proof of this theorem relies on the interleaving distance.

It is evident to mathematicians that persistent homology is closely related to spectral sequences. Zigzag persistence may turn out to be of theoretical importance to spectral sequences.

References

- Allen Hatcher, Algebraic topology. (2002) Cambridge University Press, xii+544 pp. ISBN 0-521-79160-X and ISBN 0-521-79540-0.
- Adams, Colin (2004). The Knot Book: An Elementary Introduction to the Mathematical Theory of Knots. American Mathematical Society. ISBN 0-8218-3678-1
- The Shape of Space: How to Visualize Surfaces and Three-dimensional Manifolds 2nd ed (Marcel Dekker, 1985, ISBN 0-8247-7437-X)
- Bredon, Glen E. (1993), Topology and Geometry, Graduate Texts in Mathematics 139, Springer, ISBN 0-387-97926-3, retrieved 2008-04-01.
- Edelsbrunner, Herbert; Harer, John (2010-01-01). Computational Topology: An Introduction. American Mathematical Soc. ISBN 9780821849255.
- Bott, Raoul; Tu, Loring W. (2013-04-17). Differential Forms in Algebraic Topology. Springer Science & Business Media. ISBN 9781475739510.

Permissions

All chapters in this book are published with permission under the Creative Commons Attribution Share Alike License or equivalent. Every chapter published in this book has been scrutinized by our experts. Their significance has been extensively debated. The topics covered herein carry significant information for a comprehensive understanding. They may even be implemented as practical applications or may be referred to as a beginning point for further studies.

We would like to thank the editorial team for lending their expertise to make the book truly unique. They have played a crucial role in the development of this book. Without their invaluable contributions this book wouldn't have been possible. They have made vital efforts to compile up to date information on the varied aspects of this subject to make this book a valuable addition to the collection of many professionals and students.

This book was conceptualized with the vision of imparting up-to-date and integrated information in this field. To ensure the same, a matchless editorial board was set up. Every individual on the board went through rigorous rounds of assessment to prove their worth. After which they invested a large part of their time researching and compiling the most relevant data for our readers.

The editorial board has been involved in producing this book since its inception. They have spent rigorous hours researching and exploring the diverse topics which have resulted in the successful publishing of this book. They have passed on their knowledge of decades through this book. To expedite this challenging task, the publisher supported the team at every step. A small team of assistant editors was also appointed to further simplify the editing procedure and attain best results for the readers.

Apart from the editorial board, the designing team has also invested a significant amount of their time in understanding the subject and creating the most relevant covers. They scrutinized every image to scout for the most suitable representation of the subject and create an appropriate cover for the book.

The publishing team has been an ardent support to the editorial, designing and production team. Their endless efforts to recruit the best for this project, has resulted in the accomplishment of this book. They are a veteran in the field of academics and their pool of knowledge is as vast as their experience in printing. Their expertise and guidance has proved useful at every step. Their uncompromising quality standards have made this book an exceptional effort. Their encouragement from time to time has been an inspiration for everyone.

The publisher and the editorial board hope that this book will prove to be a valuable piece of knowledge for students, practitioners and scholars across the globe.

Index